PERGAMON INTER W7- BOO- O63
of Science, Technology, Engineering and Social Studies
The 1000-volume original paperback library in aid of education, industrial training and the enjoyment of leisure
Publisher: Robert Maxwell, M.C.

INORGANIC GEOCHEMISTRY

WITHDRAWN

THE PERGAMON TEXTBOOK
INSPECTION COPY SERVICE

An inspection copy of any book published in the Pergamon International Library will gladly be sent to academic staff without obligation for their consideration for course adoption or recommendation. Copies may be retained for a period of 60 days from receipt and returned if not suitable. When a particular title is adopted or recommended for adoption for class use and the recommendation results in a sale of 12 or more copies, the inspection copy may be retained with our compliments. The Publishers will be pleased to receive suggestions for revised editions and new titles to be published in this important International Library.

Other Related Pergamon Titles of Interest

Books

AHRENS
Origin and Distribution of the Elements (Proceedings of the Second Symposium)

ATKINSON & ZUCKERMAN
Origin and Chemistry of Petroleum

CONDIE
Plate Tectonics and Crustal Evolution

* DOUGLAS & MAXWELL
Advances in Organic Geochemistry 1979

* JEFFERY & HUTCHISON
Chemical Methods of Rock Analysis, 3rd Edition

OWEN
The Geological Evolution of the British Isles

* RAMDOHR
The Ore Minerals and Their Intergrowths, 2nd Edition (2 vols.)

* RICKARD & WICKMAN
Chemistry and Geochemistry of Solutions at High Temperatures and Pressures

* RIDGE
Annotated Bibliographies of Mineral Deposits in Africa, Asia (exclusive of the USSR) and Australasia

ROCKS OF THE WORLD

* SAND & MUMPTON
Natural Zeolites (Occurrence, Properties, Use)

SIMPSON
Rocks and Minerals

SPRY
Metamorphic Textures

* VAN OLPHEN & FRIPIAT
Data Handbook for Clay Materials and other Non-metallic Minerals

WHITTAKER
Crystallography: An Introduction for Earth Science (and other Solid State) Students

Journals

Computers & Geosciences

Geochimica et Cosmochimica Acta

Journal of Structural Geology

Organic Geochemistry

Physics and Chemistry of the Earth

The terms of our inspection copy service apply to all the above books except those marked with an asterisk (*). Full details of all books and journals and a free specimen copy of any Pergamon journal available on request from your nearest Pergamon office.

INORGANIC GEOCHEMISTRY

by

PAUL HENDERSON

Head of Rock and Mineral Chemistry
Department of Mineralogy
British Museum (Natural History), London

PERGAMON PRESS

OXFORD · NEW YORK · TORONTO · SYDNEY · PARIS · FRANKFURT

U.K.	Pergamon Press Ltd., Headington Hill Hall, Oxford OX3 OBW, England
U.S.A.	Pergamon Press Inc., Maxwell House, Fairview Park, Elmsford, New York 10523, U.S.A.
CANADA	Pergamon Press Canada Ltd., Suite 104, 150 Consumers Rd., Willowdale, Ontario M2J 1P9, Canada
AUSTRALIA	Pergamon Press (Aust.) Pty. Ltd., P.O. Box 544, Potts Point, N.S.W. 2011, Australia
FRANCE	Pergamon Press SARL, 24 rue des Ecoles, 75240 Paris, Cedex 05, France
FEDERAL REPUBLIC OF GERMANY	Pergamon Press GmbH, 6242 Kronberg-Taunus, Hammerweg 6, Federal Republic of Germany

Copyright © 1982 Paul Henderson

All Rights Reserved. No part of this publication may be reproduced, stored in a retrieval system or transmitted in any form or by any means: electronic, electrostatic, magnetic tape, mechanical, photocopying, recording or otherwise, without permission in writing from the publishers.

First edition 1982

Library of Congress Cataloging in Publication Data
Henderson, Paul.
Inorganic geochemistry.

(Pergamon international library of science, technology, engineering, and social studies)
Includes bibliographical references.
1. Geochemistry. I. Title. II. Series.
QE515.H43   1982   551.9   81-13959

British Library Cataloguing in Publication Data
Henderson, Paul
Inorganic geochemistry.—(Pergamon international library)
1. Geochemistry
I. Title
551.9   QE515

ISBN 0–08–020448–1 (Hardcover)
ISBN 0–08–020447–3 (Flexicover)

Printed in Great Britain by A. Wheaton & Co. Ltd. Exeter

To

ELIZABETH, GIDEON AND LAURA

Foreword

It is not an easy task to write a text which presents a corpus of existing knowledge in a discipline and yet which also keeps close enough to the advancing front of research that dogmas are seen not to be secure. The task is particularly difficult when the range of factual knowledge is vast, as it is in geochemistry, and when the text is intended, as this one is, as a first introduction to the subject beyond the essentials of undergraduate geology and chemistry.

Although V. M. Goldschmidt, who laid the foundations of the discipline, conceived of geochemistry on as broad a basis as it has developed since his death just after World War II, it might not have been foreseen how many special subdisciplines it has spawned. As a consequence it is difficult to deal adequately with all the developments which have taken place during the last 35 years within the scope of a single text. Dr. Henderson has nevertheless managed to review all of these developments to some degree and to achieve a satisfactory balance overall. Students wishing to pursue any of the sub-disciplines will find an introduction to each, along with indications where more specialized accounts may be found, expressed in clear and concise language and with some attention given to current research interests.

A number of geochemistry texts have appeared in recent years, but few if any present as harmonious a balance of different topics as does this text, for which I am privileged by the kindness of the author and publisher to write this foreword.

DENIS M. SHAW

Contents

7. THERMODYNAMIC CONTROLS OF ELEMENT DISTRIBUTION   155

8. KINETIC CONTROLS OF ELEMENT DISTRIBUTION   184

Introduction

GEOCHEMISTRY has become a subject of such diversity that it is no longer practicable to cover all its aspects within the scope of one relatively small volume. Here the aim has been to provide an understanding of the principal chemical controls on element distribution in the inorganic systems of the Earth and its related bodies.

The book is divided into three parts. The first gives accounts of theories of element formation and of the chemical constitutions and nature of the Earth, Moon and meteorites, in order to define the compositional constraints within which the chemical processes operate. The second part deals with those chemical factors (namely, crystal and liquid structures, thermodynamics and kinetics) which control element distribution in minerals and rocks but with an emphasis on igneous and metamorphic systems, and shows how an understanding of these factors can be used to help solve petrological and geochemical problems. There are also short reviews on the observed distribution of the elements and of some geochemical uses of isotopes. The third part discusses the geochemistry of natural waters, including the topic of rock–water interaction. The book does not deal with phase equilibrium relationships since the subject is so often covered in texts in petrology, nor with the various methods of isotope geochronology.

The book is intended for senior undergraduates in the Earth sciences and for students doing postgraduate courses in geochemistry, or starting geochemical, petrological or mineralogical research. It is hoped that it might be useful also to the more seasoned researcher and to the university lecturer. A knowledge of approximately first-year undergraduate chemistry is presumed.

A substantial part of this book stems from lecture courses I have given at the Universities of Glasgow, London and Oxford. I have been much encouraged during the writing of the book by the interest of former students and many colleagues and by the constructive comments of those who have read and critically reviewed particular chapters. For this I thank Ed Scott, Noel Gale, Bob Hutchison, Eric Whittaker, Chris Scarfe, Brian Burley, Des McConnell, Jack Nolan and Roy Lowry. I am indebted especially to Denis Shaw and Clive Bishop for reading the entire manuscript and suggesting many improvements. I am, of course, responsible for any remaining errors or stylistic deviations. I thank also Peter Henn of Pergamon Press Ltd. for his continual encouragement and patience.

PAUL HENDERSON

Units and Symbols

THE International System of Units (abbreviated, SI) has not yet been (nor is perhaps likely to be) adopted fully by most scientists. Some units outside SI are much used in the geochemical research literature, it being neither convenient nor desirable to discard them. Many of the SI units are used in this book (e.g. joules instead of calories) but the angstrom, atmosphere, bar and degree Celsius have been retained. Appendix I lists the relevant SI units, together with their symbols, and gives a selection of conversion factors and fundamental constants. Some other commonly used symbols are given below.

A	nuclide mass number $(Z + N)$
$\overset{\circ}{A}$	angstrom $(10^{-10}m)$
a	year
a_i	activity of component i
aq	aqueous
atm	atmosphere
C	concentration
$^\circ C$	degree Celsius
CFSE	crystal field stabilization energy
D	diffusion coefficient
D_O	frequency factor
d	day
E	electromotive force
E	activation energy of reaction or viscous flow
E_H	electrode potential relative to hydrogen half-cell
e^+	positron
e or e^-	electron
e^x or exp x	exponential of x
eV	electron volt
F	Faraday constant
f_i	fugacity of component i
G	Gibbs function
g	gas
H	enthalpy

h	hour
J	joule
J	flux of material per unit area
K	equilibrium constant
K	thermodynamic temperature
K_D	partition coefficient
k_h	Henry's law constant
k	partition coefficient
l	liquid
log	logarithm to the base 10
ln	natural logarithm
M	molar mass
m_B	molality of solute B (i.e. amount of B divided by mass of solvent)
N	neutron number
N_A	Avogadro constant
n	neutron
n_e	number of electrons
OSPE	octahedral site preference energy
P	pressure
p	proton
P_B	partial pressure of substance B in a gas mixture
ppm	parts per million
Q	activation energy of diffusion
R	gas constant
REE	rare-earth element
r	radius of atom or ion or crystal nucleus
S	entropy
s	solid
s	second
T	temperature
$t_{\frac{1}{2}}$	half-life
t	tonne
t	time, or age of mineral or rock
V	volume
X_i	ionic fraction of ith ion in a structural site
x_B	mole fraction of substance B
Z	atomic number (proton number)
α	alpha particle
γ	gamma ray
γ_i	activity coefficient of component i
η	viscosity

λ decay constant of a radioactive nuclide
μ_B chemical potential of substance B
ν neutrino
ρ density
[B]   concentration of substance B

PART ONE

THE COMPOSITIONAL FRAMEWORK

1

Meteorites

1.1 Introduction

The availability of meteoritic material has been of considerable consequence in our study of the solar system (Table 1.1) and in the development of our knowledge of the geochemical behaviour of the elements. Meteorites might originate from comets or, more probably, from the asteroid belt located between Mars and Jupiter, and since the asteroids are likely to be the surviving remnants of fragmented planetesimals, meteorites provide a potential means of ascertaining the nature of some planetary interiors, including that of the Earth. Certain meteorites are considered to represent the most primitive material in the solar system because they have undergone little or no chemical fractionation or change since their formation. Therefore, they offer clues to the composition of the solar system and their nature provides evidence concerning nucleosynthesis and of the validity of calculations on the condensation sequence of minerals from the primitive solar nebula.

Chemical analysis of the free metal as well as the silicate and sulphide phases that exist in many meteorites has direct relevance to the quantitative estimation of element distribution amongst these phases within the Earth, while isotope studies of meteorites have led to the determination of the Earth's age. Hence, meteorites have an important part to play in defining the compositional parameters with which geochemistry is concerned.

This chapter commences with the definition and classification of meteorites. The mineralogy and composition of each principal group is then discussed; later sections deal with meteorite cooling rates, isotope studies, and origin.

1.2 Definition and classification

Meteorites may be simply defined as solid extraterrestrial material that survives passage through the Earth's atmosphere and reaches the Earth's surface as a recoverable object (or objects) (Wasson, 1974). They range from microscopic particles to objects of large volumes with masses of many thousands of kilograms, and they also show a wide variety of shapes (Mason,

3

TABLE 1.1. *The solar system*

	Mercury	Venus	Earth	Mars	Jupiter	Saturn	Uranus	Neptune	Pluto†
Mean solar distance × 10⁶ km	57.9	108:2	149.6	227.9	778.3	1427	2869.6	4496.6	5900
Revolution period	88 days	224.7 days	365.26 days	687 days	11.86 a	29.46 a	84.01 a	164.8 a	247.7 a
Rotation period	58.65 days	−243 days*	23 h 56 m 4s	24 h 37 m 23 s	9 h 50 m 30 s	10 h 14 m	−11 h	16 h	6 days 9 h
Equatorial diam., km	4880	12,104	12,756	6787	142,800	120,000	51,800	49,500	2890
Mass (relative to Earth)	0.055	0.815	1 (5.98×10^{27} grams)	0.108	317.9	95.2	14.6	17.2	0.0024
Volume (relative to Earth)	0.06	0.88	1 (1.08×10^{12} km³)	0.15	1316	755	67	57	0.012
Density (g/cm³)	5.4	5.2	5.5	3.9	1.3	0.7	1.2	1.7	1.14
Atmosphere (main components)	None	CO_2, Ar, N_2	N_2, O_2	CO_2, N_2, Ar	H_2, He	H, He	H, He, CH_4	H, He, CH_4	None (?)
Surface gravity (relative to Earth)	0.37	0.88	1 (982 cm/sec)	0.38	2.64	1.15	1.17	1.18	?
Known satellites	0	0	1	2	16	21	5	3	1

* Minus sign indicates retrograde rotation.
Table is based, with some modifications and additions, on data given in *The Solar System. A Scientific American Book.* Freeman, 1975.
† Some data for Pluto are taken from a set of papers on this planet in *Icarus*, vol. 44, 1980.

1962; Wasson, 1974). It is estimated that about 500 meteorites reach the Earth's surface each year with a total mass of between 3×10^6 and 3×10^7 tonnes. However, only five or six meteorites are recovered after observed falls each year.

There is a wide variety in the types of meteorites, ranging from those composed almost entirely of metal to those composed almost entirely of silicates; their classification is a difficult problem and any scheme is subject to criticism on one count or another. Meteorites may be divided conveniently into four principal categories on the basis of their metal content:

Chondrites ⎱ Stones	around 10% metal
Achondrites ⎰	around 1% metal or less
Stony-irons	around 50% metal
Irons	greater than 90% metal

However, while such a broad subdivision of meteorites is useful and often used in writings on the subject, it tends to separate some meteorites for which there is good evidence of a genetic relationship and vice versa (Wasson and Wai, 1970; Wasson, 1974). The classification by Mason (1962), based upon these four categories, is given in Table 1.2 with modifications to names of the chondrites and irons so as to follow current usage by most researchers. A further discussion of the classification chondrites is given below (Section 1.3.1).

Table 1.2 also gives the number of 'falls' and 'finds'. A 'fall' is a meteorite that has been seen to fall and subsequently collected, while a 'find' is one which was not seen to fall but is recognized as a meteorite from its structure, mineralogical and chemical compositions. Falls are important in that they provide a sounder basis for the statistical assessment of the frequency of meteorite infall and of the relative proportions of the different types than do finds; also, they are fresher. Finds are likely to be a biased sample of meteorite infall as irons are more readily recognized as meteorites and may be more resistant to weathering than are many stones. This bias is reflected in the different frequencies of falls and finds of the four main meteorite groups (Table 1.2). However, the various meteorite groups are represented in areas conducive to preservation (e.g. Western Australia, Antarctica) in the same proportions as for falls.

1.3 Mineralogy and composition

1.3.1 *Chondrites*

Numerically these are the most important of all meteorite groups. They are called chondrites because they contain chondrules—spheroidal silicate bodies

TABLE 1.2. *Classification of meteorites*

	Falls	Finds	Fall frequency (%)	Find frequency (%)
STONES	*752*	*576*	*95.4*	*53.4*
CHONDRITES	685*	558*	86.9	51.7
Enstatite chondrites	11	8	1.4	
H (high-Fe) chondrites [olivine–bronzite chondrite]	254	263	32.2	
L (low-Fe) chondrites [olivine–hypersthene chondrite]	291	224	36.9	
LL (low-Fe; low metal) chondrites [amphoterites]	54	19	6.9	
Carbonaceous chondrites	33	8	4.2	
ACHONDRITES	*67**	*18**	8.5	*1.7*
Ca-rich: Aubrites [enstatite achondrites]	8	1	1.0	
Diogenites [hypersthene achondrites]	9	1	1.1	
Ureilites [olivine–pigeonite achondrites]	4	4	0.5	
Chassignite [olivine achondrites]	1	0	0.1	
Ca-poor: Angrites [augite achondrites]	1	0	0.1	
Nakhlites [diopside–olivine achondrites]	1	2	0.1	
Eucrites [pyroxene–plagioclase achondrites]	24	6	3.1	
Howardites [pyroxene–plagioclase achondrites]	17	3	2.2	
STONY-IRONS	*10*	*64*	*1.3*	*5.9*
Pallasites [olivine stony-irons]	4	43	0.5	
Mesosiderites [pyroxene–plagioclase stony-irons]	6	21	0.8	
IRONS	*26†*	*439*	*3.3*	*40.7*
I AB [mostly coarse octahedrites]	6	80	0.8	
II AB [hexahedrites and coarsest octahedrites]	5	46	0.6	
III AB [medium octahedrites]	5	141	0.6	
IV A [fine octahedrites]	3	37	0.4	
IV B [ataxites]	0	11	0	
Other small groups	3	51	0.4	
Anomalous irons	4	73	0.5	

* Totals include some unclassified chondrites or achondrites.
† Wasson (1974) suggests that the true number of falls may be 44.
Table based on Mason (1962), Wasson (1974) and Hutchison *et al.* (1977).

up to a few millimetres in diameter—which are set in a fine-grained matrix, also of silicate material but with some metal and sulphide. Some (type C1—see below) carbonaceous chondrites do not contain chondrules but on mineralogical and chemical grounds they are classified as chondrites. Chondrules may be of various types, often consisting of olivine or pyroxene or these minerals together, but some are glassy. Descriptions of chondrules can be found in Van Schmus (1969). The prevailing view appears to be that many chondrules are rapidly crystallized or glassy droplets, produced by melting of pre-existing silicate material, and subsequently subjected to varying degrees of thermal metamorphism. However, one important problem relating to their origin is that these chondrules occasionally occur together with 'lithic' chondrules, i.e. ones with textures similar to those of igneous rocks.

Olivine, pyroxene, plagioclase feldspar and kamacite ($Fe_{0.93-0.96}Ni_{0.07-0.04}$; element proportions given in weight fractions) are common minerals in most chondrites. Taenite ($Fe_{0.8}Ni_{0.2}$), apatite, daubreelite ($FeCr_2S_4$), schreibersite ($(Fe, Ni)_3P$), chromite and native copper are some of the minor minerals (more detailed listings may be found in Keil, 1969). Analyses of the chondrites show (Table 1.3) that there is little or no compositional overlap between the different groups. The distinct chemical nature of each group as revealed, for example, in a plot of weight percentage of iron in the silicate phases versus the combined weight percentage of iron present in the metal and sulphide phases (Fig. 1.1) provides a means of assigning individual meteorites to their appropriate group.

However, the subdivision of the chondrites on chemical criteria alone fails to take account of important differences in mineralogy and texture within each group. To help overcome this failing Van Schmus and Wood (1967) drew up a new classification based on the petrological nature as well as the chemical nature of the chondrites. They argued that within some of the groups there is evidence of variation in the degree of chemical equilibration. For example, an absence of primary glass and the presence of distinct feldspar grains are characteristics of the more recrystallized chondrites which are also the most equilibrated. Six petrological types were designated, each with particular characteristics some of which are summarized here:

Type 1. No chondrules. High volatile content.

Type 2. Olivine and pyroxene crystals of variable composition within any one meteorite. A large matrix to chondrule ratio. The presence of igneous glass. The sulphide phase in chemical disequilibrium with the metal phase.

Type 3. Olivine, pyroxene and igneous glass as in type 2, but sulphide phase in equilibrium with metal phase. Well-defined chondrules. Significant carbon content.

Type 4. A transitional type but still with well-defined chondrules. Only

TABLE 1.3. *Compositions of some chondrite meteorites and average compositions of some achondrites*

	Carbonaceous chondrite, type I (Orgueil)	Carbonaceous chondrite, type II (Murray)	Enstatite chondrite (Hvittis)	H-group chondrite (Quenggouk)	L-group chondrite (Linum)	LL-group chondrite (Manbhoom)	Average aubrite	Average diogenite	Average ureilite	Average eucrite	Average howardite
Fe	0	0	20.04	17.89	6.89	0.70	2.29	0.79	3.94	0.80	0.33
Ni	0	0	1.96	1.75	1.24	0.04	0.17	0.03	0.11	–	0.11
Co	0	0	0.07	0.08		0.81	–	–	0.05	–	0.05
FeS	15.07	7.67	7.27	4.96	7.31	5.18	1.25	1.12	1.66	0.41	0.73
SiO_2	22.56	28.69	41.53	36.40	39.41	40.52	54.01	52.11	40.83	48.17	49.75
TiO_2	0.07	0.09	–	0.13	0.14	0.09	0.06	0.19	0.15	0.51	0.11
Al_2O_3	1.65	2.19	1.55	2.70	3.73	3.00	0.67	1.18	0.54	13.91	8.71
MnO	0.19	0.21	–	0.31	0.31	0.32	0.14	0.32	0.40	0.46	0.78
FeO	11.39	21.08	0.34	8.93	13.95	19.45	0.97	16.05	12.16	15.99	13.26
MgO	15.81	19.77	23.23	23.39	23.31	25.65	35.92	25.85	37.43	7.10	16.10
CaO	1.22	1.92	0.74	1.66	1.68	1.51	0.91	1.41	0.87	10.94	6.53
Na_2O	0.74	0.22	1.26	0.96	0.70	1.07	1.32	0.004	0.11	0.67	0.95
K_2O	0.07	0.04	0.32	0.08	–	0.13	0.10	0.001	0.04	0.13	0.28
P_2O_5	0.28	0.32	0.18	0.08	0.12	0.35	0.22	0.01	0.08	0.11	0.07
H_2O	19.89	12.42	–	0.30	0.07	0.44	1.14	0.14	–	0.44	0.25
Cr_2O_3	0.36	0.44	0.56	0.52	0.27	0.64	0.06	0.80	0.85	0.39	0.42
NiO	1.23	1.50	–	–	0	–	0.26	–	–	–	–
CoO	0.06	0.08	–	–	0	0.26	–	–	–	–	–
C	3.10	2.78	0.86	0.1	–	–	–	–	2.23	–	–
Others*	6.96	0.62	–	–	–	–	0.51	–	–	–	–
Sum	100.65	100.04	99.91	100.28	99.13	100.16	100.00	100.00(5)	101.45	100.03	101.05†

* In the type I and type II carbonaceous chondrites, the figures refer to weight loss on ignition, which is equivalent to the content of complex organic matter. In the other meteorites, the figures are for amount of CaS quoted in the analysis.

† Total includes an additional 2.62% Fe_2O_3.

Table is based on data given in Wiik (1956); Mason and Wiik (1964); Keil (1969) and Hutchison *et al.* (1981).

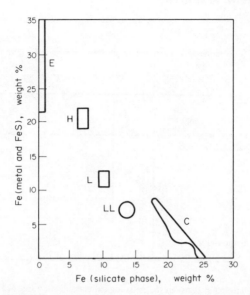

FIG. 1.1. Classification of chondrites on basis of iron content in metal and sulphide phases, and of oxidized iron in silicates. Areas depict approximate, composition fields for the different groups: E = enstatite chondrites; H = high-Fe, low-metal chondrites; C = carbonaceous chondrites. (After Keil, 1969)

slight variation in composition of olivine and pyroxene. Low carbon content. Low-calcium clinopyroxene about 20% of total pyroxene.

Type 5. Olivine and pyroxene of uniform composition. Pyroxene is nearly all orthopyroxene. Chondrules discernible but diffuse.

Type 6. Recrystallized texture. Good development of interstitial plagioclase crystals (about $An_{11}Or_6Ab_{83}$). Chondrules very diffuse. No glass. Very little carbon content.

There is probably a genetic relationship between some of the types, e.g. types 3 to 6 probably represent a metamorphic sequence with type 6 being the most metamorphosed (Wasson, 1974). However, a genetic relationship between types 1 and 2 is unlikely.

Figure 1.2 relates the Van Schmus and Wood classification to that given in Table 1.2. The areas with dashed lines have no meteorites with the appropriate characteristics. There is no synonym for the C4 chondrites shown in Fig. 1.2, and no carbonaceous chondrites of types 5 and 6 are known. In some later work, Van Schmus and Hayes (1974) noted that on the basis of abundances of some non-volatile elements (Si, Ca, Al, Ti, Cu and Fe) the four distinct petrographic types of carbonaceous chondrites could be grouped into two chemical divisions—a fact that is sometimes used in classification.

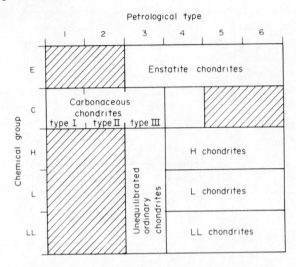

FIG. 1.2. Relationship between the petrological classification (according to Van Schmus and Wood, 1967) and chemical groups of the chondrites. There are no recorded meteorites with the characteristics defined by the shaded areas. E = enstatite chondrites; H = high-Fe chondrites; L = low-Fe chondrites; LL = low-Fe, low-metal chondrites; C = carbonaceous chondrites.

The carbonaceous chondrites, Cl, etc., are texturally distinct from all the other groups. They are particularly interesting not only because of the absence of chondrules in Cl but also because of the presence of volatile constituents which indicate that they are unfractionated and primitive. The Cl chondrites contain a large amount of chlorite and some have water-soluble salts such as magnesium sulphate and a magnetic spinel (DuFresne and Anders, 1962). Minor minerals include elemental sulphur, dolomite, olivine and graphite as well as some unpyrolysed organic compounds (Nagy, 1975). The indications are that these meteorites must have existed always at temperatures below about 300°C since their formation.

One particularly important carbonaceous chondrite fell in northern Mexico in 1969. This is the large meteorite called 'Allende', a C3 chondrite. The matrix ($\sim 60\%$) is dark grey and consists predominantly of iron-rich olivine; there are two types of chondrules, one Mg-rich ($\sim 30\%$) and the other Ca,Al-rich ($\sim 5\%$) and there are Ca,Al-rich aggregates ($\sim 5\%$) (Mason, 1975). The Ca,Al-rich chondrules are unusual in their large size (up to 25 mm diameter), mineralogy and composition. They consist of a Ti–Al-rich pyroxene, melilite, spinel, with minor plagiochase, and perovskite. Although these chondrules are deficient in alkalis they contain greater amounts, by about 14 times, of the rare-earth elements compared with the average for chondrites. A number of authors (e.g. Grossman, 1972; Gray *et al.*, 1973) consider that the chondrules represent the earliest known condensate material from the solar nebula.

In Allende, as with other carbonaceous chondrites (especially Cl), element concentrations are almost identical to those of condensable elements observed in the sun (Fig. 1.3). The mineralogical and chemical nature of the carbonaceous chondrites has led to their composition being used in estimating relative abundances of the condensable elements in the solar system (Chapter 2).

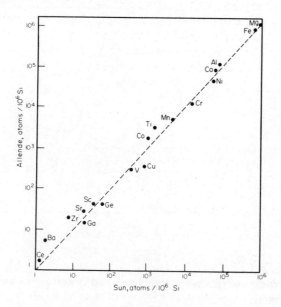

FIG. 1.3. Relationship between concentrations of non-volatile elements in the carbonaceous chondrite, Allende, and in the sun, normalized to 10^6 atoms of silicon. (After Mason, 1975, with permission.)

1.3.2 Achondrites

These have lower contents of metallic nickel–iron and coarser textures than most chondrites. Most achondrites have a brecciated structure but a few contain good igneous textures as is shown, for example, by Moore County which is an eucrite with layering characteristic of an igneous cumulate. Some eucrites have a vesicular structure that is similar to structures seen in terrestrial lavas and could indicate an extrusive origin. Many achondrites resemble terrestrial igneous rocks, and thereby are more difficult to recognise as meteorites.

The mineralogy is varied, but pyroxenes and plagioclase feldspar are important constituents, and some types are olivine-rich. Minor minerals include kamacite, taenite, troilite, schreibersite, chromite, magnetite and cristobalite. Average chemical compositions of some achondrites are given in Table 1.3 and it can be observed that, as in the chondrites, the chemical nature including the degree of oxidation of each group is quite distinct (Fig. 1.4).

The mineralogy and textures of many achondrites suggest a magmatic origin, with subjection to shock at a later stage to give a brecciated structure.

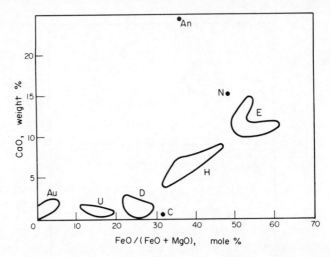

FIG. 1.4. Classification of achondrites on basis of CaO content (wt %) and FeO/(FeO + MgO) mole ratio. Areas depict approximate, composition fields for the different groups: Au = aubrites; U = ureilites; D = diogenites; H = howardites; E = eucrites. One analysis point only for each of following: C = chassignite; N = nakhlite; An = angrite. (After Keil, 1969)

1.3.3 Irons

Currently, the classification of these meteorites (Table 1.2) is based on variations in the contents of nickel, gallium and germanium. Lovering, followed by Wasson and his co-workers (e.g. Lovering *et al.*, 1957; Scott and Wasson, 1975), showed that the meteorite compositions fall into discrete fields on diagrams of the type shown in Fig. 1.5. However, this chemical classification shows some correspondence with the earlier structural classification based on the type and structure of the iron as follows:

Hexahedrites: entirely kamacite; often as large, single cubic crystals.
Octahedrites: show Widmanstätten structure which is revealed on etched surfaces. This structure results from development, on cooling of the

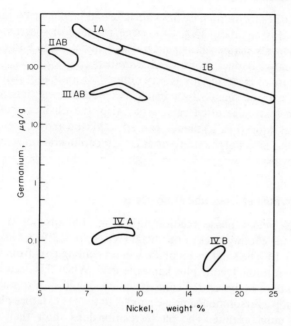

Fig. 1.5. Logarithmic plot of Ga content versus Ni content in various iron meteorite groups. (After Scott and Wasson, 1975)

Fe–Ni phase, of four sets of kamacite plates oriented along the planes of a regular octahedron (Goldstein and Ogilvie, 1965). (The structure may be used to determine the cooling rate of a meteorite, see Section 1.4.) Further classificatory subdivisions of the octahedrites may be made according to the width of the kamacite plates (see Wasson, 1974, p. 35).

Ataxites: fine-grained intergrowth of kamacite and taenite. They are Ni-rich.

The approximate relationship between the chemical and structural classifications is given in Table 1.2.

Some iron meteorites contain nodules up to several centimetres in diameter which consist of troilite, graphite, pyroxene, olivine and sphalerite. Some of the hexahedrites show evidence of shock in that they contain fine mechanical twinning lamellae which have formed along the {211} planes of the kamacite.

1.3.4 Stony-irons

This group is relatively small and most of the specimens are finds. They show a coarse texture with about equal proportion (by weight) of silicate and metal phases.

In the pallasites the iron phase has kamacite and taenite in Widmanstätten intergrowth. Olivine (about Fa_{10-20}) as large, rounded or sometimes angular crystals is the only major silicate mineral and there are, in addition, minor amounts of native copper, troilite and schreibersite.

The structure of mesosiderites is very different from that of pallasites. The metal phase may be finely disseminated or sometimes is present as large nuggets. The texture is often cataclastic. Also, the olivines show a wider compositional range (Fa_{9-48}) than in pallasites (Mason and Jarosewich, 1973). One interesting aspect of the stony-irons is their cooling rate in comparison to that of the irons.

1.4 Cooling rates of irons and stony-irons

The Fe–Ni metal phase relationships (Fig. 1.6) provide a means of determining the cooling rates of the metal (Wood, 1964, 1968; Goldstein and Short, 1967 a, b). The solid metal phase will on cooling pass from the taenite stability field into the taenite plus kamacite field. When this occurs kamacite crystals with their body-centred cubic structure should start to nucleate. The crystals develop as exsolution lamellae parallel to the {111} lattice planes of the face-centred cubic taenite. The phase relationships show that as cooling continues the amount of kamacite must increase at the expense of the taenite. This requires the diffusion of nickel within the metal phase away from the growing kamacite into the remaining taenite.

The temperature dependence of this diffusion and the faster diffusion rate of nickel in kamacite compared to that in taenite leads to disequilibrium nickel distributions with a characteristic M-shape, shown in an analysis profile in Fig. 1.7. In practice the taenite crystals at low temperatures are usually so far from their equilibrium concentration of nickel that the interiors of the crystal transform into a very fine-grained mixture of taenite and kamacite, called plessite.

Besides the constraints of the phase relationships, the shape of the 'M-profile' depends upon:

(a) Diffusion coefficients of nickel in taenite and kamacite as a function of temperature.
(b) The initial bulk nickel content of the metal.
(c) Nucleation temperature of the kamacite.
(d) Cooling rate.

Given a particular M-profile, the cooling rate may be determined if the diffusion coefficients ((a) above) are known and the nucleation temperature established. Application of this method to the determination of cooling rates of irons has yielded a wide range of values. Some group I irons have cooled at rates between 1 to $10^4 \, °C/Ma$, and some group IIA irons at about $2°C/Ma$.

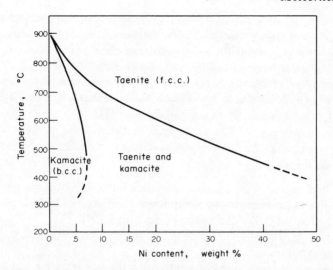

FIG. 1.6. Ni–Fe phase diagram, at approximately 10^5Pa (b.c.c. = body-centred cubic; f.c.c. = face-centred cubic).

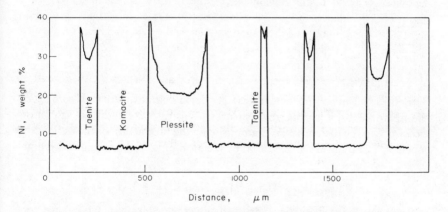

FIG. 1.7. An example of the type of Ni distribution in Widmanstätten structure, as determined by electron probe microanalysis. (Actual Ni content profiles in various iron meteorites, together with microscopic views of Widmanstätten structure can be seen in Wood, 1964).

Groups IA and IIIB have slow cooling rates of about 1 to 10°C/Ma (Scott and Wasson, 1975; Smith, 1979). However, some pallasites also cool slowly, at rates of about 1°C/Ma. Buseck and Goldstein (1969) suggest that these meteorites came from deep within the parent body and that most irons formed from different parent bodies or came from isolated metal pools within the silicate mantle of the pallasite parent. Wood (1967) and Taylor and Heymann (1971) applied the method to some chondritic metal and obtained cooling rates for fourteen ordinary chondrites of between 1 and $10°C/Ma$.

Meteorites that have been subjected to shock metamorphism, which caused severe reheating, often record very rapid cooling rates and, exceptionally in the the case of two chondrites, as fast as 1 and $100°C/day$ (as determined by related methods of estimation, Smith and Goldstein, 1977).

1.5 Isotope studies

A study of the radioactive isotopes and their stable daughter isotopes can, in theory, give us information on the following five events (Anders, 1963):

(a) The date of fall of the meteorite on to the Earth.
(b) The date of breakup of the meteorite parent body; the so-called 'cosmic-ray exposure age'.
(c) The age of last cooling of a meteorite: the 'gas-retention age'.
(d) The age of formation.
(e) Formation of the elements of the solar system; the 'age of nucleo-synthesis'.

These will be discussed in turn.

1.5.1 Date of meteorite fall

Before the meteorite penetrates the Earth's atmosphere it is subject to bombardment by cosmic-ray particles. This process induces 'spallation' reactions in the meteorite, an example being the production of the radioactive nuclide ^{36}Cl (half-life $= 3 \times 10^5$ a) by reaction of the ^{56}Fe nuclide with high-energy protons:

$$^{56}Fe + {}^1H \rightarrow {}^{36}Cl + {}^3H + {}^3He + 2{}^4He + 3{}^1H + 4n.$$

Once the meteorite has fallen, it is effectively shielded from such bombardment by the Earth's atmosphere. Hence, the measurement of the amounts and proportions of spallation induced ('cosmogenic') nuclides and their daughter products allows the determination of the date of fall.

1.5.2 Cosmic-ray exposure age

The mean absorption depth of primary cosmic rays in meteorite matter is of the order of 1 m. Therefore, the inner parts of a parent body will be shielded from spallation reactions until the body breaks up. As in Section 1.5.1 above, it is possible to determine the age of break-up from a study of stable and unstable cosmogenic nuclides in meteorites but this requires the assumption that the cosmic flux was constant with time. Meteorites that came from the surface of the parent body will give anomalous break-up dates. There are other uncertainties in the interpretation of the data but the results indicate very distinct differences in the cosmic-ray exposure ages of irons and stones. The former range from less than 200 to 1500 Ma, while the latter are mostly between 0 and 20 Ma. The differences, and the fact that some meteorite groups show a clustering in their exposure ages (e.g. H chondrite ages cluster around 4 Ma), reflect different major phases of disruption in the source region of meteorites. Because the production of cosmogenic nuclides falls off with depth in a body, it is also possible to determine the approximate pre-atmospheric size of a meteorite from the distribution pattern of cosmogenic nuclides (such as ^{3}He) within it.

1.5.3 Gas-retention age

At temperatures higher than 200 to 300°C any gases that may have been produced within a meteorite by the decay of a radioactive nuclide (e.g. ^{40}Ar from ^{40}K) may diffuse through the meteorite and be lost. The measurement of the concentration of such daughter and parent isotopes will give the date at which loss of gaseous nuclide ceased. If a meteorite has not suffered a heating event after it first solidified then its 'gas-retention age' will be the same as its 'formation age'.

Gas-retention ages determined from the proportions of ^{40}K and ^{40}Ar for a large number of meteorites range from 0.3 Ga to 4.6 Ga. The different ages result from thermal 'metamorphic' events which may have originated from impact heating at different times.

More recently the application of a new dating technique to the study of retention ages has revealed important evidence which helps our understanding of the formation and subsequent history of the solar system. The technique involves the irradiation of a piece of the meteorite together with a rock standard of known age, with a neutron flux. This induces nuclear reactions so that some ^{39}K in the meteorite will be converted to ^{39}Ar through the reaction:

$$^{39}K + n \rightarrow {}^{39}Ar + p$$

After irradiation the ^{39}Ar/^{40}Ar ratio may be determined, and as this is directly related to the ^{40}K/^{40}Ar for the meteorite, an age may be calculated. However, more information may be obtained if the meteorite material, after irradiation, is subjected to controlled heating steps whereby the ^{39}Ar/^{40}Ar ratio of the thermally released argon may be determined for different temperatures. In effect this allows the investigation of variations in the ratios of the daughter ^{40}Ar to parent ^{40}K within a particular meteorite. (An example of a fractional release pattern is shown in Fig. 3.2.)

A number of meteorites show no variation in the argon ratio by this method and they have the oldest ages of about 4.5 Ga. It is considered that these bodies have had a relatively simple and undisturbed history since their formation and cooling soon after the formation of the solar system. Others show little variation but have different ages that are all less than 4.5 Ga, indicative of a shock metamorphic event at the recorded 'age'. Still others show complicated variations in the argon ratio with heating and so have a variation in 'age'. Such a situation may result from incomplete outgassing of the argon during a thermal event in the history of the meteorite or for other reasons which are not yet understood.

So far this technique has been applied mainly to chondrites. However, a nakhlite gave a low gas-retention age of 1.3 Ga and showed a simple argon release pattern.

1.5.4 The age of formation

The measurement of the proportions of daughter and parent isotope also allows the age of formation of the meteorite to be calculated, provided that none of the daughter isotope has been lost (which might occur if the daughter isotope is a gas, see above). The principles of this method are discussed briefly in Chapter 9 of this book and so only the results for meteorites will be discussed here. The 'age' that is determined is the time at which the process of chemical fractionation of the relevant isotopes into different phases of the cooling meteorite ceased.

Isotopes of Rb and Sr have been the most widely used in this method although much has been done using U and Pb isotopes. Isotopes of Sm and Nd and more recently, of Lu and Hf are proving useful (e.g. Patchett and Tatsumoto, 1980).

A possible dating method for iron meteorites using the radioactive isotope of naturally occurring rhenium—^{187}Re—which decays to ^{187}Os with a half-life of about 43 Ga has not been successfully applied because of uncertainties in the value of the half-life and the similarity in the geochemical behaviour of rhenium and osmium with resultant clustering of points on an isochron.

Most of the data for all meteorites give formation ages close to 4.55 Ga. Two results—one for an iron meteorite and another for an achondrite—yield ages of about 3.8 Ga and the nakhlite, Nakhla, yields a precise age of 1.24 Ga (Gale et al., 1975) which is taken to indicate that significant element fractionation occurred at these times. The 4.55 Ga age may represent the time of initial solidification of all meteorites.

1.5.5 *Age of nucleosynthesis*

The formation age gives the time of solidification of a meteorite. So what is the age of the material that gave rise to meteorites? Since nucleosynthetic processes producing the heavy elements in any one nebula probably take place over a considerable period of time, it is impossible to give a precise answer to the question. However, an estimate of the time interval between the ending of nucleosynthesis and the consolidation of meteorites (the 'lapse time') may be made through the detection in meteorite material of daughter isotopes of short-lived radionuclides, which are now extinct. Radionuclides with half-lives that are too short for their preservation to the present day but are long enough for some of the nuclide to have survived the time interval between nucleosynthesis and meteorite condensation have proved to be the most useful. ^{129}Xe, the daughter product of extinct ^{129}I (half-life $\simeq$ 16 Ma), is used in this way and it has been shown that the lapse time is of the order of 100 Ma. (The C3 meteorite, Allende, has a large excess of ^{129}Xe.)

1.6 Origin

The presence of three distinct meteorite types—stones, stony-irons and irons—inevitably led people to believe that meteorites originated from some planetary body which had fractionated into a metal-rich core within a silicate envelope. The disruption of such a body would lead to the creation of the different meteorites, with the stony-irons coming from the metal core–silicate mantle interface and the various stones coming from different regions of the silicate-rich mantle. There is, however, a lot of evidence which does not fit the 'one parent body' hypothesis. One of the most important facts is the difference in age distribution shown not only by the various meteorite types but also between the meteorite groups. Furthermore, the compositional hiatuses within the chondrites and the irons suggest that each group formed in separate parent bodies. It is also easier to reconcile the evidence of cooling rates with this suggestion. On the basis of the contents of Ga, Ge and Ni alone, most iron meteorites can be assigned to one of twelve different genetic groups each of which probably derived from a different parent body (Scott and Wasson, 1975).

Differences in the isotopic ratios of oxygen in all classes of meteorites also point to origins involving many parent bodies. Six categories can be defined, each with distinct ^{18}O/^{16}O and ^{17}O/^{16}O ratios (Clayton *et al.*, 1976):

1. A 'terrestrial' group consisting of terrestrial and lunar rocks; enstatite chondrites; achondrites (other than ureilites) and stony-irons.
2. Type L and LL chondrites.
3. Type H chondrites.

4. Anhydrous minerals of C2, C3 and C4 chondrites.
5. Hydrous matrix minerals of C2 chondrites.
6. Ureilites.

The objects of one category cannot be derived by fractionation or differentiation from source materials of any other category. However, the compositions can be accounted for by suitable combinations of mixing of isotopically distinct primordial compositions, together with isotopic fractionation. (A discussion of stable isotopes is given in Chapter 9.)

If we require many parent bodies for meteorites then the asteroid belt between Mars and Jupiter provides a possible source. Here there are numerous planetesimals with diameters ranging up to about 1020 km (Ceres) and many have orbits which cut the orbit of Mars and some that of the Earth. Hence, the capture of some of these bodies by the Earth's gravitational field is a reasonable possibility. Recorded data, although rather limited, on meteorite orbits are consistent with this view. The cooling-rate data and work on pressures of formation of some iron meteorites (Schwarcz *et al.*, 1975) require the radii of the parent bodies of iron meteorites to be greater than 70 km and probably of about 180 km. Some of the objects in the asteroid belt have such radii but there is also the likelihood that slightly larger bodies have been disrupted by mutual collisions.

The early nature and formation of the asteroid bodies is more enigmatic. Many of the achondrites appear to have been derived through magmatic processes similar to those that produced terrestrial igneous rocks, whereas the nature of chondrites—especially the presence of chondrules—cannot be explained readily in this way. The direct condensation of solid material from a low density, hot, partially-ionized gas under conditions of marked thermal disequilibrium is considered by some (e.g. Arrhenius and Alfvén,1971; Blander and Abdel-Gawad, 1969) to be a likely mechanism that could account for the existence of chondrules as well as other textural, mineralogical and compositional characteristics. Carbonaceous chondrites (C1) with their high content of volatiles and their absence of chemical equilibrium could then be material that directly condensed from the solar nebula and which has not been subjected to later re-equilibration. Other meteorites may represent varying degrees of chemical and physical reworking, including some melting and fractional crystallization of this initial condensate. Many researchers prefer the hypothesis that the origin of chondrules involved the rapid crystallization of supercooled silicate liquid (see Nelson *et al.*, 1972). However, it is possible that some subcooled liquids condensed directly from the solar nebula, or alternatively the liquids formed after vaporization of meteoritic material on collision (Larimer, 1967).

If it is true that at least some meteorites are a simple accretion of nebular material, without much subsequent reworking and re-equilibration, they offer important clues about the sequence of nebular condensation and may be used to test the validity of theoretical models of condensation. From use of the ideal gas law, and of thermodynamic data for gaseous species, together with mass balance and other considerations, it is possible to calculate the equilibrium distribution of the major elements between vapour and solid for a cooling gas of cosmic composition (at an appropriate total pressure, e.g. 10 Pa) (Lord, 1965; Larimer, 1967; Grossman, 1972; Grossman and Larimer, 1974). A typical calculated condensation sequence is:

1. Condensation of solid Os, Re, Zr at $T \gg 1680$ K, followed by Al_2O_3 and then $CaTiO_3$ and $Ca_2Al_2SiO_7$ down to about 1500 K.
2. $CaMgSi_2O_6$ appears at 1387 K, and with further cooling is accompanied by Fe–Ni–Co alloys and then, at 1370 K, Mg_2SiO_4, which later reacts with the vapour to form $MgSiO_3$, during which all remaining gaseous Si is consumed.
3. Below 1250 K, Cu, Ge and Ga condense and form solid-solutions with pre-existing, condensed metal; Na, K and Rb condense and enter into solid-solution with $CaAl_2Si_2O_8$. All alkali metals condense by 1000 K.
4. Below 750 K, metallic Fe begins to oxidize; troilite becomes stable; Pb, Bi, In and Tl condense between 600 and 400 K, while at 405 K, magnetite forms.

The Ca,Al-rich inclusions in the Allende chondrite form some of the evidence that testifies to the relevance of the condensation sequence. The fact that iron–nickel alloys have higher condensation temperatures (at the probable prevailing pressure) than magnesium silicates such as forsterite and enstatite, also supports the origin of the core and mantle of the Earth by a heterogeneous, accumulation process. However, as mentioned above, there is the likelihood of the condensation being a disequilibrium process; this would provide an explanation of some of the characteristics of meteorites, e.g. chondrules, but the general pattern of the condensation sequence would remain the same, at least at the high-temperature stages.

Prior condensation of silicate, metal and sulphide phases, before their aggregation into planetisimals, provides the materials which can be mixed to form some meteorite types, for example, the chondrites with their chondrules and metallic grains (volatile-free and formed at high temperatures) set in a volatile-rich matrix (formed at low temperature) (Larimer and Anders, 1967). The aggregation process is likely to be accompanied by heating which would give rise to prograde thermal metamorphism; thereby accounting for the observation that the petrologic types 3 to 6 of chondrites suggest a metamorphic sequence involving starting material of the same composition.

Much later impact processes, caused by the collisions of planetesimals, would produce other textures, such as veining by glass, and minerals characteristic of shock metamorphism, which are present in some meteorites.

The Ar–Ar method of dating has shown evidence for the existence of shock-reheating events in the history of some meteorites after their formation as solid bodies. Shock metamorphism also provides a mechanism for the differential loss of volatile elements.

The various hypotheses of meteorite origin are diverse but to show one viewpoint we can take that by Wasson (1974) on the origin of the ordinary (H, L and LL) chondrites, as was summarized in his book on meteorites. The stages of formation may be listed from the stages when condensation of solids from the solar nebula started:

(a) Condensation of Fe–Ni metal.

(b) Chondrule formation during condensation.

(c) Temperature at lowest point. Condensation ceased. Agglomeration of material, formed during stages a, b, and part of c, into small planetesimals.

(d) The planetesimals heated, resulting in metamorphism and loss of some volatiles.

(e) Heating declined. Accretion and mixing of planetesimal material to give larger bodies—the meteorite parents.

(f) Much later break up of the parent bodies.

(g) Debris was perturbed into Earth-crossing orbits and eventually captured by the Earth as meteorites.

It should be noted that some workers believe that events of stage (e) occurred at the same time as the agglomeration in stage (c).

1.7 Tektites

These are small glassy objects which have the appearance of the rock obsidian but with an unusual composition. Although they have never been seen to fall they have sometimes been considered to be meteorites on account of their morphology and composition. Taylor (1973) gives an account of these objects and shows that the present evidence is in favour of them having a terrestrial origin through meteorite or cometary impact.

Some suggestions for further reading

BUCHWALD, V. F. (1975) *Handbook of Iron Meteorites*. University of California Press. 3 volumes.

MASON, B. (1962) *Meteorites*. Wiley. 274 pp.

MASON, B. (1979) *Cosmochemistry*. Part 1, *Meteorites*. Chapter B. Data of Geochemistry (ed. M. FLEISHER), 6th edition. U.S.G.S. Professional Paper. 440-B-1. 132 pp.

NAGY, B. (1975) *Carbonaceous Meteorites*. Elsevier Scientific Publ. Co. 747 pp.

SEARS, D. W. (1978) *The Nature and Origin of Meteorites*. Adam Hilger Ltd. 187 pp.

SMITH, J. V. (1979) Mineralogy of the planets: a voyage in space and time. *Mineralogical Magazine*, **43**, 1–89.

WASSON, J. T. (1974) *Meteorites. Classification and Properties*. Springer-Verlag. 316 pp.

2

'Cosmic' Abundances and Nucleosynthesis

2.1 Introduction

The mineralogy, texture and chemical composition of some of the carbonaceous chondrites, as discussed in the previous chapter, all point to their primitive nature and their possible representation of unfractionated, solar-nebular condensates. Chemical analysis of these meteorites and their constituents provides important indicators of the sequence of cooling and condensation as well as of the composition of the solar nebula. Knowledge of this composition is significant not only to studies of the production of elements in stars, and of other astrophysical problems (such as establishing planetary compositions) but also to terrestrial aspects by providing a suitable reference composition against which the direction and extent of element fractionation processes can be measured.

Perhaps a more fundamental reference frame is to be found in the abundances of the elements in the Cosmos. Unfortunately, this composition is unobtainable because the Cosmos itself has undergone fractionation into various components such as stars, interstellar medium and planetary bodies, and to integrate the compositions of all these is impossible. We will also see that during the evolution of a star the relative abundances of the nuclides, which make up the stellar mass, change as a result of the production of heavier elements at the expense of the lighter ones. We will, therefore, have to be content with abundance values that only approximate to that of the overall composition of the Cosmos and which are probably more representative of our own solar system.

This more limited objective is still not easy to attain. The sources for our measurements are not homogeneous and in some cases the quality of information is reduced by serious restriction in availability and by problems of observation and interpretation. There are five possible sources:

(a) The Earth, Moon.
(b) Meteorites.

23

(c) The Sun and stars.

(d) Gaseous nebulae and the interstellar medium.

(e) Cosmic rays.

The surfaces of the Earth and Moon may be sampled and analysed but in neither case is the surface chemically uniform and so attempts to find average compositions of the crusts may contain significant degrees of error. The crust of the Earth comprises less than 1 % of its mass and the overall compositions of the mantle and core can only be obtained from indirect evidence. Similarly, the composition of the bulk of the Moon can be estimated only by indirect means. It is clear that the compositions of these two bodies are still speculative. The problems associated with the first of the five sources serve to illustrate the difficulties inherent in any attempt at a compilation of the abundances of the elements. Some of these attempts will now be discussed.

2.2 Abundance of the elements

The interstellar medium from which new stars are formed offers the best *potential* for an assessment of cosmic abundances because it is a relatively unfractionated part of the universe, but unfortunately it is virtually impossible to analyse quantitatively the bulk of this material. Abundance data can be obtained from those portions of the medium which, as a result of radiation from an intensely hot star, fluoresce and emit a bright line spectrum (e.g. Orion nebula). The results are fairly reliable for the more abundant elements and would be more reliable if it were not for certain physical factors, such as temperature stratification in the nebula, which produce uncertainties in the determinations. To overcome this difficulty Aller (1968) proposed that abundance data obtained from stars recently formed from the interstellar medium should be representative of the medium itself, and he produced a compilation based upon these stars, the Sun, and the more reliable data from the diffuse nebulae. Later, Ross and Aller (1976) determined the chemical composition of the Sun by spectral analysis of the solar atmosphere. Their results are given in Table 2.1 in which the abundances are normalized to $Si = 10^6$ 'atoms' or ions. (Note that in the original paper the abundances are normalized to $H = 10^{12}$ atoms; the conversion to $Si = 10^6$ atoms was made to allow direct comparison with Table 2.2, see below.) Some elements are absent from Table 2.1 because of serious analytical difficulties. Except for Li, Be and B, the non-volatile component of the solar atmosphere fits well with element abundances in carbonaceous chondrites.

Astrophysical determinations, together with abundance data from meteorites and the Earth's crust, generally form the basis of compilations of the relative abundances of all the elements in the solar system. Such compilations are sometimes referred to as 'cosmic abundances' but the authors

TABLE 2.1. *Solar elemental abundances*
Normalized to Si = 10^6 atoms

1	H	2.24×10^{10}	26	Fe	7.08×10^5	59	Pr	0.102
2	He	1.41×10^9	27	Co	1.78×10^3	60	Nd	0.380
3	Li	0.22	28	Ni	4.27×10^4	62	Sm	0.12
4	Be	0.032	29	Cu	2.57×10^2	63	Eu	0.1
5	B	< 2.8	30	Zn	6.31×10^2	64	Gd	0.295
6	C	9.33×10^6	31	Ga	14	66	Dy	0.257
7	N	1.95×10^6	32	Ge	70.8	68	Er	0.13
8	O	1.55×10^7	37	Rb	8.91	69	Tm	0.041
9	F	8.12×10^2	38	Sr	17.8	70	Yb	0.2
10	Ne	8.32×10^5	39	Y	2.82	71	Lu	0.13
11	Na	4.27×10^4	40	Zr	12.6	72	Hf	0.14
12	Mg	8.91×10^5	41	Nb	1.8	74	W	1.1
13	Al	7.41×10^4	42	Mo	3.24	75	Re	<0.01
14	Si	1.0×10^6	44	Ru	1.51	76	Os	0.11
15	P	7.08×10^3	45	Rh	0.562	77	Ir	0.16
16	S	3.6×10^5	46	Pd	0.71	78	Pt	1.26
17	Cl	7.1×10^3	47	Ag	0.16	79	Au	<0.13
18	Ar	2.2×10^4	48	Cd	1.59	80	Hg	<2.8
19	K	3.24×10^3	49	In	1.00	81	Tl	0.18
20	Ca	5.01×10^4	50	Sn	2.2	82	Pb	1.91
21	Sc	24.5	51	Sb	0.22	83	Bi	<1.8
22	Ti	2.51×10^3	55	Cs	< 1.8	90	Th	0.035
23	V	2.34×10^2	56	Ba	2.75	92	U	<0.09
24	Cr	1.15×10^4	57	La	0.302			
25	Mn	5.89×10^3	58	Ce	0.794			

Source: Ross and Aller (1976).

of these tables would be the first to admit that their applicability is far more restricted than the name implies. Two compilations which are of importance are those by Suess and Urey (1956) and by Cameron (1959). The Suess–Urey compilation is a semi-empirical one based partly on measured abundances and partly upon considerations of nuclear stability, while that by Cameron is based more upon theories of nucleogenesis. Quite serious criticisms have been made of various parts of these tables (see, for example, Aller, 1961) and it is clear that at the present time no compilation is satisfactory in all its detail. However, the general *pattern* of element abundances that emerges is similar in all compilations and cannot be far from the true situation.

The initial basis of the Suess–Urey compilation is the abundance pattern shown by the rare-earth elements. The concentrations of these elements in sedimentary rocks were assumed to be a representative average of the crust of the Earth. The data, given in Fig. 2.1, show two features:

(a) The abundances alternate rhythmically between elements of even and odd atomic number (Z).

(b) There is a systematic decrease in abundance with an increase in atomic number.

TABLE 2.2. *Abundances of the elements (Primordial Solar System)*
Normalized to $Si = 10^6$ *atoms*

1 H	3.18×10^{10}	29 Cu	540	58 Ce	1.18
2 He	2.21×10^9	30 Zn	1.144×10^3	59 Pr	0.149
3 Li	49.5	31 Ga	48	60 Nd	0.78
4 Be	0.81	32 Ge	115	62 Sm	0.226
5 B	0.94	33 As	6.6	63 Eu	0.085
6 C	1.18×10^7	34 Se	67.2	64 Gd	0.297
7 N	3.74×10^6	35 Br	13.5	65 Tb	0.055
8 O	2.15×10^7	36 Kr	46.8	66 Dy	0.36
9 F	2.45×10^3	37 Rb	5.88	67 Ho	0.079
10 Ne	3.44×10^6	38 Sr	26.9	68 Er	0.225
11 Na	6.0×10^4	39 Y	4.8	69 Tm	0.034
12 Mg	1.016×10^6	40 Zr	28	70 Yb	0.216
13 Al	8.5×10^4	41 Nb	1.4	71 Lu	0.036
14 Si	1.00×10^6	42 Mo	4.0	72 Hf	0.21
15 P	9.6×10^3	44 Ru	1.9	73 Ta	0.021
16 S	5.0×10^5	45 Rh	0.4	74 W	0.16
17 Cl	5.70×10^3	46 Pd	1.3	75 Re	0.053
18 Ar	1.172×10^5	47 Ag	0.45	76 Os	0.75
19 K	4.20×10^3	48 Cd	1.48	77 Ir	0.717
20 Ca	7.21×10^4	49 In	0.189	78 Pt	1.4
21 Sc	35	50 Sn	3.6	79 Au	0.202
22 Ti	2.775×10^3	51 Sb	0.316	80 Hg	0.4
23 V	262	52 Te	6.42	81 Tl	0.192
24 Cr	1.27×10^4	53 I	1.09	82 Pb	4
25 Mn	9.30×10^3	54 Xe	5.38	83 Bi	0.143
26 Fe	8.3×10^5	55 Cs	0.387	90 Th	0.058
27 Co	2.21×10^3	56 Ba	4.8	92 U	0.0262
28 Ni	4.80×10^4	57 La	0.445		

Sources: Cameron (1973), except value for B, Curtis *et al.* (1980).

In order that the compilation could be extended, the assumption was made that a pattern of this kind should hold for the other elements. Information on abundances for the lighter elements was taken from astrophysical determinations and for heavier elements from their concentrations in chondritic meteorites. Many of these data were of poor quality and so the figures were adjusted to produce a smooth variation in the abundances, as revealed by the rare-earth results. The postulate that the cosmic abundances of elements show a smooth function with atomic number detracts, at first sight, from the viability of this compilation since the postulate has no theoretical foundation. But equally there is no theoretical reason why the abundances should not be a smooth function of atomic number, and the smooth pattern can be readily interpreted in terms of nuclear processes and structure. Other aspects of elemental abundances, which will be discussed later, are consistent with the smooth trend and thereby lend support to the validity of the postulate.

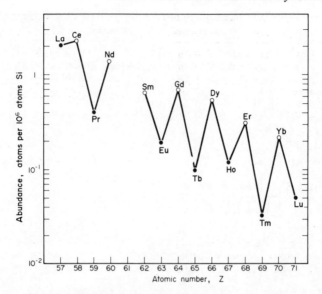

FIG. 2.1. The rare-earth element abundance pattern in sedimentary rocks as used by Suess and Urey to produce their estimate of element abundances in the solar system, see text. Open circles even Z, filled circles odd Z.

Cameron based his compilation on physical factors such as neutron-capture cross-sections of the nuclides but lack of physical data forced him to take the Suess—Urey abundance figures for a number of elements. In general his results are similar to those of Suess and Urey. Later, however, Cameron (1968, 1973) compiled new tables with greater reliance being placed on the concentrations of elements in type 1 carbonaceous chondrites (C1). In recent years there has been a considerable increase in the availability of good-quality data on meteorite composition. This development has revealed that the C1 meteorites may represent the initial composition from which the other chondritic meteorites have fractionated and has led many authors (e.g. Cameron, 1968; Goles, 1969; Anders, 1971) to suggest that these meteorites represent unfractionated samples of the non-volatile component of primordial matter from which the solar system was formed. Cameron's 1973 compilation is given in Table 2.2 and Fig. 2.2. If the pattern of this figure were to be compared with the earlier ones by Cameron (1959) and by Suess and Urey (1956) it would be seen that they are remarkably similar. Thus, the various lines of evidence that we have discussed are self-consistent and allow us to put a good measure of trust in the abundance figures.

Nothing has yet been said of the use of primary cosmic radiation as a source for inferring element abundances. Its use is restricted, as we do not fully

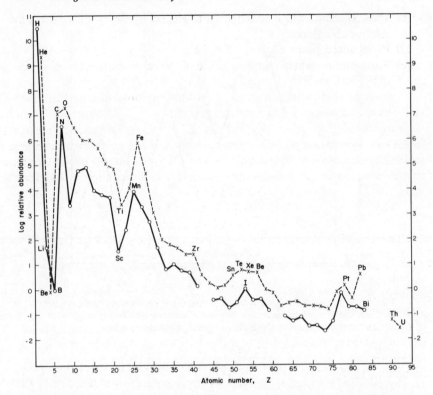

Fig. 2.2. The abundances of the elements, normalized to $Si = 10^6$ atoms, in the
solar system (Cameron, 1973; boron abundance, estimate by Curtis *et al.*, 1980).
Crosses – elements of even atomic number; circles – elements of odd atomic
number.

understand its origin and the data are limited to the more abundant elements,
but on the other hand, certain element ratios may be obtained with an
accuracy of 10 or 20 %, which is better than from many other lines of evidence
(Anders, 1971). Shapiro *et al.* (1970) have calculated the primordial cosmic-
ray composition from observations of cosmic rays at the top of the Earth's
atmosphere. They found the relative atomic abundances of the cosmic-ray
elements to be C 100; N 12; O 105; Ne 20; Mg 23; Si 20; S 3; Cr 8; and Fe 23. For
the non-volatile elements these ratios are similar to those found in the C1
meteorites.

The characteristics of the plot in Fig. 2.2 are:

(a) H and He are the most abundant elements.

(b) The abundances decrease in a roughly exponential manner with increase
in atomic number (Z) until $Z = 45$.

(c) Those elements which are adjacent to He, i.e. Li, Be, B, are very low in relative abundance.

(d) Pronounced peaks occur at O and Fe.

(e) Furthermore, when a study is made of the relative abundances of the individual isotopes (Cameron, 1968), it is seen that those isotopes with mass numbers which are multiples of 4 (i.e. multiples of α-particle mass) have enhanced abundances.

Compilations of element abundances are not only of use as the reference frame for fractionation processes but are also the source material for building theories of element formation. Any theory of nucleosynthesis must account for the observed abundances and the nature of the distribution with atomic mass.

2.3 Theories of nucleosynthesis

At the present time, the only theories which can successfully account for the abundance pattern of the elements are those based upon the formation of elements by nuclear reactions in stellar interiors. Earlier 'equilibrium' and 'non-equilibrium' theories do not account for all the observed facts and little credence is now given to them. (Aller, 1961, reviews these theories.) It was Burbidge et al. (1957) who developed the hypothesis of element synthesis by nuclear reactions with hydrogen as the sole starting material. Their work is the foundation for the current state of the theory in which the postulated sequence of events is related to stellar evolution. We start, therefore, with a brief review of this evolution. A more detailed exposition of this topic and of nucleosynthesis may be found in Clayton (1968).

2.3.1 Stellar evolution

The initial formation of a star occurs as a result of the onset of gravitational instability in the gas cloud of the interstellar medium. Energy is released as the gravitational collapse occurs and thereby the temperature of the gas, as well as its density, rises. With this increase in temperature and density, the pressure of the interior also rises and brings the unstable collapse to a halt. The increase in temperature reaches a point which is sufficient for the onset of thermonuclear reactions. These occur in the central regions of the young star and the energy supplied from the reactions is radiated, in part, from the surface.

The early phase of contraction involves a rapid rise in brightness of the star but with a relatively constant surface temperature. This phase is followed by one in which contraction continues, but at a slower rate, with a rise in surface temperature and luminosity. All this time the hydrogen at the centre of the star is being consumed in nuclear reactions and the slow contraction of the core continues. The production of heavier nuclides from light ones in the stellar interior would, on its own, result in the reduction of pressure but this is offset by the gravitational contraction which is maintained at an equilibrium rate by the concomitant rise in temperature. The further rise in temperature causes the expansion of the outer layers of the star so that the total radiation of energy may be increased through a larger surface area. The further evolution of the star is dominated by processes of this type. The core of the star is the place where nuclear reactions predominate. The successive reactions occur through the rises in temperature from the work released in gravitational contraction and by the possibility of additions of material from the outer layer.

The advanced stages of stellar evolution are complicated and need not concern us here. We do need to appreciate, however, that the evolution continues through many stages of increase in core temperature with associated core contraction and with the balance of energy being maintained by changes in the overall size and luminosity of the star. The probable nature of the thermonuclear reactions which take place in the centres of stars and the likely temperatures at which they occur will now be briefly discussed. However, the actual processes which occur in a star depends upon its mass and initial composition (i.e. whether it is a first, second or later generation star), see Clayton (1968).

2.3.2 Hydrogen and helium burning

We have already indicated that hydrogen is the most abundant element in the 'Cosmos'. Approximately 90 % of all nuclei of stellar material is hydrogen. It is, therefore, reasonable to make hydrogen the starting material in the manufacture of elements in stellar interiors. Burbidge *et al.* (1957), in their theory of nucleosynthesis, postulated that hydrogen was the sole material of the first stars, and so we require an initial nuclear reaction which produces heavier nuclides at the expense of the hydrogen. Any such process is referred to as *hydrogen burning*.

It can be shown that the most probable hydrogen-burning reaction is one which converts two atoms of hydrogen into one helium atom via the production of the nuclides ^2D and ^3He in intermediate stages, as follows:

$$^1H + {}^1H \rightarrow {}^2D + e^+ + \nu \,(\text{neutrino*})$$
$$^2D + {}^1H \rightarrow {}^3He + \gamma$$
$$^3He + {}^3He \rightarrow {}^4He + 2\,{}^1H$$

---

* Neutrino, ν, a particle with zero charge, spin $\frac{1}{2}$, and of very small or zero rest mass.

Or, as is usually written for nuclear reactions:

$$^1\text{H}(\text{p}, e^+ \nu)^2\text{D}$$
$$^2\text{D}(\text{p}, \gamma)\,^3\text{He}$$
$$^3\text{He}(^3\text{He}, 2\text{p})^4\text{He}$$

This reaction path is probable in that it only involves two particle interactions and is exothermic (for every one ^4He produced from four ^1H nuclei) to the extent of 26.73 MeV. There are in fact no other two-particle exothermic reactions which could occur in a gas made up only of protons and alpha particles that would result in a stable product (e.g. ^2He, which could be produced by the interaction of two protons, is unstable and immediately breaks up again to give two protons). The question that now arises, however, is: are other reactions possible which involve the intermediate products ^2D and ^3He, besides the ones given above? No other deuterium reaction is likely as all other possibilities involving this nuclide proceed at a negligible reaction rate, while the ^2D, ^1H reaction is extremely rapid. ^3He could, however, be consumed by the production of ^7Be through the reaction:

$$^3\text{He}(\alpha, \gamma)\,^7\text{Be}$$

and then with the consumption of the ^7Be nuclide in two possible cycles of reactions, both of which proceed at a rapid rate:

$$^7\text{Be}(e^-, \nu)^7\text{Li} \quad \text{OR} \quad ^7\text{Be}(\text{p}, \gamma)^8\text{B}$$
$$^7\text{Li}(\text{p}, \gamma)^8\text{Be} \qquad\qquad ^8\text{B} \rightarrow\, ^8\text{Be} + e^+ + \nu$$
$$^8\text{Be} \rightarrow 2\,^4\text{He} \qquad\qquad ^8\text{Be} \rightarrow 2\,^4\text{He}$$

Thus there are three possible chains for ^4He production. They may all operate simultaneously if there is sufficient ^4He to act as an 'autocatalyst'. The temperature at which hydrogen burning takes place is estimated to be about 5×10^6 K to 10^7 K.

Up to this point we have considered young stars which are made up of hydrogen as the only starting material for nuclear reactions. However, spectra from stars which have recently formed from the interstellar medium indicate that heavier elements are present and these may have originated by the recycling of part of the mass of large stars back into the interstellar medium. The fact that the interstellar medium itself does not consist entirely of hydrogen requires us to consider the effect that heavier elements may have in possible hydrogen-burning reactions. One such possibility is the CN cycle:

$$^{12}\text{C}(\text{p}, \gamma)^{13}\text{N} \longrightarrow\, ^{13}\text{C} + e^+ + \nu$$
$$^{13}\text{C}(\text{p}, \gamma)^{14}\text{N}$$
$$^{14}\text{N}(\text{p}, \gamma)^{15}\text{O} \longrightarrow\, ^{15}\text{N} + e^+ + \nu$$
$$^{15}\text{N}(\text{p}, \alpha)^{12}\text{C}$$

The effective sum of these reactions is the conversion of four protons into one ^{4}He nuclide, the ^{12}C acting as a catalyst. If any ^{16}O is present two or more reactions are possible:

$$^{16}O\,(p, \gamma)^{17}F \longrightarrow \,^{17}O + e^+ + \nu$$
$$^{17}O\,(p, \alpha)^{14}N$$

The ^{16}O may also be produced from ^{15}N produced in the CN cycle by:

$$^{15}N\,(p, \gamma)^{16}O$$

and while the frequency of this reaction in relation to that producing ^{12}C is only about 4 in every 10^4 reactions, it is none the less of some significance. A CNO bi-cycle, therefore, exists which occurs at temperatures below 10^8 K:

$$^{12}C\,(p, \gamma)^{13}N \longrightarrow \,^{13}C + e^+ + \nu$$
$$^{13}C\,(p, \gamma)^{14}N$$
$$^{14}N\,(p, \gamma)^{15}O \longrightarrow \,^{15}N + e^+ + \nu$$
$$^{15}N\,(p, \alpha)^{12}C$$

$$^{15}N\,(p, \gamma)^{16}O$$
$$^{16}O\,(p, \gamma)^{17}F \longrightarrow \,^{17}O + e^+ + \nu$$
$$^{17}O\,(p, \alpha)^{14}N$$

The processes of hydrogen burning occupy the greatest part of the active burning life of a star.

To produce the heavier elements it is now necessary for *helium burning* to occur. This starts when the hydrogen burning is complete and the core has undergone a further contraction to obtain a density of about 10^5 and a temperature in the region of 10^8 K. The helium-burning analogue of the hydrogen-burning process is the one which involves the production of ^{8}Be from α–α impacts. Although the half-life of ^{8}Be is very small (2.6×10^{-16} s) it is just sufficient to allow a small equilibrium concentration of ^{8}Be to be built up. (It is estimated that under the conditions which prevail at this stage of stellar evolution there is one ^{8}Be nucleus for every 10^9 alpha particles.) This equilibrium concentration allows the interaction of ^{8}Be and ^{4}He particles to occur:

$$^8Be\,(\alpha, \gamma)^{12}C$$

In view of the short half-life of ^{8}Be, we can consider the production of ^{12}C to be one which involves the interaction of three alpha particles. The probability of this reaction occurring is small but it must be remembered that at this stage there are large concentrations of helium nuclei of high energy and there is also a considerable time (about 10^8 years) during which the process may continue.

It is to be noted that the helium-burning process as outlined above does not involve the production of any lithium, beryllium or boron nuclei as stable end products. The low cosmic abundance of these elements is consistent with the operation of this helium burning mechanism.

With the existence of stable ^{12}C nuclei in the stellar interior, the possibility of further alpha-particle processes now arises. Such reactions could be:

$$^{12}C(\alpha, \gamma)^{16}O \qquad \text{(energy release} = 7.2\,\text{MeV)}$$
$$^{16}O(\alpha, \gamma)^{20}Ne \qquad \text{(4.7 MeV)}$$
$$^{20}Ne(\alpha, \gamma)^{24}Mg \qquad \text{(9.3 MeV)}$$

The development of heavier nuclei by this mechanism at the temperature of about $10^8\,K$ is prevented by the rapid increase of the coulomb barrier.

The precise composition of the gas at the end of helium burning is still only speculative. We can be sure that the most abundant nuclei are ^{12}C and ^{16}O but the existence and the proportions of other nuclei are uncertain. For example, ^{14}N is a product of the CNO cycle and its existence during the helium-burning processes would allow the production of ^{18}O and ^{22}Ne by an alpha-capture mechanism. As we do not know the probable amount of ^{14}N which may exist at the end of hydrogen burning, we can make no estimate of the amounts of these heavier nuclei.

2.3.3 Advanced burning and photodisintegration

Gravitational contraction and a rise in temperature of the star occur at the end of the helium-burning stage. This contraction can occur because of the reduction of the outward radiation pressure from the core when the helium burning ceases. ^{12}C and ^{16}O are the most abundant nuclides in the legacy of the last stage and with the marked rise in temperature reactions of the following kinds can occur:

$$^{12}C(^{12}C,p)^{23}Na$$
$$^{12}C(^{12}C,\alpha)^{20}Ne$$

The protons produced in the first reaction are taken up by any free ^{12}C

$$^{12}C(p,\gamma)^{13}N \rightarrow ^{13}C + e^+ + \nu$$

and the alpha particles produced in the second reaction may be taken up by ^{13}C to give free neutrons

$$^{13}C(\alpha, n)^{16}O$$

or be captured by ^{16}O, ^{20}Ne, ^{24}Mg, to give ^{20}Ne, ^{24}Mg and ^{28}Si, respectively. Reaction between ^{12}C and ^{16}O does not appear to be a significant process, probably because the temperature ($\sim 8 \times 10^8\,K$) at this stage is insufficient to overcome the coulomb barrier. $^{12}C-^{12}C$ reactions, with the lower coulomb barrier, occur at a relatively rapid rate.

The ^{16}O, remaining from the helium-burning stage, may undergo similar reactions to the ^{12}C, but at a higher temperature of about 2×10^9 K, e.g.

$$^{16}O \, (^{16}O, n)^{31}S$$
$$^{16}O \, (^{16}O, p)^{31}P$$
$$^{16}O \, (^{16}O, \alpha)^{28}Si$$

The above processes can be referred to as carbon burning and oxygen burning respectively and from them the elements Ne, Na, Mg, Al, P, S and Si are produced.

In the earlier nucleosynthetic processes the fusion reactions are exothermic as the binding energy of each product nucleus is greater than that of the reactants. Binding energies show a variation with atomic mass (Fig. 2.3) and the highest binding energy occurs at the ^{56}Fe nucleus. The figure shows that for other nuclei there is a steady decrease in binding energy with increasing mass number, and a rapid decrease with decreasing mass number. Binding energy becomes an important parameter during the carbon- and oxygen-burning stages for now the temperature is sufficiently high for the onset of photodisintegration of the earlier formed nuclei. The loosely bound protons and neutrons (collectively called nucleons) are redistributed via 'photo-disintegration rearrangement' into more tightly bound states and so energy continues to be produced. How much energy is released by this process is

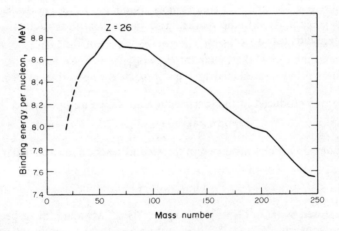

Fig. 2.3. Curve depicting the nature of the variation of binding energy (MeV) per nucleon with mass number. (After Preston, 1962)

uncertain since at this high temperature the presence of an unknown equilibrium concentration of free nucleons lowers the energy compared to a situation where all nucleons are in the bound state.

Towards the end of oxygen burning the principal nuclei are ^{28}Si, ^{31}P and ^{31}S and the temperature is of the order of 2.5×10^9 K. This is sufficient for the photodisintegration of ^{31}P (which has a lower binding energy than ^{28}Si) principally by the photoejection of protons

$$^{31}P\,(\gamma, p)\,^{30}Si$$

and then with the following:

$$^{30}Si\,(\gamma, n)\,^{29}Si$$
$$^{29}Si\,(\gamma, n)\,^{28}Si$$

The final and principal nuclide, ^{28}Si, can also undergo photodisintegration but not until the temperature is considerably higher (about 3×10^9 K).

2.3.4 The equilibrium process (e-process)

The conditions that prevail in the last stage of evolution of a star involve high temperature and density which promote a variety of nuclear reactions including gamma activated ones such as (γ, α), (γ, p) and (γ, n), with the heavier nuclei. During this stage the elements of the iron peak (Cr, Mn, Fe, Co and Ni) are produced, and the abundances can best be explained if a state of nuclear statistical equilibrium existed in the gas, between the nuclei and the free protons and neutrons. The temperature at which the e-process occurs is very difficult to predict as there is considerable neutrino flux to the surface of the star and a full understanding of the e-process has not yet been reached. The final mass fraction of the dominant nuclei depends upon temperature as is diagrammatically represented in Fig. 2.4, so it is impossible, at present, to predict the final composition at the end of the e-process.

The end of the e-process in the core of the star marks the end of any energy increase from nuclear reactions but the persistent loss of neutrinos allows the core to contract and the temperature to rise. Unless the star is very small this temperature rise will continue and lead to photodisintegration of the iron peak, a further rise of temperature and the photodisintegration of alpha particles with core collapse. The core may finally contract to a massive nucleus of neutrons to give a neutron star of low energy. Although this core has imploded, it is believed that the outer layers of the star undergo a supernova explosion and eject matter into space which will then be available for possible incorporation as part of the material of a new star.

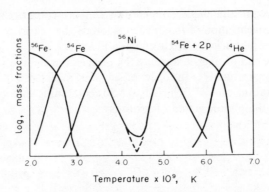

FIG. 2.4. A schematic diagram showing the abundances of the principal nuclei that result from the burning of silicon, as a function of temperature. At relatively low temperature, and slower burning, ^{56}Fe is the most stable nucleus, but at a higher temperature, ^{56}Ni becomes stable. It can be seen that there is a sensitive compositional dependence upon temperature. (After Clayton, 1968)

2.4 Production of the heavy elements

2.4.1 *Neutron-capture reactions*

The nuclear reactions that have been discussed up to this point involve the production of nuclides of mass up to and including those of the iron-group. The iron-group nuclei are apparently produced in the final stages of the evolution of a star so the question now arises as to how and when are those nuclei produced that have masses greater than those of the iron-group—the so-called 'heavy elements'? Many of them are probably produced by a non-equilibrium process involving neutron capture at a stage of stellar evolution before that giving rise to the iron-nuclei.

In any discussion of nucleosynthesis involving n-capture two points must be emphasized. First, as neutrons are uncharged there is no coulomb barrier to be overcome in a nuclear reaction and for this reason many elements can readily capture neutrons and, furthermore, will generally do so more readily if the neutron is of relatively low energy. Second, neutrons are probably not available for reaction in the very early, evolutionary stages of a star. One possible source is the reaction (Bermann *et al.*, 1969):

$$^{22}\mathrm{Ne}\,(\alpha, n)\,^{25}\mathrm{Mg}$$

^{22}Ne is readily produced from ^{14}N by the capture of two alpha particles during the hydrogen-burning stage. The ^{14}N is, of course, the main product of the

CNO bi-cycle. Other possible reactions (which also involve nuclides produced in the CNO bi-cycle) are:

$$^{13}C\,(\alpha, n)\,^{16}O$$
$$^{17}O\,(\alpha, n)\,^{20}Ne$$
and $$^{21}Ne\,(\alpha, n)\,^{24}Mg$$

When a nuclide is bombarded with neutrons it may undergo a (n, γ) nuclear reaction and so become a heavier isotope of the same element. This newly produced isotope could in turn undergo a (n, γ) type of reaction to produce another isotope and the process will continue until a radioactive isotope is produced. What happens at this stage is then determined by the extent of the neutron flux. On the one hand, if the neutron flux is very low, such that there is a greater chance of the radioactive isotope undergoing a beta-decay than of capturing another neutron, then an isobar with atomic number $Z + 1$ will tend to be produced from the decay of the radioactive isotope. On the other hand, if the flux is very great there will be a good chance that the radioactive isotope, before it has had time to undergo decay, will capture another neutron. The first path has been called the s-process because the neutron capture rate is *slow* compared to the beta decay rate, and the second path the r-process because the neutron capture rate is *rapid*.

The s-process starts just as soon as neutrons are available and, therefore, in second or later generation stars it may occur as early as the helium-burning stage. The limits of the process are imposed by the stability of the nuclides that it produces; hence, no nuclide heavier than $A = 209$ can be made by the s-process since ^{209}Bi is the heaviest stable nucleus.

The s-process has been most successful as a theory which accounts for many of the observed compositions of stars. For example, the observed occurrence of technetium in stars would be difficult to explain other than by a neutron reaction and beta-decay process:

$$^{98}Mo\,(n, \gamma)\,^{99}Mo \rightarrow\,^{99}Tc + e^- + \nu$$

(Note: The half-life of ^{99}Mo is 67 hours.) It has been established that the s-process is the cause of narrow peaks near $A = 90$, 140 and 208 in the abundance pattern of the individual nuclides.

In contrast, the r-process is not limited to the production of elements up to $Z = 83$ (Bi), since the rate of neutron capture is fast compared to the radioactive decay rate for any nuclide. A nuclide within neutron-rich matter (where the neutron density is as high as $10^{23}\,cm^{-3}$) would capture neutron after neutron until the neutron-binding energy became so low that the now neutron-rich nuclide could hold no additional neutrons. At this point it would 'wait' until it had undergone beta-decay thereby increasing the nuclear charge and allowing the further capture of neutrons as before. The waiting points will, because of neutron-binding energies, tend to be at even-N nuclei and they are the neutron-rich progenitors, which will in due course decay to stable isotopes that will often be relatively neutron-rich. The time interval between successive neutron captures in the r-process is probably of the order of 0.1 to 1 second. The process is terminated by the onset of nuclear fission which occurs near to A = 276 and the fission products are injected back into the cycle at abundance peaks about A = 130 and A = 195. The principal nuclides that are produced by the r-process are those between A = 76 and A = 204. Photodisintegration at high temperatures could also significantly reduce the effect of the r-process.

The r-process is not so well understood as the s-process, mainly because the source of a very high neutron flux is uncertain. A supernova explosion may have the right conditions for the r-process.

A study of a chart of nuclides reveals some important points that are helpful in understanding the effects of the neutron-capture processes. Firstly, it is seen (e.g. Fig. 2.5) that those elements of even atomic number generally have more stable isotopes than those of odd atomic number, and secondly, isobars of even atomic weight may have more than one stable isobar, but those of odd weight have only one.

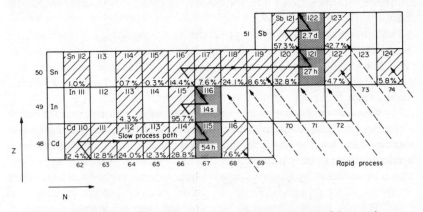

FIG. 2.5. Part of the chart of the nuclides used to show the roles of the s- and r-processes in the production of isotopes of Cd, In, Sn and Sb. The s-process path is indicated by the continuous line. The r-process decay lines enter from the lower right (see text). Shaded rectangles are naturally occurring stable isotopes; stippled rectangles are isotopes which undergo β-decay, thereby terminating the s-process path for that particular atomic number.

A section of the nuclide chart is given in Fig. 2.5 which has been described in terms of the s- and r-processes by Clayton *et al.* (1961). The s-process path is indicated by the continuous line passing through the isotopes of 110 to 115 for cadmium, of 115 for indium and 116 to 121 for tin. [113]In is a proton-rich isotope (p-nuclide) that cannot be produced by either the s- or r-process, and it is noteworthy that it is only 4 % as abundant as ^{115}In which can be produced by the s-process. The figure shows that the isotopes of tin 117, 118, 119 and 120 could be produced by the s-process and the beta-decay of neutron-rich progenitors made through the r-process. The isotope ^{116}Sn, however, could not be produced by the r-process as the isotope ^{116}Cd is stable and prevents the beta-decay of neutron-rich progenitors through to ^{116}Sn. The ^{116}Sn isotope is said to be shielded from the r-process by ^{116}Cd. The isotopes ^{122}Sn and ^{124}Sn, however, can only be produced by decay of r-process progenitors as the s-process will be stopped at the radioactive isotope ^{121}Sn. This example serves to show that the form of the distribution of stable nuclides on the nuclide chart indicates, in part, the possible production methods for the isotopes by neutron reactions. The relative abundances of the different isotopes of an element can be readily related to neutron-capture cross-sections.

It has been suggested that the production of the heavy elements could occur at the very high densities associated with the implosion of the supernova core. There is, however, much uncertainty about this, especially in that an explanation is wanting on how the heavier elements could be prevented from undergoing photodisintegration during the high-temperature stage when the core explosively re-expands.

2.4.2 Proton-capture reactions

Mention has already been made of the existence of proton-rich nuclides (e.g. ^{113}In, see Fig. 2.5) and these cannot, in general, be produced via the s- or r-processes. Burbidge *et al.* (1957) introduced the idea of a p-process which is similar to the r-process except that the flux is dominantly protons. The p-nuclei would be produced by (p, γ) reactions on heavy elements previously produced by the s- and r-processes. However, unlike neutron-capture reaction, the problem here is that fast protons are probably rare during most stages of stellar evolution because high energies would be required to give the charged protons a high speed. Hence, it is likely that only small amounts of products from any p-type reaction can result. The relatively high abundance and low coulomb barriers mean that ^{12}C, ^{14}N, ^{16}O, etc., could also be important target nuclei in a p-process (Reeves, 1971).

The source of protons during advanced burning stages is also something of a problem, although this may be from photodisintegration of elements, e.g. at temperatures above 3×10^9 K the photodisintegration of ^{28}Si proceeds rapidly with the release of p, n and α particles. The source of the protons could also be via a supernova explosion.

2.5 Production of Li, Be and B

The production of the elements Li, Be and B has not yet been accounted for in this review. Indeed, the processes by-pass these elements, and to some extent this must be a reason for the rarity of these elements in the Cosmos. There is no theory for their production that has been generally accepted. One possibility that is perhaps more acceptable than others at the present time is the production of these elements by spallation reactions (see Section 1.5.1) in a cosmic gas where the targets are principally carbon, nitrogen and oxygen (Reeves, 1971). The light nuclides produced by this mechanism could then be reduced in number by destructive proton-induced reactions of the following kind during stellar evolution:

$$^6\mathrm{Li} \, (p, \, ^3\mathrm{He})^4\mathrm{He}$$
$$^7\mathrm{Li} \, (p, \alpha)^4\mathrm{He}$$
$$^9\mathrm{Be} \, (p, \alpha)^6\mathrm{Li}$$
$$^{10}\mathrm{B} \, (p, \alpha)^7\mathrm{Be}$$

2.6 Summary statement on nucleosynthesis

The suggested stages of nucleosynthesis, which have been reviewed qualitatively above, also can be related quantitatively, in part, to the cosmic abundance pattern (Fig. 2.2). Such an exercise is beyond the scope of this book but we have seen that in general the predicted paths of stellar nuclear reactions lead to the peaks and troughs in the pattern. Thus, from the process of hydrogen burning, including the CNO bi-cycle, the elements He, C, N and O are produced (see oxygen peak in Fig. 2.2). Helium burning gives C, O, Ne and Mg but no Li, Be or B, the occurrence of these last three elements being accounted for by spallation reactions involving C, N and O targets in a cosmic gas. Carbon and oxygen burning lead to the production of Ne, Na, Mg and Si, and then the advanced burning stages and equilibrium process of nucleosynthesis give the elements of the iron group; the peak in the pattern of Fig. 2.2 being explicable in terms of nuclear binding energies. Neutron-capture processes, at rates which are slow compared to those of radioactive decay, can produce elements of atomic number up to that of Bi (83), while if the capture rates are comparatively fast, will produce a range of elements including those of $Z > 83$.

Although there are significant uncertainties about the nature of some processes during nucleosynthesis, it is satisfying to note that our understanding is sufficient to account theoretically for the observed abundance distribution of many of the elements.

Suggested further reading

CLAYTON, D. D. (1968) *Principles of Stellar Evolution and Nucleosynthesis.* McGraw Hill, New York. 612 pp.
REEVES, H. (1971) *Nuclear Reactions in Stellar Surfaces and their Relation with Stellar Evolution.* Gordon & Breach Science Publ., London 87 pp.
SELBIN, J. (1973) The origin of the chemical elements. 1. *Journal of Chemical Education*, 50, 306–310. 2. *Ibid.* **50**, 380–387.

3

The Moon

3.1 Introduction

Chemical and geophysical data about the Earth's satellite are important to the geochemist in that they impose constraints on theoretical models of the evolution of the solar system and, more particularly, of the formation and compositional relationships of the Earth–Moon system. The ancient lunar surface, undisturbed by chemical weathering processes, gives significant clues about the processes acting on the outer parts of both the Moon and the Earth during the early part of their history, and also can provide important information on the chemical compositions of their interiors. In less than a decade (1968–1976) our knowledge and understanding of the Moon underwent considerable change and expansion. The exciting exploration by man of parts of its surface and the return to Earth of lunar rock and soil allowed a reappraisal of the origins of both the Moon and the Earth. The extremely thin lunar atmosphere and the reducing nature of the chemical environment meant that the rocks which became available for study were fresh, despite their extreme age, and contained a wide scope for geochemical investigation.

The Earth–Moon distance ranges from 356,410 km (perigee) to 406,740 km (apogee); other lunar data are given in Table 3.1. The Moon's mean density ($3.34\,\mathrm{g\,cm^{-3}}$) is considerably lower than that of Earth (5.52) and is little different from that of its surface rocks (about 3.0). The Moon's moment of inertia and the similarity between the mean lunar and crustal densities impose constraints on the possible compositions and sizes of the mantle and core. A core, if present, can only be small and is likely to consist of molten iron, perhaps with a moderate admixture of FeS (Levin, 1979). Seismic and gravity data indicate that the core's radius must be less than about 350 km and that the lunar crust, rich in plagioclase and pyroxene, is, on average, approximately 60 km thick (about 50 km nearside and about 75 km on the far side) and rests on an olivine-rich mantle.

The lunar terrains may be divided into three types, each characterized by properties which are visible from Earth:

TABLE 3.1. *Lunar data*

	Absolute	Relative to Earth
Radius (av.)	1738 km	0.272
Volume	2.2×10^{10} km^3	0.020
Mass	7.35×10^{25} g	0.012
Density	3.34 g cm^{-3}	0.605
Surface gravity	162 cm s^{-2}	0.165
Escape velocity	2.35 km s^{-1}	0.21
Albedo	0.07	0.2
Structure:		
Crust	$\sim$ 60 km thick (average)	
Lithosphere	$\sim$ 1300 km thick	
Core (?)	$\leqslant$ 350 (?) km radius	

Mare: large area of low relief and very low albedo;

Upland (ejecta blanket): mild relief and slightly higher albedo than the mare;

Highland: mountainous relief; relatively high albedo.

Each of these terrains contain additional surface features, including craters, rilles, rays, scarps and ridges. These are described and their origins discussed by Taylor (1975) and in some other modern textbooks on the Moon. Photographic results from early orbiting missions showed that the far side of the Moon is lacking in the large maria which occur on the near side; the nature of the far side is suggestive of a very ancient topography.

Much of the Moon's surface is covered by a layer (the regolith) of loose rock and mineral particles, glassy fragments and minor amounts of meteorite debris. Brecciation is present to some considerable depth. For example, in the Descartes highland region, seismic data indicate that the upper 10 km is brecciated. The fragmentation has been caused mainly by meteorite impact as well as by the great temperature change (150°C) between lunar night and day acting on material of low thermal conductivity. The virtual absence of an atmosphere and of water on the Moon means that the formation of the regolith cannot have involved terrestrial-type weathering processes. Meteorite impact has not only contributed to the break up of rock but also to its distribution over wide areas.

Nine lunar missions have returned samples to Earth (Table 3.2 and Fig. 3.1). The six Apollo missions (USA) were all manned while the three unmanned Luna missions (USSR) automatically retrieved samples of the regolith. A total of over 380 kg of material has now been returned. In this chapter a brief review is given of the petrological and chemical characteristics of lunar materials. A more detailed account may be found in the textbook by Taylor (1975).

TABLE 3.2. *Lunar sample retrieval missions*

Mission	Date	Locality	Terrain
APOLLO	11 July 1969	Mare Tranquillitatis 0.7°N, 23.5°E	Mare
	12 Nov. 1969	Oceanus Procellarum 3.2°S, 23.4°W	Mare
	14 Jan. 1971	Fra Mauro 3.7°S, 17.5°W	Upland-Ejecta
	15 July 1971	Hadley Base/Apennine Front 26.1°N, 3.7°E	Nr. highlands
	16 Apr. 1972	Descartes Mtn Cayley Plain 8.9°S, 15.5°E	Highlands
	17 Dec. 1972	Taurus Littrow 20.17°N, 30.77°E	Highland
LUNA	16 Sept. 1970	Mare Fecunditatis 0.7°S, 56.3°E	Mare
	20 Feb. 1972	N of M. Fecunditatis 3.5°N, 56.5°E	Highland
	24 Aug. 1976	Mare Crisium 12.45°N, 62.12°E	Mare

3.2 Lunar samples: Mineralogical and petrological summary

A broad petrological division of the collected material gives only three categories:

(a) Crystalline igneous rocks.

(b) Breccias.

(c) Fines, regolith or soil, which comprises particles less than 1 mm in size (Klein, 1972).

In the maria, the igneous rocks are mainly basaltic in nature; often more vesicular or vuggy than terrestrial analogues and usually without flow-orientation textures. Chemical and textural evidence suggests that the parent magmas of many of these rocks were at least an order of magnitude lower in viscosity than those producing terrestrial basalts. In the highlands the grain-size of the igneous rocks shows a considerable variation from fine to coarse (gabbroic) and some specimens clearly were formed by the accumulation of crystals from a cooling silicate melt (i.e. cumulative origin).

The mare basalts have a simple mineralogy of plagioclase feldspar (usually An_{92} to An_{80} and with narrow zoned rims) and pyroxene, commonly augite, ferroaugite or pigeonite, which is frequently zoned in a complex way (Klein, 1972). Magnesian olivine and ilmenite are usually minor minerals but some olivine basalts have been found, especially in Oceanus Procellarum. There is a wide diversity in the accessory minerals including three that were first discovered on the Moon (armalcolite, $(Fe, Mg)Ti_2O_5$; pyroxferroite $(Fe, Ca)SiO_3$; and tranquillityite $Fe_8(Zr, Y)_2Ti_3Si_3O_{24}$).

Apollo 11, the first mission which returned surface material from the maria, showed that at least two basalt types exist. At Mare Tranquillitatis these are

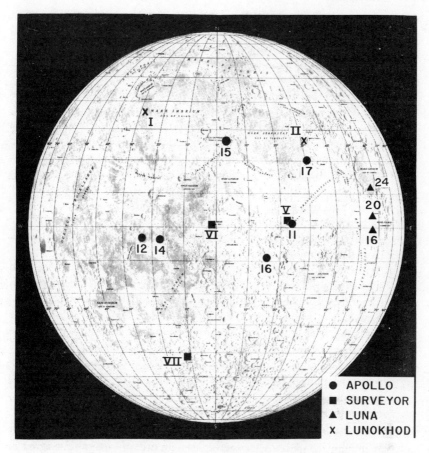

FIG. 3.1. The landing sites of the Apollo, Luna, Surveyor and Lunokhod missions. (After Taylor, 1975)

low-potassium, titanium-rich basalt and high-potassium, titanium-rich basalt. The Apollo 12 mission also returned two basalt types—one with phenocrysts of olivine, the other with clinopyroxene. By the end of the Apollo programme six general basalt types became distinguishable (Taylor, 1975; Vaniman and Papike, 1977):

(a) Olivine normative.
(b) Quartz normative.
(c) High-K, Ti basalt.
(d) Low-K, Ti basalt.
(e) Aluminous maria basalt.
(f) Very low-Ti basalts; ($< 1\%$ TiO_2)—abbrev. VLT basalt.

In all of these clinopyroxene is the dominant modal mineral, followed most often by plagioclase. Olivine, where present, rarely exceeds 5% modal proportion except in the olivine normative basalts where it can reach about 20%. Ilmenite is a common accessory in most of these basalt types, while chromite is present in the very low-Ti basalts.

Many of the crystalline rock samples returned from the highland regions are brecciated. There is much evidence to show that the highland crust has suffered extensively from shock and impact processes. Detailed investigations of the clasts present in the breccia samples, the breccias themselves, and of volcanic rocks have revealed the existence of the following highland rock types:

(a) Anorthosite; almost monomineralic; calcic plagioclase (An_{90-98}).

(b) Anorthositic gabbro; calcic plagioclase; clinopyroxene with minor olivine and opaque minerals; probably a common rock type.

(c) Gabbroic anorthosite, usually porphyritic basaltic texture (not strictly gabbroic); phenocrysts of calcic plagioclase (dominant); clinopyroxene (zoned); and minor olivine (unzoned).

(d) Spinel troctolite; plagioclase- and olivine-rich basalt; texture often suggestive of rapid crystallization; magnesian olivine—composition sometimes up to Fo_{94}; calcic plagioclase; probably a rare rock type.

(e) Low-K, Fra Mauro basalt (or high-alumina basalt); abundant in the highlands (not only in Fra Mauro); plagioclase, clinopyroxene, orthopyroxene, minor ilmenite and troilite.

(f) Medium-K, Fra Mauro basalt; plagioclase, orthopyroxene, clinopyroxene, minor ilmenite and others; sometimes called 'KREEP' basalt on account of relatively high contents of K (about 0.5%), rare-earth elements, P, and associated elements (U, Th, Zr, etc.).

(g) High-K, Fra Mauro basalt; rare rock type containing more than 1% potassium.

Rare pieces of dunite have also been recorded. (For an example of some of the many existing descriptions of these rock types, the reader is referred to Brown *et al.*, 1973.)

The highland breccias contain fragments of these different rock types as well as regolith material, all set in a fine-grained, sometimes glassy, matrix. Lithification was probably produced by shock melting and is variable in extent. Some of the breccias from the Descartes Mountains show evidence of having derived from the fragmentation of a fractionated plutonic complex. There is reasonable petrological evidence to show that some of these rocks (and possibly some from Tauros Littrow) are igneous cumulates but with the texture disrupted by impact processes (e.g. Haskin *et al.*, 1974). The breccias found in the maria contain fragments of mostly local rock types but including some that must have been derived from the highlands.

The loose material of small particle size ($\leqslant 1$ cm)—the regolith—on the

Moon's surface consists of fragments of rocks and minerals and glass. Small fragments of meteorites and meteorite minerals also occur. Like the breccias the nature of the regolith varies with locality. Regolith of the highlands has a *very* small mare component, while that of the maria has a small highland component. The glass particles are produced almost certainly by melting during impact processes.

The material of the Moon's surface can be seen to represent the end-product of the processes of magmatic differentiation, rock disruption and brecciation, and some remelting and consolidation. In broad terms the petrology is simple because there are few rock types—far fewer than on Earth. In detail, however, the rocks and their geneses are often complex, and some show the workings of chemical reactions that are not, or rarely, seen in natural magmatic systems on Earth. One example is the subsolidus reduction of some oxide minerals (Haggerty, 1972). At most of the landing sites, material has been collected which contains oxide assemblages with intergrowth relationships such as that shown by the co-existence of Cr–Al–ulvöspinel, ilmenite and metallic iron. This assemblage is interpreted as resulting from the reduction of a primary, but metastable, ulvöspinel under conditions of low oxygen fugacity* to give intergrowths of ilmenite and an ulvöspinel enriched in Mg, Cr and Al. More intense reduction gave some metallic iron and a further rise in the Cr, Al and Mg contents of the co-existing ulvöspinel.

3.3 Some age relationships

The use of isotope-dating methods in lunar specimens returned to Earth has revealed a sequence of events commencing more than 4.4 Ga ago with the differentiation of a lunar crust. For example, application of the Ar–Ar dating method with controlled heating steps (see Section 1.5.3) to 're-equilibrated' and 'relict' (or unequilibrated) plagioclase feldspars of a breccia from the Descartes Mountains gave a plateau at 3.98 Ga and an upper age of 4.5 Ga respectively, Fig. 3.2 (Jessberger *et al.*, 1974). These results indicate at least a two-stage evolution of the lunar crust; its formation (4.5 Ga ago) followed by brecciation and then an intense metamorphism at 3.98 Ga ago. The highland rocks usually have ages of between 3.9 Ga and 4.0 Ga—these reflecting, in part, the events which formed the late ringed-basins; e.g. the impact which produced the vast Imbrian basin has been dated at about 3.9 Ga. It is evident that the Moon was being bombarded by large meteorites up to 3.9 Ga ago.

The mare basalts often show by their stratigraphical relationships that they are younger than the highland rocks; this is borne out by the isotope chronology. The flooding of the maria occurred over an extended period from

---

* See Chapter 7 for a discussion on fugacity.

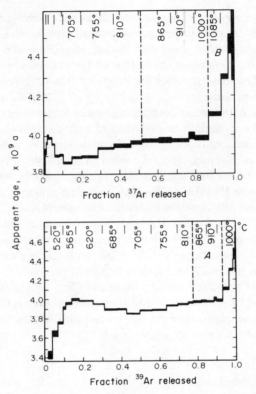

FIG. 3.2. Argon release patterns for an Apollo 16 breccia. The lower diagram (A) shows the apparent age against the fractional release of ^{39}Ar, and is dominated by Ar release from K-rich, Ca-poor accessory mineral phases. The upper graph (B), of apparent age against fractional release of ^{37}Ar, reflects the composition of Ar from Ca-rich plagioclase. (The ^{37}Ar is produced by a (n, α) reaction on ^{40}Ca.) Note the well-defined intermediate plateau at an age of 3.98 Ga. (After Jessberger *et al.*, 1974)

about 3.9 Ga to 3.1 Ga ago. Since that period the Moon's surface has been relatively quiescent, although occasional important events have occurred, e.g. the formation ages of the large craters Copernicus and Tycho are estimated to be 900 Ma and 100 Ma respectively.

3.4 Chemical characteristics

Rock analyses almost invariably show that eight elements—O, Si, Al, Fe, Ca, Mg, Ti and Na—constitute most of the lunar crust; a similar situation exists for the Earth's crust. The notable differences between the two are, however, the higher abundances of titanium in rocks from some sites and the relatively low abundances of potassium and sodium in most of the collected lunar rocks compared with terrestrial analogues. Chromium is also more abundant in the lunar specimens.

Some selected analyses of mare basalts and highland rocks are given in Table 3.3. They show the wide variations shown by some elements (e.g. Ti) in samples from different sites. Highly acidic rocks appear to be very rare or absent, although some mesostasis areas of the rocks can have high silica contents and be equivalent in composition to rhyolite, and some small grains of granitic glass have been found.

The rocks of the maria have basaltic compositions but with a significant range in Al_2O_3, as well as TiO_2, contents. The highland rocks show much higher CaO and Al_2O_3 contents than the mare basalts; this being a simple reflection of the feldspathic nature of the former. The regolith material also has wide compositional variations from site to site, again reflecting the mineralogy of the underlying terrain. (Analyses of the regolith samples are not given here. A summary of their compositions may be found in Taylor, 1975.)

There are many distinctive characteristics in the chemistry of lunar rocks that distinguish them from terrestrial analogues, and give clues about their petrogeneses and of the Moon's origin. These include:

(a) A depletion in volatile elements (such as K, Na, Pb, Hg, Tl) relative to both the Earth's surface rocks and chondrites; no water.

(b) The rocks have formed under conditions of very low oxygen fugacity.

(c) An enrichment in refractory elements compared to chondrites by about an order of magnitude.

(d) The existence of strong positive correlations between different element concentrations; some of these indicate that magmatic processes have been active.

(e) The presence of distinctive trace element distributions, including those of the rare-earth elements, which are sensitive indicators of petrogenetic processes.

(f) Some large differences in the minor element compositions of the igneous rocks, especially Ti variations.

Some of these characteristics will now be discussed.

3.4.1 Volatile elements

The general depletion of volatile elements in the Moon poses a number of problems. Sodium concentration, for example, is about an order of magnitude lower compared with equivalent terrestrial rocks but it is difficult to account for this by selective volatilization of elements during lava extrusion, especially

TABLE 3.3. Examples of the major element composition (wt %, as oxides) of some lunar rocks

	Olivine basalt Apollo 15	Quartz basalt Apollo 12	High-K basalt Apollo 11	Low-K basalt Apollo 11	Aluminous maria basalt Apollo 12	VLT basalt Apollo 17	Highland anorthosite	Highland anorthositic gabbro	Highland troctolite	Low-K Fra Mauro basalt	Medium-K Fra Mauro basalt
SiO_2	44.2	46.1	40.5	40.5	46.6	48.7	44.3	44.5	43.7	46.6	48.0
TiO_2	2.26	3.35	11.8	10.5	3.31	0.50	0.06	0.39	0.17	1.25	2.1
Al_2O_3	8.48	9.95	8.7	10.4	12.5	11.4	35.1	26.0	22.7	18.8	17.6
FeO	22.5	20.7	19.0	18.5	18.0	19.0	0.67	5.77	4.9	9.7	10.9
MnO	0.29	0.28	0.25	0.28	0.27	0.26	–	–	0.07	–	–
MgO	11.2	8.1	7.6	7.0	6.71	9.4	0.80	8.05	14.7	11.0	8.70
CaO	9.45	10.9	10.2	11.6	11.82	10.2	18.7	14.9	13.1	11.6	10.7
Na_2O	0.24	0.26	0.50	0.41	0.66	0.15	0.80	0.25	0.39	0.37	0.70
K_2O	0.03	0.071	0.29	0.096	0.07	0.04	–	–	–	0.12	0.54
P_2O_5	0.06	0.08	0.18	0.11	0.14	–	–	–	–	–	–
S	0.05	0.07	–	–	0.06	–	–	–	–	–	–
Cr_2O_3	0.70	0.46	0.37	0.25	0.37	0.60	0.02	0.06	0.09	0.26	0.18
Total	99.5	100.3	99.4	99.7	100.5	100.25	100.5	99.9	99.8	99.7	99.4

Sources: Taylor (1975); Vaniman and Papike (1977).

as the lavas are often uniform in sodium content. Furthermore, positive correlations exist between volatile and non-volatile element contents. The loss of volatiles during or soon after accretion, at a time when substantial melting of the Moon probably occurred, seems to be the favoured explanation. An alternative is that the accreting material was initially low in volatile elements.

3.4.2 Oxygen fugacity

Early in the investigation of lunar rocks it was realized that the petrogeneses had involved very reducing conditions. Many estimates have been made of the prevailing oxygen fugacity (f_{O_2}) during lunar rock formation. Sato *et al.* (1973) measured directly the fugacity between 1000 and 1200 °C with a solid-electrolyte oxygen cell, and obtained values in the range $10^{-12.3}$ to $10^{-15.7}$ atm $(10^{-0.7}$ to $10^{-2.7}$ nPa). These compare with values for terrestrial basalts of about 10^{-8} atm. The low f_{O_2} is the cause of the extremely low concentration of ferric iron and of a much higher $Eu^{2+}/(Eu^{2+} + Eu^{3+})$ ratio (about 0.7) in lunar rocks compared to terrestrial ones. The precise cause (or causes) of the reduced state of the surface and near-surface rocks is not yet established, although the high-vacuum environment may have contributed. It is possible, however, that the whole Moon is, on average, in a more oxidized state than is the Earth as the latter has a relatively large Fe–Ni core.

3.4.3 Refractory elements

Those elements considered to be early condensates during the cooling of the solar nebula (Grossman and Larimer (1974), and see Section 1.6) are enriched in the lunar surface relative to chondrites by about an order of magnitude. Elements such as Zr, Nb, the rare earths, U and Th are included in this group. The enrichment may be confined to the lunar surface layer and could have arisen from the fractionation of a predominantly molten Moon. During this event the crystallizing phases such as olivine and pyroxene would accumulate towards the interior, while the refractory elements, which are not incorporated into these phases, would be concentrated towards the surface.

It is possible that the Moon represents the accumulation of early condensates from the solar nebula without the significant addition of later, more volatile condensates. However, it would be necessary for this material to have been homogeneously accreted (see later).

3.4.4 Element correlations

Strong positive correlations exist between concentrations of elements, whether they be refractory or volatile, which are not incorporated into the major crystal phases during fractional crystallization; such correlations occur in both highland and mare rocks. A good example is shown by Rb (volatile) and Ba (refractory) in Fig. 3.3. Other examples include Cs–Ba; Zr–Nb; Th–rare earths (Taylor, 1975). Similar correlations exist in fractionated, terrestrial, igneous rocks and so it is reasonable to suggest that magmatic fractionation processes have operated in the formation of lunar surface rocks. Furthermore, the correlations occur in data for both highland and mare rocks; this fact suggests that accretion of the Moon was relatively homogeneous. It would be difficult to account for the observed correlations if the Moon had been heterogeneously accreted, unless after accretion a substantial part of the outer Moon had become molten and well homogenized.

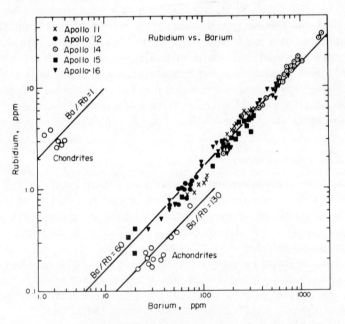

Fig. 3.3. Correlation between the volatile element, Rb, and the involatile element, Ba, for both maria and highland samples. (After Taylor, 1975)

3.4.5 Rare-earth element distributions

These distributions have proved to be particularly interesting because of the marked differences between the rare-earth element (REE) patterns of mare rocks and highland rocks. If the REE concentrations in lunar samples are normalized to REE abundances in chondrite meteorites (or 'cosmic abundances', see Chapter 2) any chemical fractionation which may have occurred is more readily discernible. Diagrams of chondrite normalized abundances versus atomic number for lunar samples are given in Figs 3.4 and 3.5. Certain features of these patterns are noteworthy:

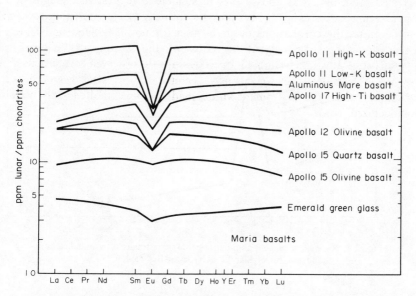

FIG. 3.4. Element abundance patterns, normalized to abundances in chondrites, for maria basalts. (After Taylor, 1975)

(a) All lunar samples (except anorthosite) are enriched in the REE relative to chondrites, some by as much as 200 times.
(b) All the mare basalts show a depletion (of varying extent) in Eu compared with the adjacent rare earths Sm and Gd. (This is referred to as a negative europium anomaly.)
(c) Many of the highland samples show an enrichment in Eu relative to Sm and Gd (positive europium anomaly).
(d) The patterns for mare basalts are not all parallel. There is relatively less of the lighter REE (La to Nd) compared with the heavier REE (Tb to Lu) in the Ti-rich basalts.

The enrichment of REE in lunar-surface samples compared with chondrites (point (a) above) may be explained partly by the selective concentration of these elements towards the surface (see Section 3.4.3). However, if these relatively high concentrations are present in much of the crust, then on average the whole Moon is about three times richer in REE than chondrites.

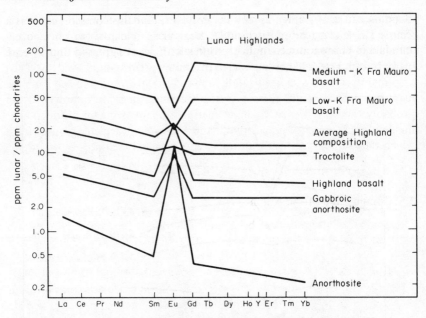

FIG. 3.5. Rare-earth element abundance patterns, normalized to abundances in chondrites, for highland rock types. (After Taylor, 1975)

The enrichment may have come about through multi-stage fractionation processes in which the primary differentiation and consolidation of the molten Moon were followed by repeated episodes of partial melting and crystal fractionation of crustal materials, to give magmas enriched in those elements which were not incorporated into the major rock-forming minerals. Variations in the slopes of the patterns (point (d) above) may also have arisen by the same means, but the smallness of these variations and the flatness of the REE normalized patterns suggest that fractional crystallization cannot have been extensive during the formation of the mare lavas; e.g. extensive fractional crystallization of clinopyroxene would leave a residual magma with a REE pattern showing enrichment in the lighter REE relative to the heavier REE (see Section 5.4).

The presence of the europium anomalies (points (b) and (c) above) are particularly intriguing. Firstly, they show that the parent magmas of both mare and highland rocks were not primary, as there is no evidence to suggest that europium had a different condensing behaviour from the other REE during the cooling of the solar nebula, i.e. the Moon is not depleted in Eu relative to the other REE. Secondly, the fact that the anomalies are opposite in nature in the highland and mare materials suggests that there may be a common cause. It seems improbable, however, that this common cause is a simple single-stage crystallization of plagioclase feldspar (which is often enriched in europium relative to the other REE, see Chapters 5 and 6) from mare basalt. For calculations show that a very substantial, and unlikely, amount of plagioclase crystallization (about 60 to 90 %) and removal from the magma would be required to give the measured negative anomaly. More probably, there has been multi-stage fractionation involving the production of highland crust enriched in europium, and a europium-depleted residue from which the mare basalts were later derived by partial melting.

3.5 Composition of the Moon

A number of attempts have been made to estimate the overall composition of the Moon. Three of these will be discussed to illustrate the kinds of approaches that may be adopted in tackling this problem. These are the work of Wänke *et al.* (1973), Ganapathy and Anders (1974) and Taylor (1975).

Wänke *et al.* were among those who showed that element concentration ratios are often constant for a large number of lunar samples over a wide concentration range even though one element may belong to the group of refractory elements and the other to the group of volatile elements. They noted particularly that the K/La ratio (at about 70) was constant for most samples. They also observed that refractory elements such as La are found in proportions close to those observed in the inclusions in the carbonaceous chondrite, Allende (see Section 1.3.1), while the ratios of many of the alkali metal abundances are similar to those in Cl chondrites. On the basis of these observations they concluded that the Moon is composed of two components—a high-temperature condensate (ϕ) equivalent to the Allende inclusions, and a chondritic component (χ). From a knowledge of the K and La abundances in these two components and the K/La abundance ratio in the Moon, the proportion of the two components may be calculated from equation (3.1):

$$(K/La) = 70 = \frac{A[K]_\chi + (1 - A)[K]_\phi}{A[La]_\chi + (1 - A)[La]_\phi} \tag{3.1}$$

where A is the chondritic fraction. Values are $[K]_\chi = 800$ ppm; $[K]_\phi = 0$; $[La]_\chi = 0.35$ ppm; and $[La]_\phi = 4.9$ ppm, which when substituted yield the value of 31 % for A. Hence, from the known proportions of chondritic and high-temperature condensate materials, an overall composition for the Moon may be calculated (Table 3.4) if the compositions of the two components are given.

Ganapathy and Anders (1974) also used the observed constancy in element

TABLE 3.4. *Composition of the Moon*

Element	Wänke et al. (1973)	Ganapathy and Anders (1974)	Taylor (1975)
Li ppm	–	8.7	–
C ppm	–	9.9	–
O %	–	41.42	–
F ppm	–	30	–
Na ppm	1770	900	–
Mg %	8.9	17.37	18.7
Al %	12.0	5.83	4.3
Si %	15.0	18.62	20.6
P ppm	–	538	–
S %	–	0.39	–
K ppm	250	96	100
Ca %	12.8	6.37	4.3
Sc ppm	81	40	20
Ti ppm	7000	3380	1800
V ppm	440	340	48
Cr ppm	–	1200	1330
Mn ppm	700	330	1000
Fe %	8.6	9.00	8.2
Co ppm	–	240	–
Ni ppm	–	0.51	–
Rb ppm	0.87	0.33	0.29
Sr ppm	93	60	43
Y ppm	24	10.9	7.5
Zr ppm	67	65	30
Nb ppm	4.8	3.3	2.2
Cs ppb	0.039	33	13
Ba ppm	33	16.8	17.5
La ppm	3.5	1.57	1.1
Sm ppm	–	0.86	0.75
Eu ppm	0.81	0.33	0.27
Tb ppm	–	0.22	0.18
Yb ppm	–	0.95	0.73
Lu ppm	–	0.160	0.11
Hf ppm	2.4	0.95	0.67
Th ppb	–	210	230
U ppb	0.086	59	60

ratios to determine the number and proportions of components but they started with the assumption that the Moon, as also the meteorites, resulted from the fractionation of the condensation products of the solar nebula. They reasoned that the Moon (and the Earth) must consist of three kinds of primordial condensate, as found in meteorites: (1) an early condensate, (2) metallic nickel-iron and (3) magnesium silicates. During cooling of the solar nebula these metal condensates reacted, in part, to form sulphides, and at the final stages of the cooling, a volatile-rich component condensed. Furthermore, just before or during aggregation some of the silicate and metal phases were briefly melted, with loss of their volatile elements. Thus, according to their model the Moon consists of seven components:

(a) Early condensate (0.301).
(b) Unmelted metal (−).
(c) Melted metal (0.061).
(d) Sulphide (0.011).
(e) Unmelted silicate (0.070).
(f) Melted silicate (0.557).
(g) Volatile-rich condensate, containing Cd, Hg, B, In, Tl, Pb, Bi, Cl, Br, I, H, C, N, Ar, Kr and Xe (0.0004).

The proportions of these can be estimated from certain chemical constraints; e.g. the amount of early condensate is given by the bulk U content of the planet (estimated from heat-flow data) and the amount of volatile-rich condensate can be ascertained from the Tl/U ratio (Tl being one of the volatile elements). The determined proportions from their model are given in brackets above. The composition of each component is obtained from the condensation sequence and solar-system abundances of the elements (Chapters 1 and 2). With these data it is then possible to calculate the composition of the Moon, see Table 3.4. The type of approach used by Ganapathy and Anders is proving a useful basis for calculations on the compositions of both the Moon and the Earth.

Taylor (1975) starts with the assumption that the Moon was accreted homogeneously, and then from a consideration of various data concludes that for the refractory elements, the Moon is, on average, about five times richer than the abundances in Type I carbonaceous chondrites (Cl). The geochemical constraints he uses are:

(a) The heat-flow data suggest an enrichment in U of about five times the Cl average U abundance.
(b) The orbital gamma-ray data suggest an enrichment of Th equal to about four times the abundance in Cl.
(c) The trace element abundances of highland rocks, if representative of the 60-km-thick crust, would require at least three to four times the Cl abundances to provide the amounts of the elements in the crust alone.

The concentrations of volatile elements are then computed from the observed volatile–refractory element ratios (e.g. as in Fig. 3.3). Taylor's estimate is given in Table 3.4.

These three methods of deriving the composition all use the observed constancy of selected element ratios in one way or another but they start out with very different postulates. The estimate by Wänke *et al.* (1973) is significantly different from the other two but there is still probably insufficient evidence to reject any of the three attempts. However, there is some evidence from the study of oxygen isotope ratios which imposes further constraints on the choice of theoretical model. The Earth and the whole Moon have virtually the same oxygen isotope composition, and they probably formed from the same batch of solar–nebular condensate. The oxygen isotope composition of the 'high-temperature' inclusions in the meteorite, Allende, differs markedly from that of the Moon and, therefore, the presence in the Moon of a large proportion of high-temperature condensate, as proposed in the model by Wänke *et al.*, is unlikely. The many attempts, with their different approaches, to estimate the Moon's composition will probably lead to significant revisions being made over a number of years.

Certain important features of the composition need to be emphasized. Compared to Cl meteorites the Moon has a very low abundance of volatiles, but is enriched in some of the early condensate elements (Ca, Al, U). Compared to Earth it is far poorer in iron (about four times lower), but is enriched in calcium and aluminium. Clearly, the Earth, Moon and Cl meteorites each contain very different proportions of the condensate fractions of the cooling solar nebula.

Suggested further reading

FRONDEL, J. W. (1975) *Lunar Mineralogy*. Wiley, New York, London. 323 pp.

RINGWOOD, A. E. (1979) *Origin of the Earth and Moon*. Springer-Verlag, New York, Heidelberg, Berlin. 295 pp.

SCHMITT, H. H. (1975) Apollo and the geology of the Moon. *Jl Geol. Soc. Lond.* **131**, 103–119.

TAYLOR, S. R. (1975) *Lunar Science: A Post-Apollo View*. Pergamon Press Inc., New York, Oxford. 372 pp.

4

The Earth

4.1 Introduction

The fractionation of the material of the Earth, either during or subsequent to accretion, has provided a layered, or shelled, structure with the characteristics given in Table 4.1. A knowledge of the composition of the whole Earth and of each of the major shells is important to our understanding of the formation of the planet, its subsequent history and the tectonic and petrogenetic processes operating today.

There is a consensus that our planet, Earth, formed directly by accretion of material formed in the solar nebula. The mode of accretion is, however, disputed; it being either homogeneous or heterogeneous. Homogeneous accretion occurs when the rate of accretion is slow compared with the rate of cooling of the nebula, so that chemical equilibrium is maintained between the condensates and the nebular gases; heterogeneous accretion arises when the accretion is relatively rapid so that newly formed condensates are isolated from, and cannot equilibrate with, the earlier condensates and the remaining nebular gases. Theoretical models of heterogeneous accretion (e.g. Clarke *et al.*,1972) sometimes have an early high-temperature accumulation of Fe–Ni alloy forming the Earth's proto-core, followed by accretion, at lower temperatures, of the outer parts. Theoretical considerations of the condensation sequence, however, indicate that silicates and oxides, both rich in calcium and aluminium, condense before Fe–Ni metal (Section 1.6). Anderson and Hanks (1972) have suggested therefore that the nucleus of the proto-earth consisted of such high-temperature condensates and was rich in Ca, Al, Ti, Th, U and the rare-earth elements. After formation of the nucleus, iron condensed next, then magnesium silicates, followed by condensates rich in K and the volatiles, including water. Other authors (e.g. Ringwood, 1966a) consider that homogeneous accretion is more probable, and that chemical reduction, loss of volatiles, melting and differentiation occurred simultaneously and as a direct result of the accretion process.

Theoretical models of heterogeneous or homogeneous accretion are not without their attendant difficulties. It is easier to account for the difference in

TABLE 4.1. *The shells of the Earth*

	Mean thickness (km)	Mean density (g cm⁻³)	Volume (10²⁰ m³)	Mass (10²³ kg)	Mass fraction (%)	Region	Depth to boundaries (km)	Feature of region
ATMOSPHERE	3.75	—	—	0.000051	~0.00008	A	—	Heterogeneous
HYDROSPHERE	~17	1.02	0.0138	0.0141	0.024		—	
CRUST		2.8	0.1021	0.286	0.48		0	
							33* (continental)	
MANTLE	2883	4.5	9.00	40.16	67.1	B	413	Heterogeneous
						C	984	Transition region
						D	2898	Probably homogeneous
CORE	3473 radius	11.0	1.76	19.36	32.4	E	4982	Homogeneous fluid
						F	5121	Transition layer
						G	6371	Inner core-solid?

* Crustal thickness under the continents is variable and averages about 33 km. The crust is 5 to 8 km thick under the oceans.

bulk compositions of the Earth and the Moon, as well as the diversity of meteorite types and features, on the basis of heterogeneous rather than homogeneous accretion. Some recent, theoretical computations on the condensation sequence, however, support an homogeneous accretion. Consideration of the equilibria of a large number of compounds, each consisting of one or more of the elements Na, K, F, Cl, Br and P, shows that only homogeneous accretion can account for the presence of certain minerals (e.g. schreibersite, $(Fe, Ni)_3 P$) and the absence of others (e.g. alkali sulphides and ammonium phosphates) in meteorites, and also for the observed abundances of the above elements in the Earth (Fegley *et al.*, 1980).

In view of the uncertainties over the mode of condensation, it is not surprising that there is also uncertainty about the amount of time required for accretion. It may have been rapid—about 10^5 a— or considerably longer. The cessation of the process, or at least the time when the Earth effectively became a closed system, occurred about 4.6×10^9 a ago—'the age of the Earth' (York and Farquhar, 1972). At present, some of the oldest known rocks are those in the Godthaab district, West Greenland, with an average age for a number of samples of 3.75×10^9 a (Black *et al.*, 1971; Moorbath *et al.*, 1972).

The evidence for the layered structure of the Earth is mainly geophysical and is reviewed in many textbooks on geophysics (e.g. Jacobs *et al.*, 1974). Some of the basic data are given in Table 4.1 and Figs. 4.1 and 4.2. Pressure, density and temperature variations within the Earth have been estimated with the aid of theoretical models of its structure. Those data in Table 4.1 and Fig. 4.1 are based upon a model by Bullen (see Jacobs *et al.*, 1974).

4.2 Composition of the whole Earth

Only a very small fraction of the mass of the Earth is available for direct study and analysis—the crust, hydrosphere and atmosphere together make up less than 1 % by mass, and so it is necessary to use indirect methods to estimate the Earth's composition. Two possible starting-points are available. One is to attempt to determine the compositions of each of the major subdivisions of the interior and to integrate them together according to their estimated mass fractions. The second is to postulate the parentage of the Earth's material and then, by determining the composition of the parent, to arrive at a composition for the whole Earth. Both approaches are constrained by the results of geophysical experiments, but the former approach would still produce many possible compositions unless further constraints, such as the nature of possible parent materials, are imposed.

It has always appeared as a reasonable assumption that the Sun, planets and meteorites of the solar system are related in composition since they were very

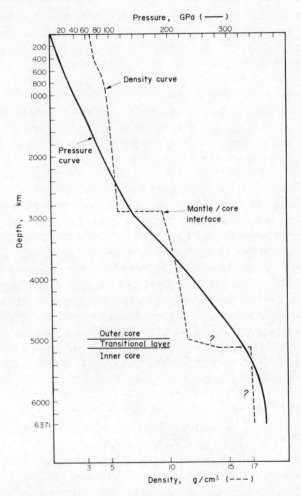

FIG. 4.1. Pressure and density variation within the Earth according to a model by Bullen.

probably derived via the same set of nucleosynthetic events. It is justifiable, therefore, to postulate that the abundances of the non-volatile elements are in the same proportion in the Earth as they are in the Sun or meteorites, or as in the 'cosmic abundance' compilations (discussed in Chapter 2). A number of estimates have been made on this basis, two examples being by Mason (1966) and Ringwood (1966a). Mason assumed that the mantle and crust have the same composition as the 'average' chondrite; that the iron alloy of the core has the composition of nickel–iron in chondrites, plus the average amount (5.3%)

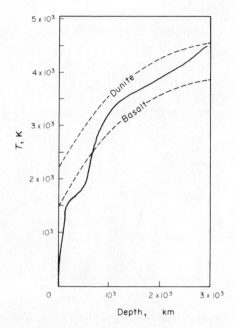

Fig. 4.2. Temperature variation (continuous line) with depth in the mantle as estimated by Tozer. Dashed curves are melting temperatures for dunite and basalt. (After Jacobs *et al.*, 1974)

of troilite found in these meteorites. His estimate, along with that by Ringwood, is given in Table 4.2. Ringwood assumed that the Earth is made up of material with the composition of C1 chondrites.

The validity of using the compositions of chondrites as an indication of the Earth's composition needs to be tested. Larimer (1971) developed the approach initiated by Hurley (1968a, b) who had attempted to define the upper and lower limits for the concentrations of potassium and rubidium in the whole Earth, without resorting to meteorite compositions. The method may be illustrated for potassium. An absolute lower limit for the terrestrial abundance of this element may be fixed if it is assumed that *all* of the Earth's ^{40}Ar is in the atmosphere and has been derived through radioactive decay of ^{40}K. With a knowledge of: the age of the Earth, the decay constants for ^{40}K, the proportion of ^{40}K in naturally occurring potassium at the present day and the amount of argon in the atmosphere, the lower limit may be computed. It is found to be about 520×10^{15} t, corresponding to about 90 ppm. This assumes that there has been no loss of argon from the atmosphere. An upper limit is set by giving a conservative estimate of the amount of ^{40}Ar released from the solid

TABLE 4.2. *Estimated com-*
positions of the Earth (wt %)

	By Mason (1966)	By Ringwood (1966a)
Fe	34.63	31
O	29.53	30
Si	15.20	18
Mg	12.70	16
Ni	2.39	1.7
S	1.93	—
Ca	1.13	1.8
Al	1.09	1.4
Na	0.57	0.9
Cr	0.26	—
Mn	0.22	—
Co	0.13	—
P	0.10	—
K	0.07	—
Ti	0.05	—

Earth and computing the potassium content as before. The loss of 50% of the Earth's argon to the atmosphere was considered by Hurley to be a suitably conservative figure. This puts the upper limit for potassium at 1040×10^{15} t, or about 175 ppm. (There is, however, little evidence to support the 50% or any other retention figure, so until this evidence is forthcoming the upper limit for potassium must remain something of an open question.)

A similar approach may be adopted for other elements and is especially suitable for radioactive elements and their daughters. The estimates so obtained may then be compared to abundances in different meteorites. Larimer did this for twenty-two elements and showed the chondrite model of the Earth to be acceptable, except that chondrites are overabundant in Na, K and Rb. For volatile elements such as Bi, Hg, Pb and Tl the terrestrial abundances fall below or within chondritic abundances but some are above the level in eucrites (achondrites). Also, the eucrites are consistently too high in their abundances of the refractory and alkali elements. Thus a chondrite model is preferable to an achondrite one for the Earth's composition. Larimer's estimates of the terrestrial abundances in ordinary chondrites (H, L and LL chondrites) and eucrites are given in Table 4.3. The only points of comparison that can be made with the estimates by Mason and by Ringwood are for sodium and calcium and it can be seen that agreement is poor (Tables 4.2 and 4.3). However, with the uncertainties that still exist over the origin of meteorites it is important to try to assess the composition of the Earth by as many independent means as possible.

TABLE 4.3. *Element abundances in the Earth and selected meteorites*

	Ordinary chondrites	Earth	Eucrites
Na %	0.68	0.07	0.3
K ppm	850	130	360
Rb ppm	2.8	0.42	0.24
Cs ppm	4 to 619	9	12
Ca %	1.21	0.95	7.49
Sr ppm	11	12	85
Ba ppm	4.1	5.3	35
Sc ppm	8.0	16.5	27
Y ppm	2.0	4.2	23
La ppm	0.24	0.52	3.7
Th ppb	43	35	440
U ppb	12	12	130

Based on Larimer (1971).

Probably one of the more successful approaches at the present time is the one by Ganapathy and Anders (1974) who argued that the Earth and the Moon each consist of at least seven different components of the same type that comprise the ordinary chondrites; see discussion in Section 3.5. Known geochemical constraints, similar in type to those used in computing the Moon's composition (Section 3.5), enabled the authors to estimate the proportions of the seven components in the Earth, as follows: (a) early condensate (0.092), (b) melted metal (0.240), (c) unmelted metal (0.071), (d) troilite (0.050) (e) melted silicate (0.418), (f) unmelted silicate (0.114), (g) volatile-rich material (0.015). In contrast to the Moon, the Earth is estimated to be much richer in both the metal and volatile-rich components but poorer in the early condensate fraction (cf. Section 3.5). Their estimate of element abundances is given in Table 4.4, and Fig. 4.3. The diagram reflects the assumption that each trace element is associated exclusively with one component and, therefore, that cosmochemically similar elements are enriched or depleted by the same factor. The estimate may be contrasted with one (Table 4.4) by Smith (1977, 1979) who set out to test the model by Ganapathy and Anders, by using many sources of data about the compositions of the individual shells of the Earth. He assumed the mantle to consist dominantly of garnet peridotite and the core to be of a composition represented by iron meteorites admixed with troilite. It is interesting to note that despite the different approaches, the two estimates show reasonable agreement (except perhaps for sodium and potassium) and allow some credence to be placed in them. They are, at the present time, the best practical assessments of a bulk composition that we have.

TABLE 4.4. *Abundances of elements in the Earth according to estimates by Ganapathy and Anders (1974), G, and Smith (1979), S, μg/g, unless otherwise indicated)*

		G	S			G	S
1	H	78	66	45	Rh	0.32	
3	Li	2.7		46	Pd	1.00	
4	Be ng/g	56		47	Ag ng/g	80	
5	B ng/g	470		48	Cd ng/g	21	
6	C	350	220	49	In ng/g	2.7	
7	N	9.1	19	50	Sn	0.71	
8	O%	28.50	31.3	51	Sb ng/g	64	
9	F	53		52	Te	0.94	
11	Na	1580	850	53	I ng/g	17	
12	Mg%	13.21	13.7	55	Cs ng/g	59	
13	Al%	1.77	1.83	56	Ba	5.1	
14	Si%	14.34	15.1	57	La	0.48	
15	P	2150	1830	58	Ce	1.28	
16	S%	1.84	2.91	59	Pr	0.162	
17	Cl	25	45	60	Nd	0.87	
19	K	170	130	62	Sm	0.26	
20	Ca%	1.93	2.28	63	Eu	0.100	
21	Sc	12.1		64	Gd	0.37	
22	Ti	1030	928	65	Tb	0.067	
23	V	103		66	Dy	0.45	
24	Cr	4780	4160	67	Ho	0.101	
25	Mn	590	470	68	Er	0.29	
26	Fe%	35.87	31.7	69	Tm	0.044	
27	Co	940		70	Yb	0.29	
28	Ni%	2.04	1.72	71	Lu	0.049	
29	Cu	57		72	Hf	0.29	
30	Zn	93	93	73	Ta ng/g	29	
31	Ga	5.5		74	W	0.250	
32	Ge	13.8		75	Re ng/g	76	
33	As	3.6		76	Os	1.10	
34	Se	6.1		77	Ir	1.06	
35	Br ng/g	134		78	Pt	2.1	
37	Rb	0.58		79	Au	0.29	
38	Sr	18.2		80	Hg ng/g	9.9	
39	Y	3.29		81	Tl ng/g	4.9	
40	Zr	19.7		82	Pb	0.13	
41	Nb	1.00		83	Bi ng/g	3.7	
42	Mo	2.96		90	Th ng/g	65	
44	Ru	1.48		92	U ng/g	18	20

4.3 Composition of the core

The idea that the core consists of an alloy of iron and nickel has a long history but is still accepted today, albeit with some modifications. Some of the core, the outer part, is likely to be fluid in nature, since no shear waves penetrate the mantle–core interface. Furthermore, the alloy cannot be pure

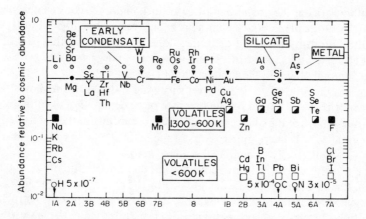

FIG. 4.3. Model composition of the Earth, relative to cosmic abundances. It is based on the assumption that each component contains its cosmic complement of trace elements. The ordinate is divided according to the chemical groups of the periodic table. (After Ganapathy and Anders, 1974)

Fe–Ni for this has too high a density and too small a bulk sound speed to be compatible with geophysical data. A lighter element needs to be added to the estimated composition. Sulphur and oxygen have both been proposed (Murthy and Hall, 1970; Anderson *et al.*, 1972; Ringwood, 1977).

About 10% by weight of sulphur would be required in the core to produce the correct density, and there is sufficient sulphur in C1 chondrites to make this feasible. Ringwood (1966 a, b), however, considers that a large proportion of the sulphur of the solar nebula dust was lost, together with other volatiles, as a result of heating and melting during the primary accretion process of the Earth. He favours oxygen as the lighter alloying element (Ringwood, 1977), but the lack of data on the physical properties of Fe–Ni alloys containing dissolved FeO at the high temperatures and high pressures of the core precludes a reliable estimation of the amount of oxygen.

Geophysical data are consistent with either of these alternative alloying materials, but, at the present time, there are many more proponents for the inclusion of sulphur than there are for oxygen.

4.4 Nature of the mantle

Since the general consensus of opinion is that the core consists of a metallic alloy, so it is that the mantle is essentially made up of silicate, but with significant variations in nature with depth. The uppermost part, part of the lithosphere, is probably both laterally and vertically heterogeneous; it is about

70 km thick under the oceans and twice as thick under the continental shields. From the nature of material having a probable deep-seated origin—such as the nodules found in kimberlite diatremes—the mineralogy of this part of the mantle is likely to consist of olivine, pyroxene and garnet, and with some spinel, amphibole and phlogopite as minor phases.

Beneath the lithosphere is a low-velocity zone at depths from about 100 to 150 km. It is characterized by the presence of some liquid derived by partial melting ($\sim 1\%$?) of the silicate material, thus accounting for the low velocity of the seismic waves. These upper layers of the mantle—down to the transition zone—are collectively known as the upper mantle.

The transition zone is from about 400 km depth to about 900 or 1000 km depth. Its top is thought to be demarcated by the change in magnesium–iron orthosilicate from the olivine structure to the spinel structure. Other structural transitions occur in this zone. The lower mantle, from the base of the transition zone to the core, is probably homogeneous and is considered (e.g. Anderson *et al.*, 1972) to contain more iron than the upper mantle and probably consists of a mixture of $(Mg, Fe)SiO_3$ and $(Mg, Fe)O$ minerals. Structural transformations of minerals are also likely to occur with depth in the lower mantle.

The nature and composition of the upper mantle has been the subject of intensive research over many years principally because it is a probable source region for magmas which intrude the crust or flow out upon the Earth's surface. For example, the presence of diamonds in kimberlites indicates that these rocks must have existed at pressures corresponding to depths in the Earth of 120 to 200 km. Also, Green *et al.* (1967) and Green and Ringwood (1967) claim that the three types of basalt magma—high alumina, alkali, and tholeiite—could all be produced in the upper mantle but under different load pressures and with different directions of fractionation of the magmatic liquids during their ascent to the location of igneous activity.

To understand the production and evolution of different magmas it is necessary, therefore, to know the composition and mineralogy of the upper mantle and the appropriate phase relationships at different pressures and temperatures. Here only brief comments will be made on the possible nature of the upper mantle and its role in magma production. Fuller accounts of this field of study may be found in Carmichael *et al.* (1974), Wyllie (1971) and Yoder (1976).

The geophysical and petrological constraints on the rock types of the upper mantle preclude most except eclogite and peridotite. Eclogite, however, has too high a density (usually about 3.4–3.6) to be predominant in the upper mantle. Furthermore, an olivine–pyroxene peridotite is the commonest type of xenolith in magmas that are postulated to have risen from the mantle. This is a reflection, in part, of the limited stability of the garnet-bearing rocks, such as eclogite, at high temperatures and low pressures but it has also been taken as evidence that the major part of the upper mantle is close to peridotite in nature and composition. Most peridotite xenoliths themselves are unlikely to represent the true composition, because they generally contain too little Ca, K, Sr, Ba and certain other elements to yield basalt on partial melting. More probably the xenoliths represent a residuum after the removal of the basaltic fluid.

This kind of reasoning led Ringwood (1966, 1979) to postulate that the upper mantle beneath a peridotitic part of the lithosphere has the composition of a mixture of 3 parts peridotite and 1 part basalt—termed 'pyrolite'. On fractional melting it would yield 20–40 % of typical basalt magma, and leave a peridotite or dunite residuum. The estimated composition of the upper mantle on a 'pyrolite model' is given in Table 4.5 together with two estimates of mantle composition (Mason, 1966; Ringwood, 1966 a and b) derived from the compositions of chondrites. Ringwood has emphasized that the choice of the 3 : 1 ratio is somewhat arbitrary and significant variations are possible within the framework of the pyrolite model. Heterogeneity in parts of the upper mantle (Fig. 4.4) is also indicated.

The 'pyrolite model' is only one of a number of possibilities. It does not account adequately for the production of volatiles (water, chlorine, sulphur,

TABLE 4.5. *Estimated composition of the mantle (wt %)*

	From meteorites		'Pyrolite' Ringwood (1966a)	From lherzolite[a] Hutchison (1974)
	Mason[b] (1966)	Ringwood (1966a)		
SiO_2	48.1	43.2	45.2	45.0
MgO	31.1	38.1	37.5	39.0
FeO	12.7	9.2	8.0[c]	8.0
Al_2O_3	3.1	3.9	3.5	3.5
CaO	2.3	3.7	3.1	3.25
Na_2O	1.1	1.8	0.57	0.28
Cr_2O_3	0.55	—	0.43	0.41
MnO	0.42	—	0.14	0.11
P_2O_5	0.34	—	0.06	—
K_2O	0.12	—	0.13	0.04
TiO_2	0.12	—	0.17	0.09
NiO	—	—	—	0.25

[a] Estimate for undepleted upper mantle.
[b] Estimate for mantle plus crust.
[c] $Fe_2O_3 = 0.46$, not included.

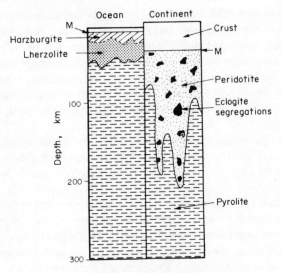

FIG. 4.4. Chemically zoned model for the upper mantle. (After Ringwood, 1979)

etc.) that are so much part of volcanic activity, and the degree of significance in the upper mantle of the rock types eclogite and lherzolite is still an open question. Some of these rocks contain a few percent of hydrous minerals, such as phlogopite, which accommodate a number of volatile elements. An estimate (Hutchison, 1974) of the composition of undepleted upper mantle material, based upon spinel lherzolite xenoliths in basalt, is given in Table 4.5. This gives much lower values for the abundances of Na_2O, K_2O and TiO_2, but in other respects is similar to the estimates by Ringwood or Mason.

4.5 Composition of the crust

The crust is the outermost zone of the solid Earth and its boundary with the mantle is at that depth where there is a sudden increase in the velocities of compressional and shear seismic waves—the Mohorovicic discontinuity (Moho). The depth of the Moho is, however, variable—see Table 4.1. It should be noted that some authors include the atmosphere, hydrosphere and biosphere within the term 'crust'. The lithic crust alone has been called the lithosphere but current usage has the outermost layer of the mantle and the lithic crust together comprising the 'lithosphere'. The lithosphere is essentially rigid, and rests on a more plastic layer, the asthenosphere, capable of slow deformation and flow.

The crust may be divided into two categories—the oceanic crust and the continental crust. It will be immediately recognized that because of the wide variety of rock types and the diversity in their modes of occurrence, an accurate estimate of the average composition of the continental crust is hard to obtain. The deeper layers of the oceanic crust are poorly known and so, despite the greater uniformity of the oceanic crust compared with the continental crust, an estimated composition is also liable to significant error.

Parker (1967) carried out a careful review of previous attempts to estimate crustal composition. Four approaches had been used to obtain an overall composition for the crust:

(a) Averaging of available rock analyses.

(b) As (a) but weighted in proportion to the rocks' occurrences.

(c) Abundance of elements based on crustal models.

(d) An indirect assessment, e.g. by estimating the mass balance of mafic and felsic rocks to give, through weathering, etc., the observed average composition of sediments (or selected sediment).

Methods (a) and (b) have more often been applied to continental crust than to the whole crust, and clearly method (d) is restricted to the continental crust. Method (c) is an extension of method (b) and still requires assumptions about the nature of the deeper crust.

Clarke and Washington (1924) used method (a) to produce an estimate for the continental crust (Table 4.6). Although obvious objections can be and indeed were raised about the approach we will see that their estimate has marked similarities to those produced by more sophisticated methods.

Method (b), although more quantitative than method (a) by taking into account rock distribution, still suffers from our lack of knowledge of the nature of the deeper crustal layers. In attempting to overcome the serious deficiency of any estimate based on assumptions concerning the deeper crust, Pakiser and Robinson (1967) used seismic data to determine the proportions of mafic and silicic rocks at depth. Their study was confined to the crust of the United States where they were able to distinguish two distinct crustal layers on the basis of seismic wave velocities. The upper layer, with the slower seismic velocity, correlates with rocks of silicic composition, while the lower layer, of faster wave velocity, correlates with rocks of basic (or mafic) composition. These data, together with average rock compositions, allowed them to compute an estimate for continental crustal composition (Table 4.6).

Earlier, Poldervaart (1955) had calculated the average composition of four structural regions of the Earth—the deep oceanic region; the sub-oceanic region (i.e. continental shelf and geosynclinal regions); the continental shield region; and the younger folded-belts region. His estimate for the composition of the Earth's crust, based on a model of crustal structure (in which he was

TABLE 4.6. *Estimates of the chemical composition of the crust*

Oxide	Clarke and Washington (1924) Entire crust	Poldervaart (1955)			Ronov and Yaroshevskiy (1976)				Goldschmidt (1954) Continental crust	Taylor (1964) Continental crust	Pakiser and Robinson (1967) Continental crust
		Oceanic crust	Continental crust	Entire crust	Oceanic crust	Continental crust	Subcontinental crust	Entire crust			
SiO_2	59.1	46.6	59.4	55.2	49.4	59.3	58.8	57.1	59.2	60.3	57.9
TiO_2	1.0	2.9	1.2	1.6	1.4	0.7	0.8	0.9	0.8	1.0	1.2
Al_2O_3	15.2	15.0	15.6	15.3	15.4	15.0	14.9	15.0	15.8	15.6	15.2
Fe_2O_3	3.1	3.8	2.3	2.8	2.7	2.4	2.4	2.5	3.4	*	2.3
FeO	3.7	8.0	5.0	5.8	7.6	5.6	5.8	6.0	3.6	7.2*	5.5
MnO	0.1	0.2	0.1	0.2	0.3	0.1	0.1	0.2	0.1	0.1	0.2
MgO	3.4	7.8	4.2	5.2	7.6	4.9	5.1	5.5	3.3	3.9	5.3
CaO	5.1	11.9	6.6	8.8	12.5	7.2	7.4	8.4	3.1	5.8	7.1
Na_2O	3.7	2.5	3.1	2.9	2.6	2.5	2.5	2.5	2.1	3.2	3.0
K_2O	3.1	1.0	2.3	1.9	0.3	2.1	2.0	1.7	3.9	2.5	2.1
H_2O	1.3	—	—	—	—	—	—	—	3.0	—	—
P_2O_5	0.3	0.3	0.2	0.3	0.2	0.2	0.2	0.2	0.2	0.2	0.3

All estimates on a water-free and CO_2-free basis except those by Clarke and Washington, and Goldschmidt.

* All Fe expressed as FeO.

forced to make a number of assumptions, e.g. that the underlying rock of the ocean floor is olivine basalt), is given in Table 4.6. Ronov and Yaroshevskiy (1976) adopted a similar approach but with the advantage of having considerably greater and more reliable data. Their computed compositions for three structural regions—oceanic crust, sub-continental crust, continental crust—as well as for the entire crust are given in Table 4.6. Comparisons with the estimates by Poldervaart show good agreement except in the cases of Ti in all regions and of K in the oceanic crust.

Many of the estimates of crustal composition are based on average igneous rock compositions despite the fact that large areas of continent may be composed of metamorphic rock. The rationale for this has been that the compositions of metamorphic rocks, whether derived from pre-existing igneous or sedimentary material, are similar to the broad types of 'average granite' or 'average basalt'. Evidence is accumulating to suggest that this is an oversimplification. For example, it has been shown (Holland and Lambert, 1972) from geochemical studies of the Canadian Shield and the Precambrian rocks of Scotland that for these regions an estimate incorporating a large proportion of granulite facies rock will be more representative of their crustal composition than one based entirely on igneous rock compositions, for granulites are known to differ in composition from broad types of igneous rocks.

A very original method of estimation, by Goldschmidt (1933, 1954), uses the idea that glacial clays are compositionally representative of the crust from which they were derived, in that glacial erosion is essentially a mechanical process. An average of many analyses of glacial clays from Norway is remarkably similar to the Clarke and Washington estimate, Table 4.6; the lower sodium and calcium values being attributable to slight leaching.

Where there is a marked difference between the trace element abundances in granitic rocks and basaltic rocks then it is possible to assess the proportions of the two rock types—granite and basalt—which have contributed to derived sediments, provided that the elements used are not labile during weathering. The rare-earth elements (REE) have been used in this way (Taylor, 1964) to show that a mix of 1 : 1 mafic to felsic igneous rock would produce the REE distribution in 'average sediment'. This igneous rock mix might thereby represent the composition of average continental crust, Table 4.6. A mix of one part felsic to one part mafic rock does not give a perfect match for all the REE but, as more data become available and with improved methods of trace element analysis, such an approach is to be encouraged. The method also provides an estimate of the concentrations of trace elements in the continental crust, Table 4.7.

Examination of Table 4.6 shows the similarity of the various estimates for

TABLE 4.7. *Abundances (μg/g) of minor and trace elements in the continental crust (from Taylor, 1964)*

3	Li	20	37	Rb	90	64	Gd	5.4
4	Be	2.8	38	Sr	375	65	Tb	0.9
5	B	10	39	Y	33	66	Dy	3
6	C	200	40	Zr	165	67	Ho	1.2
7	N	20	41	Nb	20	68	Er	2.8
9	F	625	42	Mo	1.5	69	Tm	0.48
16	S	260	47	Ag	0.07	70	Yb	3
17	Cl	130	48	Cd	0.2	71	Lu	0.5
21	Sc	22	49	In	0.1	72	Hf	3
23	V	135	50	Sn	2	73	Ta	2
24	Cr	100	51	Sb	0.2	74	W	1.5
27	Co	25	53	I	0.5	79	Au	0.004
28	Ni	75	55	Cs	3	80	Hg	0.08
29	Cu	55	56	Ba	425	81	Tl	0.45
30	Zn	70	57	La	30	82	Pb	12.5
31	Ga	15	58	Ce	60	83	Bi	0.17
32	Ge	1.5	59	Pr	8.2	90	Th	9.6
33	As	1.8	60	Nd	28	92	U	2.7
34	Se	0.05	62	Sm	6			
35	Br	2.5	63	Eu	1.2			

the major elements despite the different methods of approach. It seems reasonable to suppose that we have now an accurate appraisal of the composition of the crust, at least as far as the major elements are concerned.

4.6 The atmosphere and hydrosphere

Part Three of this book discusses some aspects of aqueous geochemistry; the compositions of ocean water, ground waters, etc., are given there. More than 70 % of the Earth's surface is covered by water, and the volume of ocean water is about $1.36 \times 10^9 \, km^3$, and so the importance of the hydrosphere in geochemical processes on the Earth's surface is obvious. Also, there has been an increasing awareness in recent years of the significance of the interactions of ground waters or ocean water with rocks both during and after magmatic activity. Much research is in progress in investigating this geochemical process and its possible importance in ore-formation. This aspect is also discussed in later chapters.

Some data of the hydrosphere and atmosphere are to be found in Tables 4.1, 4.8 and 4.9 (see also Part Three).

4.7 Geochemical differentiation

The previous sections have shown that the Earth is highly differentiated. Probably only three elements—Fe, Ni, S (or possibly O)—comprise most of

TABLE 4.8. *Structure of the atmosphere*

Region	Height (km)	Temperature change with height
Thermosphere	$\sim 80–700$	Increase, up to 1500 K, but variable
Mesosphere	50–80	Decrease, variable minimum
Stratosphere	$\sim 12–50$	Increase, up to ~ 270 K
Tropopause	12 (mean)	Mean temperature: 210 K
Troposphere	0–12	Decrease from 288 K mean

TABLE 4.9. *Composition of dry air (at sea level) (% volume)*

N	78.08	Ne	0.0018
O	20.95	He	0.00052
Ar	0.93	Kr	0.00011
CO_2	0.031	CH_4	0.0002

Also traces of: NO_2; H_2; Xe; SO_2.

the core, and only nine elements—O, Si, Al, Fe, Mg, Ca, Na, K and Ti (on a water-free basis)—make up more than 99 % of the crust. For the whole Earth, many elements, including Na, K, Al, U and the halogens, are enriched towards the surface while others, such as Mg, S, Fe, Cr, Co and Ni, are enriched in deeper layers. These facts can be appreciated from Fig. 4.5, which is a plot of the abundances of selected element concentrations in the continental crust (Tables 4.6 and 4.7) normalized to their 'cosmic' abundances (Table 2.2), when compared with Fig. 4.3. (Part Two of this book gives some explanations of the observed distribution.) The debate still continues at to whether the differentiation occurred during accretion from the solar nebula or subsequent to accretion during an internal heating and melting stage in the Earth's evolution, or possibly by a combination of both.

The continuing presence of volcanism (including the emanation of the inert gases) and tectonic activity means that geochemical differentiation of the Earth is still operating. For example, volcanic activity discharges about 1 km³ a⁻¹ of material, some of which is undoubtedly remelted crustal material while some is from the upper mantle. This process will modify the composition of the continental crust if, as appears probable, the amount of sediment lost to the ocean basins is not of the same magnitude.

Another persistent change in the chemistry of the Earth with time is the reduction in abundances of unstable isotopes through radioactive decay, Fig. 4.6. The amount of ^{40}K in the Earth was nearly twelve times greater 4.5 Ga ago compared with its abundance today. If we assume that there is about 130 ppm K in the Earth (Smith, 1979; see Section 4.2 and Table 4.4) then it can be readily

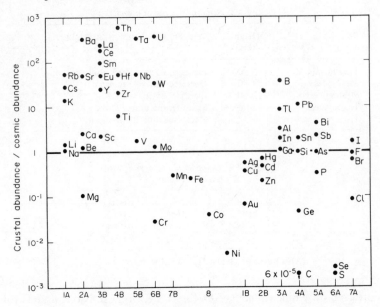

FIG. 4.5. Composition of the crust, normalized to cosmic abundances of the elements and to $Si = 10^6$ atoms. The diagram is in the same format as Fig. 4.3, with which it may be compared

shown that the present-day heat output from all the ^{40}K in the Earth is about 8 $\times 10^{21}$ J a^{-1} (2×10^{21} cals a^{-1}). A twelvefold increase in this value, together with associated increases in the heat output from other radioisotopes, including ^{232}Th, ^{235}U and ^{238}U, has important implications concerning the early heating and possible melting of the Earth.

In an attempt to characterize broadly the geochemical behaviour of elements during major differentiation processes, Goldschmidt in 1932 classified elements into various groups on the basis of their distribution among the phases in meteorites:

Siderophile elements: those which tend to concentrate in metallic iron. Typical examples: Ni; Co; Pt.

Chalcophile: those concentrating in the sulphide phase. Typical elements: Se; As; Zn; Cd.

Lithophile: those concentrating in the silicate phases. Typical elements: Al; Na; K; Ca.

Atmophile: gaseous elements occurring in the uncombined state: N; Ar; etc.

Goldschmidt also included a biophile group. A more detailed discussion and listing of the elements is given in Goldschmidt's book (1954). Obviously the

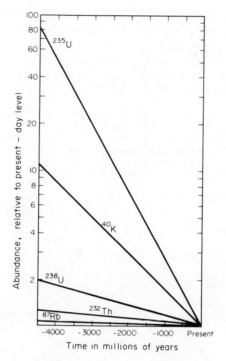

FIG. 4.6. Relative changes (log scale) in the abundances of some radioactive
isotopes in the Earth over 4500 million years.

groupings apply to the chemical conditions prevailing at the time of
differentiation of the phases, and have no wider chemical implications. Some
elements fall into two or more categories, e.g. nickel shows both chalcophile
and siderophile behaviour, but the classification is still widely used in
meteoritics and planetary studies and it has a particular relevance to work on
metallic ore-genesis when sulphide, silicate and oxide minerals can be involved.

Suggested further reading

MCELHINNEY, M. W., editor (1979) *The Earth: Its Origin, Structure and Evolution.* Academic Press.
597 pp.

RINGWOOD, A. E. (1979) *Origin of the Earth and Moon.* Springer-Verlag, New York, Heidelberg,
Berlin. 295 pp.

SMITH, J. V. (1979) Mineralogy of the planets: a voyage in space and time. *Mineralogical Magazine,*
43, 1–89.

WINDLEY, B. F., editor (1976) *The Early History of the Earth.* J. Wiley & Sons, London, New York,
Sydney, Toronto. 619 pp.

PART TWO

MAGMATIC AND METAMORPHIC SYSTEMS

5

The Data of Element Distribution

5.1 Introduction

The beginning of the twentieth century saw an important landmark in the science of geochemistry with the publication in 1908 of F. W. Clarke's *Data of Geochemistry*. This work not only gave a summary but also a critical appraisal of the already vast number of available geochemical data. Included in its scope were the chemical relationships of natural waters, volcanic gases, rocks, minerals and ores, as well as petroleum and coal. At that time, geochemistry had evolved into being much more than a descriptive science but it was soon to become even more diversified, placing a greater reliance on the sciences of chemistry and physics as well as mathematics. Today, we endeavour through the use of pertinent geochemical observations to describe and understand the formative processes of the multifarious materials of both the Earth and the solar system. To this end it is useful to have an overview of the recorded behaviour of the elements within different geochemical systems.

Part Two of this book is concerned with igneous and metamorphic rock systems, for which there is a substantial body of analytical data, much of it reviewed and summarized in the treaties on geochemistry listed in Appendix VII. The reader is referred to such works for listings of the average compositions of rock types and of the concentration levels of individual elements in the commoner minerals and rocks. In this chapter selected aspects of those observations of particular relevance to modern geochemistry and petrology are introduced. The following three chapters examine the controlling processes of element distribution, while the last chapter of Part Two discusses the relevance of isotope studies to petrogenesis and ore-formation.

5.2 The presentation of analytical data

5.2.1 Rock or mineral analyses

The concentrations of all the major elements of any rock or mineral are usually presented in terms of the weight percent of each oxide (as in Table 5.1).

Occasionally it is more appropriate to list the mole percent of each oxide. In the case of mineral sulphides or halides, the analyses are best quoted as weight percent of the individual elements. Mineral analyses are given also as the number of cations (or atoms) of each element for a fixed number of anions, usually the oxide, sulphide, or halide ion as appropriate.

There is no generally accepted definition of a trace element but in practice the term is applied to an element whose abundance, in terms of the commonly used units, is 1000 ppm (i.e. 1000 $\mu g/g$) or less. Molar abundances may be required for certain purposes such as in the thermodynamic study of element partition (see Chapter 7).

Graphical means are rarely used for the presentation of individual rock or mineral analyses; there being too many elements to allow simple handling. One notable exception is in the presentation of rare-earth element (REE) data, where it has become customary to give a graphical plot of the abundance of each REE normalized to (i.e. divided by) its concentration in chondritic meteorites against atomic number. The advantages of such a diagram are that the abundance variation between elements of odd and even atomic number is smoothed (see Chapter 2, Section 2.2) and the extent of fractionation between the REE of lower and higher atomic number is clearly shown. These plots are called Masuda–Coryell diagrams after the workers who suggested them (Masuda, 1962; Coryell *et al.*, 1963); an example is given in Fig. 5.1 for two minerals from a rock. On occasion, some authors have plotted the chondrite-normalized abundance against the ionic radius of each REE ion but this practice has little to commend it as the radius is a function of co-ordination number and ionic charge, the former varying from mineral to mineral and the latter for some REE ions occurring in more than one state within the same rock system (see Chapter 6, Section 6.5).

A minor problem associated with chondrite-normalized abundance plots is in the choice of the normalizing values. More than five different sets of REE concentrations in different groupings of chondrites have been used. Although the differences amongst the various sets are small they are none the less sufficient to make direct comparison between normalized results more difficult. It is to be hoped that some standard will be agreed soon, but in the meantime authors should be urged to cite the source of the chosen normalizing values.

5.2.2 Rock suites or series

Variations in element abundances between one rock and the next within one rock series may be an indicator of the past operation of a particular petrogenetic process. Knowledge of the general behaviour of any element

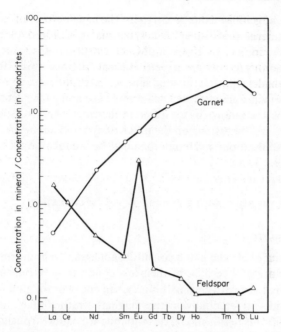

FIG. 5.1. A Masuda–Coryell diagram or chondrite-normalized plot of the rare-earth element concentrations in feldspar and garnet from a garnet-bearing anorthosite. The normalized concentration is on a log scale (ordinate) plotted against atomic number (abscissa). The high abundance of Eu relative to the other REE in the feldspar is referred to as a 'positive europium anomaly'. Europium anomalies can exist in other minerals and rocks; see Chapter 3, section 3.4.5 and Figs. 3.4 and 3.5 for a discussion and examples of chondrite-normalized REE diagrams for lunar rocks.

during different types of petrogenetic processes could be useful in elucidating the origin of a rock suite. Plots of the abundance variation of one element against that of another (inter-element correlation) give a ready visual means of assessing the broad similarities or differences in the behaviour of elements. For igneous rock systems a variation diagram is a simple display of the chemical differences and trends shown by a related set of rocks, especially in situations where the magmatic composition evolved by crystal–liquid fractionation processes.

A number of different parameters have been used for these diagrams such as the plotting of oxides (weight percent) against SiO_2 (so-called Harker diagram) or against the magnesium–iron ratio (e.g. $MgO/(MgO + FeO)$). More complex indices (e.g. the solidification index, S.I. $= 100\,MgO/(MgO + FeO + Fe_2O_3 + Na_2O + K_2O)$) have been devised for application over a wide range of igneous rock types. Triangular variation diagrams are also a common means of display, usually consisting of plots of the proportions of MgO, FeO and $Na_2O + K_2O$, or of CaO, Na_2O and K_2O in the rocks of a suite. Variation diagrams are a petrological aid by showing the chemical variation within, say, a fractionated igneous intrusion, a suite of lavas, or a progressively metasomatized rock mass, and in so doing they demonstrate the degree of coherence shown by the different elements. This last aspect is used in the next section. Discussions on the different types of variation diagrams, as well as their applications and limitations, may be found in Cox *et al.* (1979) and Wilcox (1979).

5.3 Element variation during crystal–liquid fractionation

5.3.1 General trends

The majority of elements that constitute a magma do not form minerals of their own principally because of their low concentration, but they enter, to varying degrees, a relatively small number of rock-forming minerals. Many elements reside preferentially in the magmatic rather than the crystalline phases and, for this reason alone, their abundances show strong inter-element correlations. In general, strong correlations are not observed amongst all the elements of any one group of the periodic table, although particular pairs may exhibit very close association (e.g. Al–Ga; Zr–Hf). As is shown in later chapters, it is factors such as ionic radius that determine an element's distribution and there are wide differences in these in any one periodic group. Thus, there is no simple pattern to the geochemical behaviour of all the elements but the following broad categories for many trace and minor elements in igneous rocks derived by crystal–liquid processes can be given. (Details of the major element variation shown by the common rock types may be found in most standard texts on igneous petrology.)

(a) Many of the transition elements, Sc to Cu, are enriched in the mafic or basic fractions and show fair to strong inter-element correlations.

(b) Many of the elements of atomic number 39 (Y) and greater, but excluding the platinum group elements, tend to be partitioned preferentially into the melt phase during fractional crystallization of a basic magma.

(c) Where a magmatic sulphide fraction is developed the elements Cu, Fe, Ni, Co, Ag, Au, Se, Te, In, Tl and Re tend to be associated.

(d) Any immiscible aqueous phase tends to be enriched in Na, K, Ca, Mg, Cl, S (as sulphate), Li, B, P, C (as bicarbonate or carbonate), Zn and W.

The enrichment of elements in pegmatites and volcanic emanations are discussed briefly in following sections.

The four categories are rather sweeping generalizations, and some exceptions undoubtedly can be found in any one rock suite. However, they give a useful preliminary guide to element distribution which is examined more closely in the following sections.

5.3.2 Pegmatites

Pegmatites, which represent the products of the last stage of fractional crystallization of a magma or the first stage of fractional melting by anatexis, commonly contain a number of elements at very much greater concentrations than the average abundances in the crust. Many of the elements which tend to remain in the magma during fractional crystallization are finally precipitated as relatively unusual and complex mineral or mineral assemblages.

Pegmatites are rich in silica, alkalis, alumina, water, etc. As a group they are significantly enriched, relative to crustal abundances, in the following elements: Li, Be, B, F, Sc, Ga, Ge, Rb, Y, Zr, Nb, Mo, In, Sn, Cs, REE, Hf, Ta, W, Re, Pb, Bi, Th and U. However, it should be noted that the source of the pegmatite is significant in affecting the degree of enrichment of some elements. For example, granite pegmatites are far less rich in Zr (and Hf) than those of nepheline syenites, especially agpaitic ones. This is probably a result of the early crystallization of zircon from granite magmas (cf. syenitic magmas) so preventing significant enrichment of Zr (and Hf) in residual magmas.

5.3.3 The Skaergaard Intrusion, East Greenland

The Skaergaard intrusion as first described by Wager and Deer (1939) and reviewed and summarized by Wager and Brown (1968) has provided one of the few opportunities to study the systematic variation in compositions of *both* the liquid and the solid phases during the fractional crystallization of a tholeiitic magma. The detailed mapping, petrological and geochemical work on this plutonic complex, carried out by Wager and his co-workers has given us valuable insight into the geochemical behaviour of many elements. One particularly important aspect of the work enabled the construction of an unusual, but more fundamental, variation diagram. This gives the changes in element concentrations of solid and remaining liquid, with proportion of solidified melts during progressive crystallization of a closed magma system.

One igneous complex alone cannot provide a data base for accurately predicting the geochemical behaviour of elements in other rock suites. Variations in certain parameters can lead to changes in the sequence of mineral crystallization and hence in the compositional variations of solids and liquids. One example is the effect of the partial pressure of oxygen; if this is sufficiently high, magnetite will crystallize early (i.e. at high temperature), whereas if the conditions are reducing, it tends to crystallize relatively late, as in the case of the Skaergaard intrusion. Furthermore, there are many sources of error in computations of the kind made for the Skaergaard rocks, and some interpretative aspects may be open to doubt. However, the general trends of compositional changes observed in the Skaergaard pluton seem to be sufficiently well established to merit a significant place in a discussion of element distribution.

The funnel-shaped Skaergaard layered intrusion was produced from one basic magma which was emplaced rapidly in the upper crust. (The volume was estimated initially to be about 500 km^3 but recent geophysical work (Blank and Gettings, 1973) suggests that the figure should be smaller.) Gravity accumulation of the crystallizing phases on to the floor of the magma chamber was the principal petrogenetic process leading to the sequence of rocks from those rich in magnesian olivine and calcic plagioclase at the lowest exposed horizons to those bearing fayalite, iron-rich pyroxene and more sodic plagioclases at the top. The composition of the initial magma is considered to be represented by that of the chilled margin (Table 5.1a). Knowledge of the compositions of the initial magma and of the rocks throughout the exposed layered series (Table 5.1a), together with the results of mapping and the use of reasonable assumptions about the nature of the unexposed part of the intrusion, enabled Wager (1960) to calculate the compositions of the magma at successive stages of crystal fractionation. The calculations have been extended to a number of trace elements (see Wager and Brown, 1968).

The compositional variations in successive solid and liquid fractions are given in Tables 5.1a and 5.1b. Of the major element oxides, silica shows a slight decrease in both fractions before a marked increase at the final stages of fractionation. Al_2O_3, MgO and CaO all show a steady decrease in the liquid while Na_2O and K_2O increase. The intrusion is characterized by iron enrichment in the later stages of its fractionation. The trends of TiO_2 and P_2O_5 were affected strongly by the incoming of new cumulus minerals—titanomagnetite (at about 80% solidified) and apatite (at about 97% solidified).

The trends for the trace elements compiled by Wager and Brown (1968) with some additions given in later works (e.g. Brooks (1969), Dissanayake and Vincent (1972, 1975), Vincent and Nightingale (1974)) can be put into three categories:

(a) Enrichment in the early solid products of fractionation, with a concomitant decrease in the liquid fraction, shown by Cr, Co and Ni.
(b) Enrichment in the liquid during fractionation and in the final crystallized products: Li, S, Cu, Zn, Ga, Rb, Y, Zr, Ag, Cd, In, Cs, Ba, rare earths, Hf, Ta, W, Hg and U. The example of Li is shown in Fig. 5.2.

TABLE 5.1a. *Composition of average rocks throughout the Skaergaard intrusion*

Percentage solidified	0	35	76	88.2	95.7	98.2	99.3
SiO_2	50.8	48.7	45.7	45.7	45.2	43.8	44.2
Al_2O_3	19.5	18.0	16.0	15.3	13.3	11.0	8.1
Fe_2O_3	—	0.7	1.8	3.3	3.5	3.6	4.2
FeO	4.5	6.8	10.2	13.3	16.4	20.2	25.4
MgO	9.3	9.5	9.3	6.7	4.8	2.8	0.3
CaO	13.0	12.0	10.7	10.2	9.7	9.3	8.8
Na_2O	2.12	2.20	2.43	2.85	3.17	3.08	2.62
K_2O	0.21	0.23	0.23	0.25	0.28	0.32	0.40
TiO_2	0.55	0.71	0.88	2.95	2.67	2.50	2.28
P_2O_5	0.04	0.03	0.07	0.08	0.08	2.17	1.54
MnO	0.16	0.13	0.08	0.16	0.16	0.27	0.48
Total	100.18	99.00	97.39	100.79	99.26	99.04	98.32

TABLE 5.1b. *Estimated composition of liquids of the Skaergaard intrusion*

Percentage solidified	0	35	76	88.2	95.7	98.2	99.3
SiO_2	48.1	47.3	46.8	46.9	47.5	49.8	55.0
Al_2O_3	17.2	16.4	15.2	14.1	12.5	11.8	11.8
Fe_2O_3	1.3	2.0	3.3	3.6	3.8	4.3	4.5
FeO	8.4	10.0	13.0	15.0	17.3	17.8	14.5
MgO	8.6	8.1	5.8	4.0	2.5	1.3	0.8
CaO	11.4	10.9	10.0	9.5	9.5	8.0	7.0
Na_2O	2.37	2.45	2.80	3.00	3.20	3.40	3.65
K_2O	0.25	0.29	0.33	0.40	0.65	1.00	1.40
TiO_2	1.17	1.42	2.37	2.65	2.20	1.80	1.05
P_2O_5	0.10	0.13	0.28	0.50	1.25	1.35	0.40
MnO	0.16	0.16	0.16	0.20	0.28	0.35	0.25

(c) No clear trend, or with slight enrichment in the middle fraction: Sc, V, As, Sr, Sb, Au and Tl.

Insufficient or no data are available for those elements not included either in the above categories or in Table 5.1.

A variation diagram constructed by Williams (1959) based on earlier data (Wager and Mitchell, 1951) gives a clear indication of the relative rates of change in element concentration in the Skaergaard magma during fractionation. The ratio (R) of an element's abundance in the remaining liquid at any given stage of solidification to its abundance in the initial liquid is plotted against the percentage solidified. Figures 5.3 (a) and (b) are for some divalent

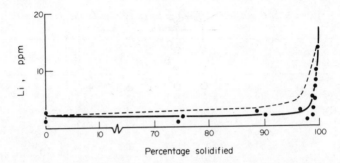

FIG. 5.2. Variation trend for lithium in the rocks of the Skaergaard layered series (continuous line) and in the liquid (dotted line). Filled circles represent Li analyses of layered-series rocks. (After Wager and Brown, 1968)

and trivalent cations respectively and show the rapid removal of Cr, Ni and Co, while Fe, Ba and La, amongst others, are enriched.

We return in later chapters to give explanations of these observations. It is sufficient at this point to state that the types and proportions of the various minerals formed during fractional crystallization are the dominant controls on the trends. It is, therefore, important to have information on the partition of elements between minerals and their supernatant magma or melt.

5.4 Partition coefficients

5.4.1 General principles

The extent to which an element is incorporated in a mineral crystallizing from a magma can be expressed conveniently by means of a *partition coefficient* (k). This is the concentration ratio (by weight) of an element (M) in the mineral to that in the magma at chemical equilibrium (5.1):

$$k_M = [M]_{\text{mineral}}/[M]_{\text{magma}}. \tag{5.1}$$

The value of k_M can be sensitive to variations in temperature, pressure and composition of both the mineral and magma at the time of crystallization. For a trace element, k is often a constant over a wide range in the element's concentration provided that T, P and major element concentrations remain fixed, but in a few instances the range appears to be very restricted (see below, and Chapter 7 for further discussion).

The partition coefficient in expression (5.1) is of limited use in the thermodynamic treatment of element distribution in rocks and minerals but it does have wide applicability in petrogenetic modelling (see Chapter 8) and as a

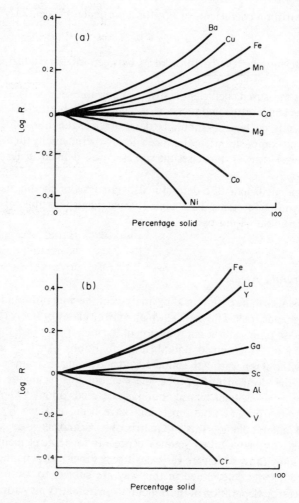

FIG. 5.3. Relative changes in element concentration in the magma during crystal fractionation of the Skaergaard intrusion. *R* is the ratio of concentrations of the element in the magma at successive amounts of solidification to the element's initial concentration: (a) divalent cations; (b) trivalent cations. (After Williams, 1959)

semi-quantitative (sometimes quantitative) guide to element distribution. In the application to petrogenetic modelling it is often necessary to use a bulk partition coefficient $(\bar{k})$. This may be calculated when the modal proportions (by weight) of the crystallizing phases are known. If minerals Q, R and S crystallize together in the modal proportion α, β and γ (where $\alpha + \beta + \gamma = 1$), the bulk partition coefficient for $\bar{k}_M$, for a particular element, M, is given by:

$$\bar{k}_M = \alpha k_M^Q + \beta k_M^R + \gamma k_M^S. \tag{5.2}$$

This coefficient gives the partition of M between total crystallizing solid and liquid.

Any one measured coefficient is strictly applicable only to the system under study at the temperature and pressure of element equilibration between the two phases. Ideally, coefficients should be determined as functions of temperature, pressure and composition; this has been done experimentally for a few systems (see below). However, most results are not well defined in terms of these parameters.

Partition coefficients are determined by analysis of the phases in natural rocks or synthetic systems. They can be measured also for mineral–mineral, or immiscible liquid–liquid pairs.

5.4.2 Natural systems

Coefficients can be determined by analysis of the mineral(s) and matrix of a suitable igneous rock. The matrix, which may be glassy or finely crystalline, is considered to represent the composition of the melt from which the mineral(s) crystallized under assumed equilibrium conditions. Analysis can be performed, without separating the mineral and matrix phases from each other, by means of the electron microprobe analyser but this technique is normally applicable to an element whose abundance exceeds about 500 ppm in both phases. At the present time most data have been obtained by analysis of separated mineral phenocrysts and matrix of basic and acid extrusive rocks. A wide variety of analytical techniques, especially for trace elements, has been used including those which are very sensitive (e.g. neutron activation analysis).

The coefficients determined in this way are subject to more than simple analytical error. Separation of the mineral from its matrix may not be complete and submicroscopic inclusions might be present in the phenocrysts. It is also difficult to assess the temperature and pressure of crystallization and whether or not the mineral–matrix pair formed under equilibrium conditions. Suitable care over the preparation and analysis of well-selected material can sometimes reduce these possible errors to acceptable levels.

Tables 5.2a, b and c are a compilation of partition data obtained from crystal–matrix pairs. Only those published papers specifically concerned with establishing coefficient values have been used. For some elements (e.g. REE) there are many published results, while for others (e.g. Zr) there are none or very few despite the relative importance of some of these elements in geochemistry. The results are grouped broadly into those for basalts and

TABLE 5.2a. *Mineral/matrix partition coefficients of elements in basaltic and andesitic rocks*

	Olivine Range	Olivine Average	Orthopyroxene Range	Orthopyroxene Average	Clinopyroxene Range	Clinopyroxene Average	Amphibole Range	Amphibole Average	Plagioclase Range	Plagioclase Average	Phlogopite (one determination)	Garnet Range	Garnet Average
3 Li	0.012–0.024	0.02			0.26	*	0.12, 0.19	*					
11 Na	4.5–8.0	6.1	0.075	*	0.27	*	0.72, 0.80	*		*			
12 Mg	0.012	*	4.7	*	2–5	*	2.8, 4.8	*					
13 Al			0.3	*									
19 K	0.0002–0.008	0.007	0.01, 0.02	*	0.21–0.6	0.3	0.33–0.86	0.60	1.6–2.1	1.9			
20 Ca	0.01–0.04	0.03	0.2, 0.4		0.002–0.27	0.03	1.1, 3.0		0.02–0.36	0.17	2.7		
21 Sc	0.14–0.22	0.17	1.2		1.7–3.2	2.7	2.2, 4.2					2.6–5.4	4
22 Ti					0.8	*		*					
23 V	0.03	*	0.6	*	0.8–1.9	1.3							
24 Cr	1.1–3.1	2.1	10		4.7–20	8.4	0.10, 1.6	*				0.06, 0.29	*
25 Mn	0.8–1.8	1.2	1.4		0.6–1.3	0.9							
26 Fe²⁺	1.2–3.0	1.9	1.8		0.6–1.0	0.8							
26 Fe³⁺			0.75		0.37–0.86	0.53							
27 Co	2.8–5.2	3.8	2, 4		0.7–2.8	1.2	1.4, 1.8	*		*			
28 Ni	8–19	14	5		1.4–4.4	2.6							
30 Zn	0.7				0.41	0.40	0.42, 0.69						
31 Ga	0.04	*	0.7		0.30–0.58	0.40							
37 Rb	0.0002–0.011	0.006	0.02, 0.03	0.006	0.001–0.28	0.04	0.04–0.4	0.25	0.85–1.3	1.0	3.1		
38 Sr	0.0001–0.02	0.01	0.01, 0.02		0.07–0.43	0.14	0.19–1.02	0.57	0.03–0.50	0.10	0.08		
49 In	0.06	*	0.45		1.7		2.2, 3.1		1.3–2.9	1.8			
55 Cs	0.0004				0.002–0.018	0.01	0.05, 0.20		0.025	*			
56 Ba	0.0001–0.011	0.006	0.012, 0.014		0.002–0.39	0.07	0.10–0.44	0.31	0.05–0.59	0.23	1.1		
57 La					0.08	*	0.17–0.44	0.27	0.14	0.14		0.03, 0.08	*
58 Ce	0.009	*	0.003–0.04	0.02	0.17–0.65	0.34	0.09–0.54	0.34	0.06–0.30	0.14	0.03	0.05	*
60 Nd	0.007, 0.010		0.03, 0.06	*	0.32–1.3	0.6	0.19	*	0.02–0.20	0.08	0.03		
62 Sm	0.003–0.015	0.009	0.01–0.10	0.05	0.43–1.8	0.9	0.34–1.46	0.91	0.02–0.20	0.08	0.03	0.07–1.25	0.6
63 Eu	0.006, 0.010		0.02, 0.08	*	0.48–2.0	0.9	0.36–1.49	1.01	0.06–0.73	0.32	0.03	0.3–1.5	0.9
64 Gd	0.012				0.82, 0.88		0.51–1.7	1.1	0.03–0.21	0.10		2.1, 5.2	*
65 Tb					1.0	*	1.02–2.0	1.4	0.03	*		4.1, 7.1	*
66 Dy	0.009, 0.014		0.05		0.56–1.46	1.1	0.64		0.01–0.20	0.09			
67 Ho			0.12, 0.29										
68 Er	0.009, 0.017		0.16, 0.46		0.53–1.3	1.0	0.48		0.02–0.24	0.08	0.03	13, 24	*
69 Tm					1.1	*							
70 Yb			0.11–0.67	0.34	0.48–1.3	1.0	0.46–1.42	0.97	0.006–0.30	0.07	0.03	4–53	30
71 Lu			0.11		0.67, 1.0	*	0.44–1.31	0.89	0.03–0.24	0.08	0.04	5–57	35
72 Hf												0.3, 0.6	*
82 Pb									0.10–0.67	0.26			
90 Th					0.003–0.05	0.02			0.009	*			
92 U	0.0024, 0.0027	*			0.004–0.08	0.04							

* See footnote to Table 5.2c.   Data sources—see Table 5.2c.

TABLE 5.2b. *Mineral/matrix partition coefficients of elements in dacitic and rhyolitic rocks*

	Orthopyroxene		Clinopyroxene		Amphibole		Plagioclase	
	Range	Average	Range	Average	Range	Average	Range	Average
3 Li	0.16, 0.21	*					0.27–066	
11 Na	0.03–0.1	0.06	0.09–0.14	0.11	0.08	*	1.2–3.1	1.5
12 Mg	50–160	100					0.03–0.5	0.3
19 K	0.0006, 0.002	*					0.08–0.33	0.19
20 Ca	0.55–2.3	1.4	8–12	10	1.8	*	1.8–4.9	3.4
21 Sc	6.0–7.7	6.9	18–28	22	20	*	0.01–0.20	0.07
23 V	4.4–10	6						
24 Cr	0.6–3	1.6	90	*			0.03–0.7	0.2
25 Mn	29, 34	*					0.03–0.31	0.18
27 Co	2.1–3.6	2.9	6.0–11	8	45	*	0.04–0.5	0.15
30 Zn	0.8–1.0	0.9			7	*	0.26–0.60	0.38
37 Rb	0.0005–0.29	0.09	0.09	*			0.02–0.46	0.09
38 Sr	0.009, 0.05	*					1.5–8.8	6†
39 Y	0.9	*						
49 In								
55 Cs	0.3						0.01–0.13	0.05
56 Ba	0.003–0.03	0.02	0.02, 0.06	*	0.035	*	0.30–0.92	0.50
57 La	0.50–0.90	0.65	0.5–0.8	0.6	0.85	*	0.24–0.49	0.32
58 Ce	0.08–1.03	0.46	0.6–1.2	0.9	0.43–1.8	1.2	0.11–0.40	0.24
60 Nd	0.11–1.20	0.62	1.4–2.9	2.1	1.0–4.3	3.2	0.06–0.29	0.19
62 Sm	0.13–1.6	0.7	2.1–3.3	2.7	1.6–8.2	5.4	0.05–0.22	0.13
63 Eu	0.11–1.0	0.5	1.6–2.3	1.9	1.2–5.9	3.6	0.82–4.2	2.0
64 Gd	0.17–2.2	1.1	2.0–4.8	3.1			0.11–0.24	0.16
65 Tb	0.8–1.6	1.2	2.0–4.1	3.0	3.0	*	0.10–0.22	0.15
66 Dy	0.26–1.3	0.7	2.2–4.0	3.3	2.3–14	9	0.04–0.45	0.13
67 Ho								
68 Er	0.43–0.73	0.61			2.4–11	8	0.03–0.08	0.05
69 Tm	0.8–1.9	1.4			5	*	0.1–0.2	0.1
70 Yb	0.73–1.2	1.0	1.6–2.8	2.1	1.9–9	6.2	0.02–0.30	0.08
71 Lu	0.76–1.4	1.1	2.0–2.6	2.3	1.8–6	4.5	0.03–0.11	0.06
72 Hf	0.04, 0.2	*	0.20–0.55	0.34	6	*	0.02–0.17	0.07
73 Ta	0.2–0.7	0.5	0.1–0.8	0.4	0.3	*	0.02–0.09	0.05
82 Pb							0.29–0.78	0.45
90 Th	0.13–0.18	0.15	0.01–0.25	0.13	0.22	*	0.01–0.09	0.04
92 U	0.09	*	0.03	*	0.40	*	0.01–0.07	0.03

andesites and those for rhyolites and dacites. Data were selected only to the extent of rejecting the very occasional extreme value. The range and average of the values are given except where one or two results only are available; in those cases the values are listed in the 'range' column of Table 5.2.

Some aspects of Table 5.2 are worthy of note:

(a) There is a large range of coefficient values amongst different elements and different minerals.

(b) There can be a wide scatter of results for the partition of any one element in a particular mineral–matrix pair while for others the values are more consistent. Some of the scatter is a result of analytical error but probably a large part arises from the dependence of coefficients on T, P and major element composition.

TABLE 5.2b (*cont.*)

	Alkali Feldspar		Biotite (1 or 2 determinations)	Garnet		Magnetite	
	Range	Average		Range	Average	Range	Average
3 Li			0.39				
11 Na			0.10				
12 Mg	0.05–0.33	0.21	22				
19 K	0.64–2.2	1.4	2.6, 5.6	0.02	*		
20 Ca	0.10–3.8	1.9	0.62			0.3–0.5	0.4
21 Sc			11.3	10.2–20.2	16.0	3.3–4.5	3.9
23 V							
24 Cr			19			5–20	11
25 Mn	0.03–0.52	0.18	6.0				
27 Co			29	1.7–3.6	2.6	19–35	28
30 Zn			20			10–14	12
37 Rb	0.11–0.80	0.38	3.3, 3.5	0.009	*		
38 Sr	3.6–26	9.4	0.12, 0.36	0.02	*		
39 Y							
49 In			3.9				
55 Cs			2.4				
56 Ba	2.7–12.9	6.6	6.4, 8.7	0.017	*	0.05–0.08	0.07
57 La			0.32	0.28–0.54	0.39	0.24–0.88	0.53
58 Ce	0.04	*	0.04, 0.38	0.35–0.93	0.62	0.28–1.15	0.61
60 Nd	0.03	*	0.04	0.53–0.73	0.63	0.35–1.80	0.88
62 Sm	0.02	*	0.06, 0.39	0.76–5.5	2.2	0.39–1.85	0.93
63 Eu	1.13	*	0.15, 0.33	0.17–1.37	0.7	0.28–0.96	0.58
64 Gd			0.08, 0.44	5.3–13.6	7.7		
65 Tb			0.39	7.2–19.6	12	0.36–1.50	0.84
66 Dy	0.006	*	0.10	29	*	0.3–1.4	0.8
67 Ho				18.4–35	28		
68 Er	0.006	*	0.16	43	*		
69 Tm						0.5–1.2	0.8
70 Yb	0.012	*	0.18, 0.67	26–67	43	0.2–0.6	0.4
71 Lu			0.74	24–64	38	0.2–0.6	0.4
72 Hf			2.10			0.2–0.6	0.3
73 Ta						0.8–1.8	1.3
82 Pb	0.84–1.4	1.0					
90 Th			0.31			0.04–0.20	0.11
92 U				0.14	*		

* See footnote to Table 5.2c.

† Dupuy (1972) also reports four values, range 15.2–40, average 27.

(c) The coefficients are in keeping with the observations on element distribution in the Skaergaard intrusion and other rock series discussed above.

(d) Partition coefficients of any one element for a particular mineral are generally greater in the more siliceous rocks.

Elements which have k values $\ll 1$ for the common crystallizing minerals during extended stages of magma fractionation are termed *incompatible*, as they are not incorporated into the crystals. Examples of some incompatible elements during the early and middle stages of fractionation of a basaltic magma are Ba, REE (excluding Eu) and U. They become enriched in the late fractionation products. The term *hygromagmatophile element* has become

TABLE 5.2c. *Minteral/matrix partition coefficients for rare-earth elements in some accessory minerals of dacitic and rhyolitic rocks*

		Allanite (1 determin- ation)	Apatite		Zircon	
			Range	Average	Range	Average
57	La	820				
58	Ce	635	17–53	31	2.3–7.4	4.2
60	Nd	460	21–81	50	2.0–6.5	3.6
62	Sm	205	21–90	54	2.6–6.5	4.3
63	Eu	80	15–21	27	1.1–5.2	3.4
64	Gd	130	22	*		
65	Tb	71				
66	Dy		17–69	42	38–54	48
68	Er		14–51	31	120–150	140
70	Yb	8.9	9–37	21	240–300	280
71	Lu	7.7	8–30	17	280–390	345

* Where only one or two determinations are available, an average is not given; the values are quoted in the range column.

Sources of data for Tables 5.2a to 5.2c:
  Brooks *et al.* (1981); Dale and Henderson (1972); De Pieri and Quareni (1978); Dudas *et al.* (1971); Dupuy (1972); Ewart and Taylor (1969); Goodman (1972); Hakli and Wright (1967); Hart and Brooks (1974); Henderson (1979); Henderson and Dale (1969); Higuchi and Nagasawa (1969); Irving and Frey (1978); Leeman (1979); Leeman and Scheidegger (1977); Matsui *et al.* (1977); Nagasawa (1970); Nagasawa and Wakita (1968); Nagasawa and Schnetzler (1971); Noble and Hedge (1970); Onuma *et al.* (1968); Philpotts and Schnetzler (1970); Schnetzler and Philpotts (1968, 1970); Schock (1977).

synonymous with incompatible element; it was introduced by Treuil and Varet (1973) and applied initially to the elements Ta, Nb, La, Th and Hf, all of which have low crystal/liquid partition coefficients for the common minerals (also see Table 5.2). Another term—*large ion lithophile*, abbreviated LIL—is applied to lithophile elements (e.g. Ba, La, U) which have ionic radii larger than the common rock-forming elements. Hence the coverage of the terms 'incompatible', 'hygromagmatophile' and LIL overlap quite significantly (see Chapter 6 for a further discussion).

Table 5.2 may be used to calculate the partition coefficient for an element in a mineral–mineral pair, by dividing the average coefficient for one mineral by that for the other.

5.4.3 Experimental systems

The dependence of partition coefficients on temperature, pressure and composition may be investigated quantitatively using experimental methods. The experiments can use either synthetic or natural starting materials which have been doped with the element(s) of interest. The principal limitation of this approach is that the experimental rock system cannot match exactly the natural one, but in some instances it can come quite close to it. Basaltic compositions are frequently studied but some work has been done on acidic rock systems. Synthetic compositions in the diopside–albite–anorthite, forsterite–anorthite–diopside or forsterite–anorthite–silica systems are frequently chosen.

Irving (1978) gave an excellent review of the experimental studies of crystal/liquid trace element partition; the general findings are discussed below, the reader being referred to the paper for further details. Experimental work has centred on a restricted number of elements, especially the transition elements Sc to Cu, REE, Sr, Rb and Ba, although a wide range of minerals has been studied including the important rock-formers: feldspars; olivine; orthopyroxene; clinopyroxene; amphiboles; and garnets.

Temperature dependence

Partition coefficients are dependent to varying degrees upon temperature in most systems. In some they are strongly dependent, e.g. Sr and Ba in plagioclase feldspar; Cr in orthopyroxene; and some rare earths in garnet. As an example, results for Sr partition between feldspar and basaltic melt as a function of temperature are given in Fig. 5.4.

The dependence of k on temperature can usually be expressed in the form of equation (5.3). (This is similar to the integrated van't Hoff equation relating the equilibrium constant and temperature of reaction.)

$$\ln k = \frac{A\,10^4}{T} + B \tag{5.3}$$

where A and B are constants and T is in kelvins. Experimental results show that the constant A is usually $> +1$ and the constant B is rarely zero or close to zero, but is often a quite large negative number (see Fig. 5.4). Thus, in general the partition coefficient decreases with increasing temperature, but in a few recorded instances (e.g. Sm partition between garnet and liquid, (Wood, 1976)) it increases.

The extent to which some changes in melt composition are affecting the values of A and B is not yet known (see below). The experimental results have not reached the stage where it would be appropriate to give a listing of A and B values for different elements and different mineral–liquid pairs. However, experimental results are in keeping with data obtained for natural rock systems (see preceding section).

Pressure dependence

There have been very few experiments investigating the dependence of k on pressure, although first results indicate that the dependence could be significant.

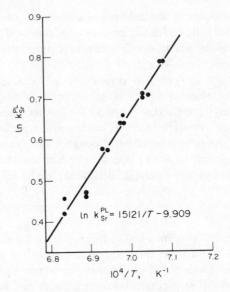

FIG. 5.4. Strontium partition coefficient (weight fraction) as a function of inverse temperature (K^{-1}) between plagioclase feldspar and basaltic melt. (After Sun *et al.*, 1974)

Compositional dependence

In studies of element partition involving both complex silicate melts and minerals which are of solid–solution series, it is virtually impossible to separate directly the contribution to changes in k values from temperature and from compositional factors. However, in some systems it is clear that composition has a significant effect on element distribution. The partition of trace amounts of Ni between olivine and melt is dependent upon MgO content of the liquid. There is also a strong compositional dependence of REE distribution in garnet–melt systems, and Watson (1977) has demonstrated large changes in the partition of Mn between olivine and silicate liquid, at constant temperature, as a function of the Si/O atomic ratio in the melt (k_{Mn} varies from about 0.8 at Si/O = 0.27 to about 1.6 at Si/O = 0.36, at 1350° C).

It has been shown in Section 5.3.2 that the partition coefficient of an element for a mineral–melt pair tends to be higher in acidic rocks than in basic ones. This is partly a result of the lower temperature of crystallization in the former but it is also a result of the differences in composition. This compositional effect is well demonstrated by results on element partition between two immiscible melts. Watson (1976) and Ryerson and Hess (1978) have investigated the partition of P, Ti, Cr, Mn, Sr, Zr, Ta, Cs, Ba, La, Sm, Dy and Lu between immiscible basic and acidic melts in iron-rich silicate systems (at P = 1 atm; T = 965 to 1180° C). All elements except Cs showed a preference for the basic composition melt, with k values (basic/acid) being: P 10–15; Sr and Ba ~ 1.5; others 2.3 to 4.3. Cs showed a preference for the acidic melt by a factor of about 3.

There is conflicting evidence on whether or not the partition of a trace element is affected by variations in its own concentration. There is, for example, experimental evidence to show that Sm partition between plagioclase feldspar and melt is unaffected by variations in the Sm concentration of the liquid, from about 1 ppm to about 5 % Sm. On the other hand, the partition of Sm between garnet and liquid is dependent upon Sm concentration below about 100 ppm in garnet (Harrison and Wood, 1980). At the present time, the concentration ranges over which coefficients remain constant are not well known. This topic is discussed further in Chapter 7.

The partition of elements which show variable oxidation states (e.g. Fe and Eu) will be affected by changes in the oxygen fugacity of the melt, as the proportions of the oxidation states change. Changes in major element composition can also affect the oxidation state of ions (Morris and Haskin, 1974). A discussion of fugacity is given in Chapter 7.

5.5 Volcanic emanations and sublimates

The volatile components lost from a magma during its formation, movement, cooling and fractionation, are represented, at least in part, by volcanic emanations and sublimate deposits. White and Waring (1963) made an excellent compilation of all the reliable data, up to that time, on the compositions of volcanic gases and of sublimates. These authors emphasized that the composition of a fumarolic gas depends, among other factors, upon the extent of mixing and reaction of the gas with air and meteoric water, as well as upon possible reaction with rocks situated between the places of gas evolution and gas collection.

Carbon dioxide is usually the dominant 'active' component (i.e. excluding water, O_2, N_2 and Ar) of collected fumarolic gases. The other common gases are CO, SO_2, H_2S and HCl. H_2 is variable but can be substantial. CO_2 normally exceeds CO in abundance but this is not always so. CO and CH_4 are common when conditions are strongly reducing; H_2 and H_2S are important components when conditions are in the range from moderately oxidizing to strongly reducing. HF, SO_3, B, Br and I can also occur.

In a recent review on some basaltic and andesitic gases, Anderson (1975) states that the compositions of volcanic gases are compatible with solubility and melt composition for H_2O, Cl and possibly sulphur gases.

Sublimate deposits can contain a wide variety of compounds but the commoner ones are sulphates and chlorides of Na, K, Ca, Mg, Al and Fe, together with ammonium chloride and native sulphur. Fluorides are abundant in some encrustations. The compositions of sublimates are modified by contributions from acid attack on the surrounding volcanic rocks, but in general many elements are found in higher concentrations in the sublimates than in the surrounding lavas. They include: Li, Be, Ti, V, Cr, Co, Ni, Zn, Ga, As, Se, Zr, Mo, Ag, Cd, Sn, Sb, Te, Ba, Pb and Bi. B and Cu sometimes occur.

5.6 Metamorphism and metasomatism

Rocks that have been metamorphosed under strictly isochemical conditions will have the chemical compositions and variations of their precursors although there will be some local redistribution of elements through the development of new minerals, metamorphic segregations, etc., during the transformation process. There is a little evidence for the mobilization of some elements (e.g. Pb, Rb, Tl and Au) during regional metamorphism, especially at high grade, but there is still a serious lack of data on this topic. (See Chapter 8 (Section 8.4) for a discussion of transformations including closed-system metamorphism.)

Many metamorphic changes are not isochemical but involve the introduction of elements from an external source into the rock mass, and the removal of other elements—the process of metasomatism. The compositional changes are brought about by chemically active fluids, and often take place under constant volume conditions so that there may be little disturbance of rock texture. Water and carbon dioxide are mobile during metamorphism and may be lost by dehydration or decarbonation reactions (e.g. reaction (5.4):

$$CaMg(CO_3)_2 + 2SiO_2 \rightleftharpoons CaMgSi_2O_6 + 2CO_2 \qquad (5.4)$$
$$\text{dolomite} \qquad \text{quartz} \quad \text{diopside}$$

or gained, e.g. production of kaolinite from alkali-feldspar (5.5):

$$2\,KAlSi_3O_8 + 2H_2O \rightleftharpoons Al_2Si_2O_5(OH)_4 + K_2O + 4SiO_2 \qquad (5.5)$$
$$\text{feldspar} \qquad\qquad \text{kaolinite}$$

Metamorphism of near-surface rocks is likely to lead to hydration and oxidation by reaction with the atmosphere and hydrosphere.

Much of the earlier work on metasomatism was descriptive and enabled the identification (Rankama and Sahama, 1950) of five types, each with their particular element associations:

   (a) Alkali metasomatism. Alkali metals.
   (b) Lime metasomatism. Ca.

(c) Fe–Mg–silicate metasomatism. Fe, Mg, SiO_2.

(d) Boron metasomatism. B, Li, F, Cl, S, Si, Sn.

(e) Carbon dioxide metasomatism. CO_2.

The addition or removal of major or trace elements during metasomatism is dependent upon the mineralogy of the metasomatized rocks and on the composition of the metasomatizing fluid, and no all-embracing statement on element mobility is possible. Cl^- and CO_3^{2-} appear to be abundant during some metasomatic reactions (e.g. scapolitization) and these constituents may, through complexing, give rise to mobility of many elements. The same can be said of F^- and SO_4^{2-}. Alkali-metasomatism can involve the introduction besides Na and K of many trace elements including Nb, Ba, REE, Ta and Re. However, the precise nature of the alkali-metasomatism is important; W and Tl are examples of elements which tend to be removed during Na-metasomatism but introduced during K-metasomatism, while Ga is associated more strongly with Na-metasomatism than with K-metasomatism. The existence of a relatively high sulphur fugacity in metasomatizing agents may lead to the release of a number of heavier metals including Hg, Ag, Sn, Sb, As, W, Mo and Au, all of which form soluble sulphide and hydrosulphide complexes. Alkaline solutions probably act as transporting media for some lighter elements, e.g. Be forms stable and soluble carbonato–beryllate complexes in alkaline solutions in the pH range of about 7 to 12 (Govorov and Stunzhas, 1963).

Attempts have been made to establish the nature of the metasomatizing fluids. For example, Johannes (1967), in an experimental study of the transport of magnesium during the metasomatic formation of magnesite, concluded that chloride-bearing solutions are more efficacious than CO_2-bearing fluids alone. Solutions or gases rich in the halogens act as transporting media for a number of elements including the alkali metals and alkaline earths. The halogens, especially chlorine, may play a significant part in the alteration of rocks, such as in the process of serpentinization. Transformation reactions of forsteritic olivine to serpentine have been written as involving only water to produce brucite as a second product:

$$2Mg_2SiO_4 + 3H_2O \rightarrow \underset{\text{serpentine}}{Mg_3Si_2O_5(OH)_4} + \underset{\text{brucite}}{Mg(OH)_2} \qquad (5.6)$$

$$\underset{\text{olivine}}{}$$

or alternatively with the removal of some silica:

$$5Mg_2SiO_4 + 4H_2O \rightarrow Mg_3Si_2O_5(OH)_4 + 4MgO + SiO_2. \qquad (5.7)$$

If there is the presence also of sufficient CO_2-bearing gases or solutions, then magnesite may develop from the magnesia released by serpentinization. Reactions (5.6) and (5.7) are almost certainly an oversimplification of the alteration process. A careful analytical microprobe study (Rucklidge, 1972) of some partially serpentinized dunite has shown that not only does trace nickel diffuse out of the unaltered material adjacent to veins of serpentine between the olivine remnants and into the central part of the veins but that high concentrations (up to 0.5%) of chlorine occur in the serpentine veins. This contrasts with an almost zero concentration in the fresh olivine. The inference is that some form of chlorine is an active agent of serpentinization, and the absence of chlorine in many analysed serpentinites is a result of the loss of this highly mobile element after alteration of the rock.

Our knowledge of element mobility during metasomatism is very limited; much more research is needed into the processes of mobilization and fixation of particular elements or element groups. One example of a study of relative mobility of elements during metasomatism is that by Martin *et al.* (1978) on the introduction of REE during fenitization of an almost pure quartzite. There is a significant increase in the concentration of REE, especially those of lower

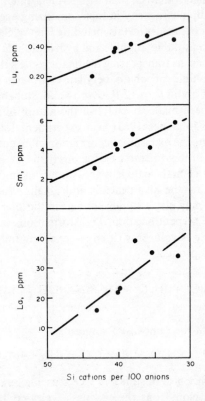

FIG. 5.5. Plot of concentrations of La, Sm and Lu versus Si cations per 100 anions in six fenite samples. (A value of 49.71 Si is representative of the quartzite before metasomatism.) (After Martin *et al.*, 1978)

atomic number, with decrease in silicon (i.e. with increase in degree of fenitization) of the rocks (Fig. 5.5). It is clear that the nature of any alteration of a rock must be established before trace-element concentrations may be used in petrogenetic interpretations. Conversely, when more data on element mobility and fixation during metasomatism are available, it should be possible to use patterns of trace-element abundances as indicators of the nature and possible extent of alteration.

Suggested further reading

Appendix VII lists many of the standard reference works on geochemical data. The following two works are probably the best sources for the most up-to-date compilations:

Handbook of Geochemistry, ed. K. H. WEDEPOHL (1969 to 1978). 2 vols., vol. 2 in 5 parts. Volume I contains 12 chapters on various topics including the composition and abundance of common rocks; volume II is a systematic treatment of the chemical elements.

Data of Geochemistry, ed. M. FLEISCHER (1962 onwards, still in production). 6th edition. U.S. Geological Survey Professional Paper 440. Chapters on various geochemical topics and compilations. See Appendix VII for list of those chapters which have been published.

6

Structural Controls of Element Distribution

6.1 Introduction

The chemical fractionation processes that magmas undergo when they crystallize, either under equilibrium or disequilibrium conditions, together with the converse of these processes during partial melting of pre-existing rocks, lead to a wide diversity of magma compositions and the extensive fractionation of a number of elements. Metamorphism may not produce so much large-scale fractionation as in magmatic systems but there may be considerable redistribution of elements within small rock volumes during new mineral growth or recrystallization. To account for the observations recorded in the previous chapter it is necessary to consider the controls that structures of melts and solids exert on element distribution in both igneous and metamorphic rocks. Molecular structure is the packing and arrangement of atoms in a chemical compound and the most important parameter affecting it is the size, or sizes, of the constituent atoms. However, the factors that give rise to order in a crystal structure also include the nature of chemical bonding between the atoms. In ionic solids the charges of the ions are important as they are locally neutralized by each other within the crystal and repulsion effects between cations or between anions will help to determine the spacing of ions of the same charge sign. In covalent solids the geometrical arrangement of the localized distribution of the electrons shared between, or among, the atoms is an important factor.

Crystallization involves the exchange of elements between liquid and solid, or between solid and solid, and hence, for any element, the relative gain or loss in chemical stability in passing from one phase to the other. The structural nature of the crystalline material determines, in part, the extent of this energy change. This chapter starts by briefly discussing, with some specific examples, the nature of crystal structures that commonly occur within the mineral kingdom. A review is then made of aspects of the properties and structural form of melts that are of relevance to geochemistry. The rest of the chapter is

102

devoted to consideration of those characteristics of elements, namely their size and their type of chemical bonding within given chemical compounds, that affect element distribution between different mineral structures or between minerals and melts.

6.2 Crystal structure

6.2.1 Introduction

The chemical compositions of the crust and mantle of the Earth lead to silicates being the most abundant constituent compounds (see Chapter 4). Many minerals have a complex structure, the general nature of which is largely determined by the type of linking of the SiO_4 tetrahedra. Variation in the degree of the linking produces a range of structural types, from ones with isolated SiO_4 tetrahedra (as in the mineral olivine, $(Mg,Fe)_2SiO_4$, in which Mg^{2+} and Fe^{2+} ions are arranged so as locally to neutralize the negative charge $(4-)$ of each separate SiO_4 group) through structures with the tetrahedra linked so as to form chains (as in those minerals that comprise the two large groups of the pyroxenes and the amphiboles) or in sheets (as in the mica group) to framework structures exemplified by the feldspars and the feldspathoids. In silicates with the framework structure the SiO_4, or $(Si,Al)O_4$, tetrahedra are linked to one another by sharing the oxygen atoms in all directions.

In most natural silicates the chemical bonding between the oxygens of the SiO_4 tetrahedra and elements outside these tetrahedra is essentially ionic, but the wide diversity of structures provides a large number of possible co-ordination polyhedra for the various cations (see Table 6.1). The existence of

TABLE 6.1. *Examples of co-ordination number of cations (other than Si) in silicate minerals or mineral groups*

Mineral	Formula	Cation co-ordination number
Olivine	$(Mg,Fe)_2SiO_4$	6
Zircon	$ZrSiO_4$	8
Sphene	$CaTiSiO_4(OH,F)$	6 and 7
Pyrope (garnet)	$Mg_3Al_2Si_3O_{12}$	Mg:8; Al:6
Sillimanite	Al_2SiO_5	6
Epidote	$Ca_2Fe^{3+}Al_2(Si_3O_{12})(OH)$	Ca:9; Fe,Al:6
Beryl	$Be_3Al_2(Si_6O_{18})$	Be:4; Al:6
Amphibole group	Various	6 and 6–8
Mica group	Various	6 and 12
Orthoclase feldspar	$K(AlSi_3O_8)$	K:9
Leucite	$K(AlSi_2O_6)$	K:12

these different polyhedra is one of the principal factors resulting in the significant fractionation of many elements from each other during geochemical processes. Furthermore, different chemical bonding characteristics are found throughout the range of mineral groups which include oxides, sulphides, native metals, halides, carbonates and silicates. In the descriptions of three mineral structures which follow, the significance of the different polyhedra as well as of the nature of the bond type are exemplified.

Descriptions of the structures of commoner minerals may be found in the books by Deer *et al.* (1966) and Wells (1975) and an introductory account of bonding in crystals, together with a description of some crystal structures is given by McKie and McKie (1974) and Wells (1975).

6.2.2 *Some mineral structures*

Here the structural characteristics of three minerals are described: orthopyroxene (a chain silicate); spinel (oxide); and the disulphide, pyrite, so as to give a basis for the discussions that follow in later sections.

Orthopyroxene

The pyroxenes are single-chain silicates with the general formula $MSiO_3$, where M = Mg, Fe or Ca. Calcium-poor varieties, the orthopyroxenes, are orthorhombic and comprise a solid solution series from enstatite ($Mg_2Si_2O_6$) to orthoferrosilite ($Fe_2Si_2O_6$). In these, the chains consist of silicon–oxygen tetrahedra linked by the sharing of two oxygens of each tetrahedron with adjacent tetrahedra. Bands of the cations (Mg^{2+}, Fe^{2+}) link the chains to each other (Fig. 6.1). These cations occupy sites of six-fold co-ordination but the sites are of two types. One, designated M1, has the cations co-ordinated to six oxygens each linked to one silicon atom, and the symmetry is approximately octahedral. The other, M2, is also co-ordinated by six oxygens, four of which are each linked to one silicon atom, and the other two are bridging atoms shared by two silicon atoms. The different bonding relationships of the oxygen atoms give rise to a distorted octahedral symmetry for the M2 site. The symmetry of each site is shown in Fig. 6.1. In this particular pyroxene, of formula $Mg_{0.93}Fe_{1.07}Si_2O_6$, the average cation to oxygen bond distance in the M1 site is 2.10Å (range 2.04–2.17Å) and with a maximum deviation in the 0—M1—0 bond angle of 7.3° from the 90° expected in regular octahedral symmetry (Ghose, 1965). In M2, the average M2—0 bond distance is 2.23 Å (range 2.04–2.52 Å) and with a maximum deviation in the 0—M2—0 bond angle of 22° from 90°. The two longer M2—0 bonds (Fig. 6.1) are those involving the two bridging oxygen atoms.

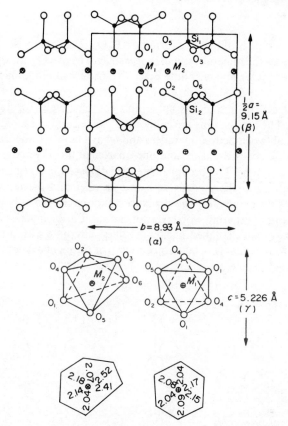

FIG. 6.1. The orthopyroxene crystal structure, projected onto (001). Oxygen co-ordination polyhedra, (100) projections, and metal–oxygen distances (Å) in each site are indicated. ⊕ M1; ⊗ M2; ● silicon; ○ oxygen. (After Burns, 1970, with permission; based on data from Ghose, 1965.)

The greater distortion and larger size of the M2 site in orthopyroxene compared with M1 site gives differences in the energies associated with each site. Thus, not only do the general structural chracteristics of ortho-pyroxene—namely, six-fold co-ordination site availability—give a con-trol to the distribution of a cation between this mineral and another, but the different energies associated with each site can create a non-random distrib-ution of the cations (Mg^{2+}, Fe^{2+}) over the available sites. This phenomenon, in which a cation shows a preference for the site it occupies because of a gain in chemical stability, is called *cation ordering* and, as will be shown later, its extent is temperature-dependent.

Spinels

Spinels are oxides of the type formula AB_2O_4. They comprise a large group which includes the minerals: spinel (*sensu stricto*) $MgAl_2O_4$; magnetite Fe_3O_4 (i.e. $Fe^{2+}Fe^{3+}_2O_4$); and chromite $FeCr_2O_4$. (A few sulphides have the spinel structure.)

The unit cell of a spinel is cubic and contains eight formula units, i.e. $A_8B_{16}O_{32}$. The oxygen ions are cubic close packed and in each unit cell there are 32 available octahedral interstices and 64 available tetrahedral interstices. Half of the 32 octahedral sites and one-quarter of the 64 tetrahedral sites are filled by the cations. The occupied sites are disposed in a regular manner so that the octahedra share edges with one another and the tetrahedra share corners with the octahedra. The structure can be envisaged as having layers of oxygen anions alternating with layers of cations. Cation layers, in which three of the four available octahedral sites are filled, alternate with others in which one of the four available octahedral and two of the eight available tetrahedral sites are filled (Fig. 6.2).

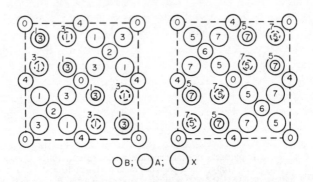

FIG. 6.2. Plan of the spinel unit cell AB_2X_4, projected on to the base of the cell. The heights of the atoms are indicated in units of $\frac{1}{8}a_0$. The lower and upper halves are shown separately. (After Greenwood, 1968)

A spinel may be 'normal' or 'inverse'. In the normal configuration the A-cations of AB_2O_4 are entirely in tetrahedral sites and the B-cations entirely in octahedral sites. In the inverse structure, half of the B-cations occupy the tetrahedral sites while the other half of the B-cations and all the A-cations are in the octahedral positions. This variation, or difference in the degree of order of the ions, may be represented by a parameter, λ, which is the fraction of tetrahedral sites occupied by B-cations. Hence, for a normal spinel $\lambda = 0$, and for an inverse spinel $\lambda = 1$. Other values of λ between 0 and 1 are possible.

Whether a spinel is inverse or normal depends upon four structural controls:
(a) The radii of the ions A and B. On radius ratio criteria, if B^{3+} is smaller than A^{2+} then the spinel will tend to be inverse.
(b) The variation in lattice energy with change in λ. A 'lattice energy' for a particular structure may be calculated from the electrostatic interactions of the constituent ions and when the energy is at a minimum value (corresponding to a particular degree of inversion, λ) the structure will be the most stable. Calculations reveal that in general the normal spinel is favoured on this criterion.
(c) The effects of crystal field energies. (Discussed in Section 6.6.)
(d) Effects of orbital overlap or ion polarization. (Discussed in Section 6.7.)

The point of balance of these four controls determines the λ-value of a spinel at thermodynamic equilibrium. Furthermore, in some spinels λ is temperature dependent. It is also possible to produce spinels with higher λ-values by rapid cooling from an elevated temperature than they would have through slow cooling. This is shown by $MgAl_2O_4$ (usually normal) which can retain small λ-values on quenching.

The configurational details of a number of spinels (not all naturally occurring) are given in Table 6.2. They provide a qualitative insight into the distribution patterns of cations over available octahedral and tetrahedral sites in a close packed array of oxygen anions. In a general way this may be applied to the exchange of cations (such as Fe^{2+}) between a silicate melt, where the cations may be occupying some of the tetrahedral sites, and a crystal where the cations may occupy only octahedral sites (as in olivine). The configurations of a number of spinels allow us to determine quantitatively the site preference energies of certain cations (see Sections 6.6 and 7.4).

Finally one must mention that the spinel structure is of particular interest because of the possibility of silicates in the mantle converting to this structure at high pressure (e.g. Mg_2SiO_4, i.e. $SiMg_2O_4$—an example of a spinel in which the oxidation states are 4 and 2 on A and B respectively, rather than 2 and 3).

Pyrite

Pyrite is a disulphide of formula FeS_2, with the iron atoms in a face-centred cubic array. The structure may be viewed as a derivative of the NaCl structure with the iron atoms on the Na positions and the centre of each S—S bond located on the Cl positions (Fig. 6.3). Sulphur atoms occupy distorted tetrahedral sites with each sulphur atom bonded to three iron atoms and another sulphur atom. Each iron atom is co-ordinated to six sulphur atoms in a slightly distorted octahedron compressed along one of the trigonal axes and which shares corners with neighbouring octahedra. The S—S bond distance is

TABLE 6.2. *Structural details of some 2, 3 spinels*
N = normal ($\lambda = 0$) I = inverse ($\lambda = 1$)
P = partially inverse ($0 < \lambda < 1$)

	MAl₂O₄		MFe₂O₄	
$MgAl_2O_4$	N		$MgFe_2O_4$	P (nearly I)
$MnAl_2O_4$	N		Fe_3O_4	I
$FeAl_2O_4$	N		$CoFe_2O_4$	I
$CoAl_2O_4$	N		$NiFe_2O_4$	I
$NiAl_2O_4$	P($>600°C$)		$CuFe_2O_4$	I
	I($<600°C$)		$ZnFe_2O_4$	N
$CuAl_2O_4$	P			
$ZnAl_2O_4$	N			
			MGa₂O₄	
			$MgGa_2O_4$	P
	MV₂O₄		$MnGa_2O_4$	P
			$FeGa_2O_4$	I
			$CoGa_2O_4$	P (nearly I)
MgV_2O_4	N		$NiGa_2O_4$	I
MnV_2O_4	N		$ZnGa_2O_4$	N
CoV_2O_4	N			
ZnV_2O_4	N			
	MCr₂O₄		MIn₂O₄	
$MgCr_2O_4$	N		$MgIn_2O_4$	I
$MnCr_2O_4$	N			
$FeCr_2O_4$	N			
$CoCr_2O_4$	N			
$NiCr_2O_4$	N			
$CuCr_2O_4$	P			
$ZnCr_2O_4$	N			

See text for comments on the temperature dependence of the degree of inversion.

2.14 Å and the average Fe—S bond distance is 2.26 Å. Vaesite, NiS_2; cattierite, CoS_2; and bravoite, $(Fe, Ni, Co)S_2$, also have the pyrite structure.

The short Fe—S bond distance in pyrite is evidence of a significant proportion of covalent bonding between the atoms. It is considered (Bither *et al.*, 1968; Burns and Vaughan, 1970) that the 3s and 3p orbitals of the sulphur atoms are sp^3 hybridized (tetrahedral arrangement) and the 4s, 4p and some of the 3d orbitals of the iron atoms are d^2sp^3 hybridized (octahedral arrangement).[*] These hybridized orbitals of one atom 'overlap' with those of another atom to form a covalent bond. This aspect is discussed further in Section 6.7.

[*] An account of valence bond theory and hybridization of orbitals may be found in Cotton and Wilkinson (1966).

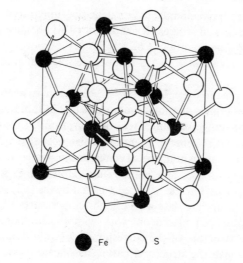

Fe ⬤   S ◯

FIG. 6.3. The structure of pyrite. (After Wuencsch, B. J., 1972. In Wedepohl, 1969 to 1978)

6.3 Silicate melts

6.3.1 General structure

A full knowledge of silicate melt structure would significantly help to advance our detailed understanding of many geochemical processes, including those involving element partition and the crystallization sequence of minerals forming from a cooling magma. Unfortunately our knowledge is still very limited. We are able to make, at best, semi-quantitative descriptions of silicate melts in terms of the linking of silicon–oxygen units and of the types of co-ordination likely to be available for cations, but these lean heavily on theoretical models rather than on a large body of appropriate experimental evidence. In this section a brief review of the relevant properties of melts is made as well as of the approaches used to establish the broad aspects of their structure. A complete description of the importance of melt structure in geochemistry would also require details of liquid metals, sulphide melts and rare magmas such as those that produce carbonatites. Much information is available on the first of these (Richardson, 1974); some on the second, but rather little on the third. Here, the account centres on silicate melts.

There is a substantial quantity of evidence to suggest that structural elements similar to those observed in the corresponding crystalline material exist in some melts. For example, the low entropy of fusion of high cristobalite indicates that the silica melt must have a structure similar to that of the crystal. Furthermore, structural studies (Richardson, 1974) on silica glass show that the Si—O bond distance is about 1.62 Å and that the co-ordination number for every silicon atom is about 4.3 (i.e. the values are very similar to those observed in cristobalite).

X-ray diffraction and thermal neutron-scattering experiments have revealed the fact that most liquids are not structureless, and in some cases it is possible to find the time-averaged arrangement of their atoms. From studies of this kind it is known that, in general, interatomic distances are slightly shorter and the average co-ordination numbers of elements are smaller in the melt than in the crystalline solid of the same composition. However, experiments on silicate melts of geological interest have been limited by the technical difficulties of obtaining the appropriate data at the high temperatures of melting; hence, much reliance has been placed on work with silicate glasses in providing evidence for the structure of the corresponding liquid. Some of the earliest attempts to establish the structural nature of silicate glasses were made by Zachariasen and his coworkers. He was one of the first people to recognize (1932) that silica-rich glasses have a quasi-crystallinity in which framework structures with silicon–oxygen linkages occur. The basic unit of structure in silicate melts, as in silicate minerals, is the $(SiO_4)^{4-}$ tetrahedron. This unit appears to be preserved over a wide range of melt compositions. For example, in the binary melt system Na_2O—SiO_2, the coordination of silicon is about four oxygens over a wide compositional range, although the coordination number of the sodium atoms changes as a function of composition. The $(SiO_4)^{4-}$ structural units may be connected together in a number of ways to give different degrees of silicon–oxygen 'polymerization': as chain polymers with x number (degree of polymerization) of structural units; or as branched structures which may or may not interconnect with themselves to give a network that is planar or an irregular three-dimensional form.

A high degree of polymerization is present in molten silica and this is reflected by the high viscosity of the melt. In this case the majority of the oxygen atoms occur as *bridging oxygens* (i.e. each oxygen atom shared by two adjacent silicon atoms) and the presence of a very low electrical conductivity suggests that the oxygen–silicon bonds are largely covalent. The addition of a small proportion of a metal oxide to a pure silica melt usually produces a marked decrease in the viscosity and a sharp increase in the conductivity showing that the melt has become less polymerized through rupturing of the bridging oxygen bonds and that the oxygen–metal bonds are, at least in many cases, ionic in nature.* The extent of the viscosity decrease is well illustrated by

* Work by Tickle (1967) on the conductance of molten alkali silicates has shown that it is the ionic species of the alkali metals that carry the current and the smaller ions (Li^+ and Na^+) travel faster than the larger potassium ion.

the addition of 2.5 mole % of K_2O to a pure silica melt (at 1700°C) where the reduction is from 2×10^6 Pa s to 2×10^2 Pa s. The extent of a viscosity change, however, is not only a function of the proportion of added metal oxide but also of the nature of the metal ion. As will be shown in Section 6.3.5, some metal ions (e.g. Ca^{2+}) exert a greater depolymerizing effect than others (e.g. Co^{2+}). Similarly, the nature of the ion has a pronounced effect on the electrical conduction (MacKenzie, 1962).

The behaviour of alumina in silicate melts is more complex than that of many metal oxides. In minerals aluminium substitutes for silicon in the SiO_4 tetrahedra (as in the feldspars) and can also occupy co-ordination positions outside these tetrahedra (as in kyanite). In silicate melts it is likely to have a similar dual role. Considerations of the structural characteristics of liquid mixtures of feldspar and silica (Flood and Knapp, 1968) indicate that for some (e.g. albite + silica composition mixtures) the aluminium exists predominantly as discrete AlO_4^{5-} tetrahedra, but in others (e.g. anorthite + silica compositions) higher polymeric units develop. In systems containing relatively more oxygen ions, aluminium would be able to enter into sixfold co-ordination but, except at high pressures, under either hydrous or anhydrous conditions aluminium is likely to be predominantly in tetrahedral co-ordination in most magmas. With increasing pressure, Al^{3+} in melts of basaltic or andesitic composition, aluminium tends to shift from four-fold to six-fold co-ordination. This shift leads to a reduction in the melt's viscosity (Waff, 1975; Kushiro *et al.*, 1976) because the melt becomes less polymerized. Viscosity decreases are not large (e.g. about a factor of 2 for an andesitic melt from 10^5 Pa to 2×10^9 Pa at 1350°C), especially in comparison to the more significant changes produced by increasing temperature (discussed below), but are nonetheless of sufficient magnitude to be of importance in their effects on the rates of crystal settling or on the ease of separation and ascent of melt from its source region in the upper mantle.

6.3.2 Network-formers and network-modifiers

Aluminium and silicon atoms, because of their propensity to form network structures in silicate melts and glasses, are referred to as *network-formers*. Also, the minor element phosphorus often acts as a network-former. Larger cations, such as Ca^{2+}, Mg^{2+}, Fe^{2+}, Na^+, K^+, which occur as major constituents in many magmas, are called *network-modifiers*—they usually disrupt and modify the framework. As discussed above for the case of aluminium, this division into network-formers and network-modifiers is not a clean one for all elements (Mysen and Virgo, 1980; Wood and Hess, 1980).

An empirical but useful indicator of the extent of polymerization is given by the ratio R, of the total number of oxygen atoms to the sum of the network-forming atoms in the magma (Lacy, 1965, 1967):

$$R = \frac{0}{Si + Al + P}. \tag{6.1}$$

R is close to 2 for melts of granitic composition and about 3 for melts of ultrabasic composition. A melt of pure silica (SiO_2), which will be highly polymerized, has an R value of 2, while one of pure forsterite composition (Mg_2SiO_4) is depolymerized and has a value of 4.

Viscosity measurements on natural magmas show that the change in viscosity with temperature usually follows an Arrhenius type relationship:

$$\eta = Ae^{E/RT} \tag{6.2}$$

where A is a constant and E is the activation energy for viscous flow (kJ mol^{-1}). Scarfe (1973) has shown that for magmas of different composition, there is a smooth variation in activation energy for viscous flow at liquidus temperatures and R-value (Fig. 6.4). As predicted, melts with the higher R-values have lower viscosities. Scarfe was able to construct a viscosity–temperature–R-grid (Fig. 6.5) so that, given the temperature and composition of any melt, it would be possible to predict its viscosity. Evidence for the existence of network-formers and network-modifiers is available also from the modelling of viscosities of magmas. Bottinga and Weill (1972) have

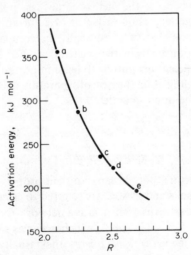

FIG. 6.4. Activation energy for viscous flow of rock melts at liquidus temperatures (~ 1200°C) at 1 atmosphere pressure, plotted against the compositional index R. (a) pantellerite; (b) basic andesite; (c) oceanic island tholeiite; (d) olivine basalt; (e) olivine melanephelinite. (After Scarfe, 1973)

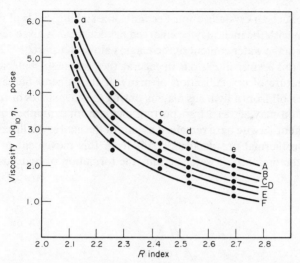

FIG. 6.5. Relationship between viscosity, temperature and composition of melts at one atmosphere pressure. Melt types as in Fig. 6.4. Temperatures, °C: A. 1150; B. 1200; C. 1250; D. 1300; E. 1350; F. 1400. (After Scarfe, 1973)

been able to predict successfully the viscosity of an anhydrous silicate melt (at one atmosphere pressure) from its chemical composition. In this model, element oxides are classified into two types:

Chain-formers: SiO_2, $KAlO_2$, $NaAlO_2$, $Ca_{1/2}AlO_2$ and $Mg_{1/2}AlO_2$.
Chain-modifiers: CaO, K_2O, Na_2O, MgO, FeO, TiO, Al_2O_3.

On this basis the calculated and measured viscosities show good agreement. Furthermore, new studies of immiscibility in silicate melt systems are beginning to indicate the structural roles of a few elements, especially the network-formers (Wood and Hess, 1980; review by Hess, 1977).

6.3.3 *Addition of water*

Whether a magma is hydrous or anhydrous is important geochemically for four principal reasons. Firstly, the addition of water to anhydrous silicate magmas, except those of ultrabasic composition, reduces their viscosity often by quite large amounts. For example, Scarfe (1973) showed that a melt of basic andesite composition at 1150°C had its viscosity reduced by a factor of 4 (1000 to 250 Pa s) when 4 % by weight of water (at $P_{H_2O} = 100$ MPa) was added. A viscosity reduction will enhance the rates of some petrogenetic processes such as mineral growth or crystal settling, and thereby affect the detailed nature of the end products of crystallization. Secondly, the stability fields of minerals in equilibrium with the melt may be enlarged or reduced at a given temperature by changes in the water content of the magma alone (Kushiro, 1972, 1975); an example of this is given in Fig. 6.6. In general, the presence of water also lowers the temperature of crystallization of many of the mineral phases. Thirdly, water driven off from a hydrous magma during the later stages of cooling and crystallization may act as a transporting medium for a number of elements which might otherwise have remained as part of the final crystallized products. Many hydrothermal ore deposits result from this movement of elements. Fourthly, the presence of water permits the formation of hydrous minerals.

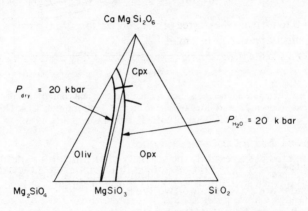

FIG. 6.6. Olivine–pyroxene liquidus boundaries with and without water, at 20 kbar (2 GPa) total pressure. (After Kushiro, 1969)

The solubility of water in many natural magmatic systems, however, is limited. Hamilton *et al.* (1964) record solubilities of 3.1 wt % at 1000 bars (100 MPa) and 9.4 wt% at 6000 bars (600 MPa) in a basaltic melt, and about 4.5 wt % at 1000 bars (100 MPa) and 10.1 wt % at 5300 bars (530 MPa) in an andesitic melt. But it must be remembered that a small per cent weight of water represents quite a large mole fraction as its molecular weight is much smaller than that of the other major element oxides. For example, 4 % by weight of water in a basaltic magma could be equivalent to a mole fraction of between 10 and 15 %. These solubilities are large enough to change significantly the structure of the melt compared with that under dry conditions. The mechanism by which this change takes place has been suggested by many workers to involve the reaction of the water with the bridging oxygens of the silicate network according to the equation:

$$H_2O + \; \equiv Si\text{—}O\text{—}Si \equiv \; \rightleftharpoons 2(\equiv Si\text{—}OH) \tag{6.3}$$

and so reduce the degree of polymerization. This reduction lowers the viscosity of the melt unless it is already depolymerized under anhydrous conditions (such as in an olivine basalt melt) when the addition of water can have no effect as a depolymerizing agent and so the viscosity change is small. Recently, the use of Raman spectroscopy in studies of the solubility of water in silicate melts at high pressure (about 2 GPa) has shown that water dissolved in those melts with a significant proportion of non-bridging oxygens, reacts to form $\equiv$ Si—OH and $M(OH)_n$ complexes (Mysen *et al.*, 1980).

It is possible, however, that water has a dual role, behaving as an acid in highly basic melts and as a base in other melts. For very basic melts:

$$H_2O + 2(\equiv Si—O)^- \rightleftharpoons \equiv Si—O—Si \equiv + 2OH^-. \tag{6.4}$$

The effect is to increase the degree of polymerization. On the basis of viscosity data this effect appears to be small.

Whether a silicate melt is basic or not, the solubility of water increases with increasing pressure but decreases with increasing temperature (see Mysen, 1977, for a review).

6.3.4 Carbon dioxide and other volatiles

Neither the solubility of other volatile components such as the halogens nor their effects on melt structure is well known, except perhaps in the case of CO_2, for which many experimental data have recently been published (e.g. Mysen, 1977; Eggler and Rosenhauer, 1978). The solubility of CO_2 in silicate melts is much less than that of water but it shows a strong rise with increase in temperature or pressure (Mysen, 1976a). This small solubility provides a means of generating a separate vapour phase during magmatic fractional crystallization, and is manifested by the CO_2-rich nature of many of the fluid inclusions found in the phenocryst phases of some volcanic rocks (Roedder, 1972).

In the past it has generally been assumed that the sole species existing in a melt is CO_3^{2-} but the work of Mysen suggests that discrete CO_2 molecules may exist in melts of jadeite, albite or nepheline composition. The generation of CO_3^{2-} probably leads to an increase in the degree of melt polymerization according to reaction (6.5) involving non-bridging oxygens:

$$2(\equiv Si—O)^- + CO_2 \rightleftharpoons \equiv Si—O—Si \equiv + CO_3^{2-} \tag{6.5}$$

but when water is present the CO_3^{2-} ion is stabilized in the melt without further polymerization (Mysen, 1977):

$$2(\equiv Si—O)^- + H_2O + CO_2 \rightleftharpoons 2(\equiv Si—OH) + CO_3^{2-}. \tag{6.6}$$

The solubility mechanism of equation (6.5) is consistent with the greater solubility of CO_2 in basic melts than in silicic ones, the former having a greater proportion of non-bridging oxygens.

The concentration of chlorine in basalts is of the order of 200 ppm, but which species (i.e. free chlorine or chloride) is present in the melt is not known. More acid igneous rocks, such as glassy peralkaline ones, can contain a few thousand ppm of chlorine but the element may be expelled during crystallization since it is not accepted into hydrous minerals when in the presence of fluorine (Carmichael *et al.*, 1974). Fuge (1977) has suggested, however, that because chlorine has a strong affinity for associated aqueous phases of a magma, it will tend to be concentrated in early formed hydrous minerals of a differentiating melt, while fluorine, by virtue of its strong affinity for the melt, will be enriched in the late hydrous minerals. In view of the strong preference for the aqueous phase it is unlikely that chlorine acts as a depolymerizing agent in silicate melts. This is in contrast to fluorine which, as the fluoride ion, may disrupt the bridging oxygens of the network and thereby reduce the melt's viscosity (Richardson, 1974):

$$\equiv Si—O—Si \equiv + F^- \rightleftharpoons \equiv Si—O^- + F—Si \equiv. \qquad (6.7)$$

It may also link with sodium or other network modifiers.

A review on fluorine in granitic rocks and melts is given by Bailey (1977) where the wide variety in concentrations of this element, even in very similar rock types, is recorded. The apparent inability of a silicate melt to retain its chlorine and possibly some of its fluorine during the final stages of crystallization has an important bearing on the processes of metasomatism and the formation of hydrothermal ore deposits.

Knowledge of the solubility of sulphur (as sulphur, sulphide or sulphate) in magmas and silicate slags is a little better. Sulphur concentration in natural basaltic melts is usually no more than 1000 ppm. Evidence from work on slags (Richardson, 1974) suggests that this occurs as sulphide, for free sulphur is insoluble, but an increase in the oxygen partial pressure of the melt reduces the amount of 'sulphide sulphur' by conversion to 'sulphate sulphur'. Katsura and Nagashima (1974) in a study of the solubility of sulphur in a basaltic magma (at 1 atmosphere pressure) showed that most of the sulphur was present as sulphide when the partial pressure of oxygen was less than 10^{-8} atm, but as dissolved sulphate when greater than 10^{-8} atm (see Table 6.3). A temperature rise also increases the solubility of sulphur (at constant partial pressures of oxygen and sulphur).

Furthermore, an increase in the silica content of a melt increases the solubility of sulphide–sulphur, suggesting that the sulphur is co-ordinated to the metal cations and not to any oxygens of the SiO_4 tetrahedra. This point has been stressed by Haughton *et al.* (1974), who showed that the maximum

TABLE 6.3. *Contents of sulphide–sulphur and sulphate–sulphur in a tholeiitic basalt at 1250°C and 1 atm pressure, and at 2.1 % SO$_2$*

		S content wt %		
$-\log P_{S_2}$	$-\log P_{O_2}$	Total S	Sulphide	Sulphate
14.0	3.65	0.006	< 0.001	0.006
10.5	5.38	0.003	< 0.001	0.003
7.7	6.79	0.001	< 0.001	0.001
3.4	8.97	0.009	0.008	0.001
3.0	10.56	0.022	0.022	< 0.001
4.0	12.20	0.11	0.11	< 0.001

solubility in a basaltic magma increased four-fold (0.05 to 0.2 wt %, at 1200°C) when the FeO content was also increased four-fold (5 to 20 wt %). It is likely that the sulphur is co-ordinated to the iron.

6.3.5 *Polymer model*

The discussion so far has described silicate melts in a qualitative way as made up of polymeric units consisting predominantly of silicon, aluminium and oxygen atoms, other cations which can modify the polymers, and free oxygen ions. We should like to know the sizes of the polymeric units, their spatial arrangement, and the nature of the interaction between them and the network-modifiers. There is no direct experimental way which can tell us the proportions of the different sized polymeric units or their degree of polymerization. Some of the indirect evidence has been presented above so that we can envisage at one compositional extreme—that of fused silica—there is just one polymeric unit with an infinite degree of polymerization, and at the other—that of molten olivine—that no polymers other than the monomer $(SiO_4)^{4-}$ exist. In between these two extremes we can expect the existence of units with differing degrees of polymerization and in differing proportions. One way of helping to establish the structural nature of silicate melts has been to apply the theoretical aspects of polymer chemistry (Flory, 1953) to them, and with the addition of general equilibrium reactions definable in thermo-dynamic terms (Toop and Samis, 1962). This approach gives us information about molecular chain lengths and weights, viscosities, and activities of ionic species in relatively simple melts (binary and ternary systems). It does not tell us about the nature, availability and occupancy of cation sites in the liquid; for that we have to turn to other theoretical and experimental evidence. In this section a brief examination of the application of the *polymer model* to silicate melts is presented. A discussion of cation co-ordination is given in the following section.

Early interpretations of the thermodynamic properties of silicate melts based upon the assumption that silica was present entirely as SiO_4^{4-} groupings were unsuccessful except for melts of a very restricted range of compositions, and it was found necessary to consider the polymerization of silica groupings in equilibria of the kind:

$$\equiv Si—O—Si \equiv + O^{2-} \rightleftharpoons 2(\equiv Si—O^-). \qquad (6.8)$$

The use of statistical polymer theory enables the calculation of the relative proportions of the various polymeric species (i.e. SiO_4^{4-}; $Si_2O_7^{6-}$; $Si_3O_{10}^{8-}$ ...) which exist in any one melt as well as the proportion of free oxygen ions (O^{2-}). The theory rests on the important assumption that the chemical reactivity of a functional group (e.g.. $\equiv SiO^-$) does not depend on the size of the molecule to which it is attached. This is the 'principle of equal reactivity' and it allows the application of a simple statistical treatment of bond distribution among units in a polymeric system. Flory (1953) discussed the experimental evidence in support of the principle and showed that it is valid for all polymeric molecules except those of very small size.

In a series of papers Masson and his co-workers have developed and extended the application of polymer to melt structures. Figure 6.7 shows some of the possible polymeric molecules that might exist in a silicate melt, ignoring for the moment the possibility of intramolecular condensations. Whiteway *et al.* (1970) have shown how we may determine from a statistical treatment the relative proportions of these species in a polymeric system and Masson *et al.* (1970) have applied this treatment to liquid silicate binary systems.

We define the extent of reaction, α, as the fraction of oxygens initially present that have become shared by more than one tetrahedral unit during polymerization. Whiteway *et al.* (1970) have derived the following expression giving the mole fraction of an x-mer (N_x) present in a silicate melt:

$$N_x = \omega_x \left(\frac{2\alpha}{3}\right)^{x-1} \left(1 - \frac{2\alpha}{3}\right)^{2x+1} \qquad (6.9)$$

and where $\omega_x = \dfrac{3x}{(2x+1)! \, x!}$.

Solutions of this equation for different α-values given in Table 6.4 show that for all values of α up to 0.5 the monomer is the most abundant of any polymeric species and that each particular polymer has its maximum abundance at a particular α-value such that the larger the polymer the greater the value of α at the maximum. It must be remembered, however, that the dimer contains twice as many silicon atoms, and the trimer three times as many as the monomer, so that the number of silicons atoms in all the higher polymeric units may exceed those in the monomer.

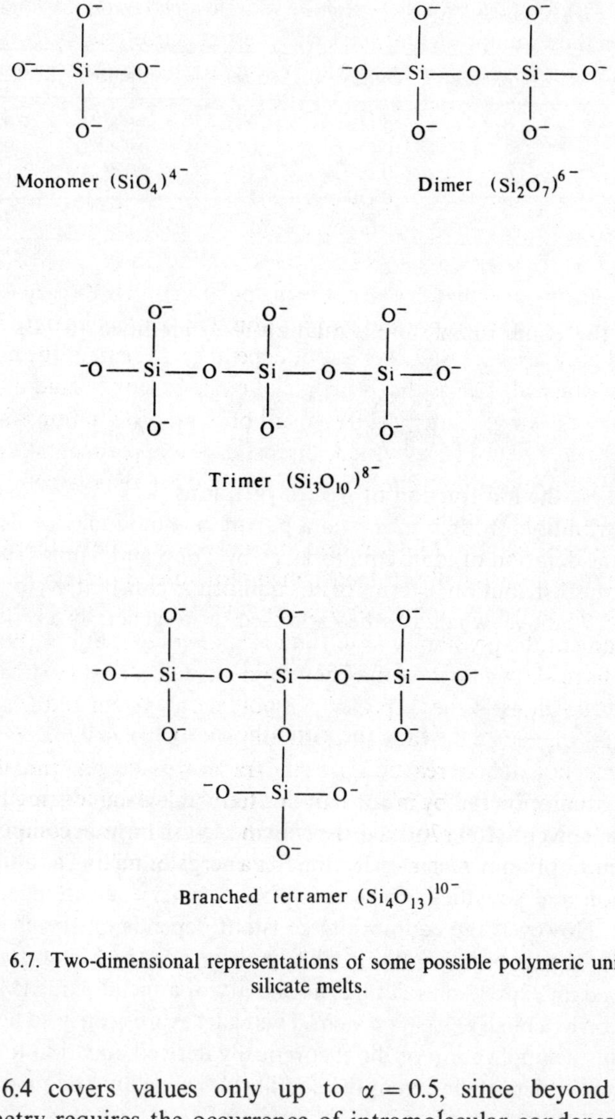

Monomer $(SiO_4)^{4-}$ Dimer $(Si_2O_7)^{6-}$

Trimer $(Si_3O_{10})^{8-}$

Branched tetramer $(Si_4O_{13})^{10-}$

FIG. 6.7. Two-dimensional representations of some possible polymeric units in silicate melts.

Table 6.4 covers values only up to $\alpha = 0.5$, since beyond that the stoichiometry requires the occurrence of intramolecular condensations, the theory being no longer applicable. This theory strictly applies to cases where the condensation reactions yield no other products besides the polymers. However, in the case of silicates a condensation reaction such as:

$$SiO_4^{4-} + SiO_4^{4-} \rightarrow Si_2O_7^{6-} + O^{2-} \tag{6.10}$$

yields free oxygen ions which accumulate in the system and help determine the

TABLE 6.4. *Mole fractions of polymeric species as a function of the extent of reaction, α*

α	Monomer	Dimer	Trimer	Tetramer	Remainder
0.1	0.813	0.142	0.033	0.009	0.003
0.2	0.651	0.196	0.078	0.036	0.039
0.3	0.512	0.197	0.101	0.059	0.131
0.4	0.394	0.170	0.123	0.064	0.249
0.5	0.296	0.132	0.078	0.053	0.441

nature of the equilibrium state. Equation (6.9) modified to take this into account gives

$$N_x = \omega_x \cdot \left(\frac{2\alpha}{3}\right)^{x-1} \left(1 - \frac{2\alpha}{3}\right)^{2x+1} (1 - N_{O^{2-}})$$ (6.11)

where $N_{O^{2-}}$ is the ion fraction of free oxygen ions.

The distribution of ionic species in a polymeric liquid may be determined also by consideration of ionic equilibria, as by Toop and Samis (1962). They expressed the distribution in terms of an equilibrium constant, K, for reactions of type (6.10) above, which can be expressed more generally as:

$$2O^- \rightleftharpoons O^0 + O^{2-}$$ (6.12)

so that $K = \dfrac{(O^0)(O^{2-})}{(O^-)^2}$, where the parentheses signify activity. (6.13)

With the assumption that K is not only constant at a given temperature but is specific for a given system and is independent of variations in compositions of that system, approximate curves for the free energy of mixing as a function of composition and activities of species in basic, binary silicate melts may be calculated. However, the equilibrium constant depends on the chain length, and account should be taken of this. This has been done by Masson *et al.* (1970), who derived an expression relating the activity of a metal oxide (MO) to the composition of a binary silicate system. This exact expression need not concern us here but the application of the theoretically derived equation to empirical data allows the determination of the equilibrium constant and provides some interesting results.

For the generalized condensation reaction:

$$SiO_4^{4-} + Si_nO_{3n+1}^{2(n+1)-} \rightleftharpoons Si_{n+1}O_{3n+4}^{2(n+2)-} + O^{2-}$$ (6.14)

let the equilibrium constant be $K_{1,n}$. Since K depends on chain length, Masson *et al.* compared the theoretical expression with the experimental curves of MO activity versus composition for the case where $n = 1$ in equation (6.14), and determined the values of $K_{1,1}$ which gave the best match with experimental results (Table 6.5).

TABLE 6.5. *Equilibrium constants*
($K_{1,1}$) of some silicate systems for
N_{SiO_2} *less than 0.5*

System	$K_{1,1}$	Temperature °C
CaO—SiO$_2$	0.0016	1600
PbO—SiO$_2$	0.196	1000
MnO—SiO$_2$	0.25	1500–1600
FeO—SiO$_2$	0.70	1257–1406
CoO—SiO$_2$	2.0	1400–1700
SnO—SiO$_2$	2.55	1100

From Masson *et al.*, 1970.

These results show the considerable effect that the cation has on the equilibrium. Melts of the CaO—SiO$_2$ system are highly depolymerized while those of the SnO—SiO$_2$ system are relatively polymerized. Despite the fact that the theory ignores intramolecular condensation there is still excellent agreement with experimental data. Therefore, such condensations cannot be extensive in many melts, at least those of systems listed in Table 6.5.

The polymer model allows us to make a number of predictions and calculations concerning the activity of species and of the proportions of various silicate groups in simple (namely, binary) melt systems. The extension of the findings to magmatic systems allows us to understand a little more of melt behaviour as a function of silica content. However, the model tells us little about immiscibility in silicate systems and nothing of the co-ordination of the metal cations. For the latter we must turn to other models of liquid structure.

6.3.6 Cation co-ordination in silicate melts

Mössbauer spectroscopy, together with optical and infra-red absorption measurements, have provided important data on the co-ordination of many elements in silicate glasses, and it has been customary to extrapolate the applicability of these conclusions from the glass studies to silicate melts. The transition elements have been excellent candidates for studies of this kind and some revealing observations have been made. For example, sodium silicate glasses that are doped with nickel show the nickel ions to be both tetrahedrally and octahedrally co-ordinated. If the temperature of the glass is raised, more of the nickel becomes tetrahedrally co-ordinated. An increase in the proportion of tetrahedrally co-ordinated nickel can also be achieved by an increase in the ionic size of the other cation. In passing through a series of alkali–silicate glasses the nickel co-ordination changes from being predominantly octahedral in a Li–silicate glass to predominantly tetrahedral in a Rb–silicate glass. Recent work (Boon and Fyfe, 1972) on spectra of ferrous iron in silicate glasses has indicated that conclusions about co-ordination number in silicate glasses cannot be extended with confidence to silicate melts. Even if a melt is quenched quickly there can still be appreciable changes in co-ordination during quenching.

However, we can obtain some clues about element co-ordination in melts from their co-ordination in glasses. Visible and infra-red absorption spectra of Ni-doped glass of olivine composition show the nickel occupying both tetrahedral and octahedral sites, yet nickel is purely octahedrally co-ordinated in crystalline olivine. It seems reasonable to suppose that some nickel in melts is tetrahedrally co-ordinated. We will see later that there are other independent experimental data that are consistent with this view.

An alternative approach to the determination of element co-ordination in melts is to take particular theoretical models of melt structure and match their predicted chemical behaviour with that observed. This approach is still in its early stages of application.

Theoretical models of liquid structure have been available for some time. Bernal was one of the earlier workers in this field. In one of his later writings (1964) on the subject he presented the results of a simple experiment involving the packing of rigid spheres in a random assembly in which the co-ordination of each sphere was determined. This was found to be on average lower (8 or 9) than in a regular close-packed array (12). Of importance to our present discussion was the identification of five types of holes within the packing (Fig. 6.8 and Table 6.6). The percentage proportion of each of these was also recorded (Table 6.6).

One interesting aspect of these results is the predominance of the number of available holes of tetrahedral co-ordination, and gives some confirmation of the lowering of cation co-ordination on melting of a crystal. Bernal also recorded the presence of a few large holes within the geometrical assembly. It can also be seen that combinations of the available polyhedra may lead to the presence of interstices of relatively high co-ordination number, e.g. as in the trigonal prism capped by half-octahedra in Fig. 6.8. Whittaker (1978), using a modified treatment of Bernal's data, showed that while tetrahedra are the most frequently occurring polyhedra in a random close-packed assembly, cavities with five vertices are the next most abundant.

This experimental approach was restricted to monatomic liquids, as spheres of only one size were used. However, it is suggested that in ionic liquids the structure is determined by the packing of the large negative ions with the smaller positive ions fitting into interstices of four-fold and five-fold co-

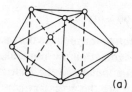

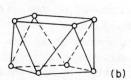
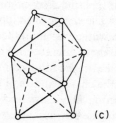

FIG. 6.8. Examples of available polyhedra in random rigid-sphere assembly. (a) Trigonal prism capped with three half octahedra. (b) Archimedean antiprism. (c) Tetragonal dodecahedron. (After Bernal, 1960)

TABLE 6.6.   *Available holes in random rigid-sphere assembly*

Polyhedron type	Number (%)	Volume (%)
Tetrahedra	73.0	48.4
Half-octahedra	20.3	26.9
Trigonal prisms	3.2	7.8
Archimedean anti-prisms	0.4	2.1
Tetragonal dodecahedra	3.1	14.8

ordination. For silicate melts this implies a close packing of the oxygens with silicon and other ions in the interstices. The preservation of the basic $(SiO_4)^{4-}$ unit in silicate melts means that the Bernal geometrical model must be of limited applicability to these systems but it does provide a starting-point for work of this kind. Its importance lies in the identification of the high proportion of available cation (including silicon) sites of low co-ordination number, and also the existence of a few large cavities, some of which probably have a high associated co-ordination number.

Further aspects of element co-ordination in silicate melts are discussed in the following sections.

6.4 Atomic substitution and isotypism

In the preceding sections we have seen that a wide variety of sites are available for element co-ordination in minerals and silicate melts. The questions that now arise are: what are the characteristics of elements that occupy these sites and what are the necessary conditions for one element to substitute for another in a given site?

The ability of an atom or ion to substitute (or *proxy*) for another in a given crystal structure rests largely on the requirement that its radius is similar to that of the replaced atom or ion. (The elements are said to be *diadochic* in that structure.) The charge of an ion is also of importance in this process—element exchange will occur more readily when charges of the ionic species are identical. These, however, are generalized statements and they are examined more closely in the following sections of this chapter.

When the charge and radii of substituting ions are similar then complete solid solution may exist between the two end-member compounds, as for example in the olivines, where substitution of Mg^{2+} by Fe^{2+} gives rise to a continuous isotypic series from Mg_2SiO_4 (forsterite) through $(Mg, Fe)_2SiO_4$ to Fe_2SiO_4 (fayalite). In the olivine series the solid solution approximates to being an ideal solution. In others—such as the alkali feldspars—the solution is non-ideal and may show a maximum or minimum melting point. Where ions differ markedly in size, then solid solution will be absent or will occur to a limited extent. *Limited solid solution* occurs when two minerals are iso-structural ('isotypic') but the difference in the size of the ions in one compared with the other is such that neither mineral can tolerate more than a few mole per cent of the other without the distortion raising the energy beyond the level where it is more stable for the two minerals to co-exist as separate phases. This is shown in the binary system $KAlSi_3O_8$—$NaAlSi_3O_8$ at low temperatures.

Besides substitutional solid solution, there also exists in minerals the phenomenon of interstitial solid solution. Here, oriented cavities or voids are sufficiently large to accommodate extraneous atoms or small molecules (e.g. as in the ring silicate, beryl).

Substitution involving different valencies of the substituted and substituting elements also occurs, and is known as altervalent substitution. In order to maintain electrical neutrality in the crystal, one (or more) of the following processes must occur:

(a) substitution of a second element of a different, and compensating, valency, at the same time;

(b) addition of an ion into an interstitial position in the structure;

(c) the development of a vacancy (e.g. $A_2^+ B^{2+} X_{cryst}^{4-} + C_{liq}^{3+} \rightarrow$
$A^+ \square C^{3+} X_{cryst}^{4-} + B_{liq}^{2+} + A_{liq}^+$).

The process (a) above could occur as:

$$A^+M^{3+}X^{4-}_{cryst} + B^{2+}_{liq} + C^{2+}_{liq} \rightarrow B^{2+}C^{2+}X^{4-}_{cryst} + A^+_{liq} + M^{3+}_{liq}$$

or it could involve substitution in the anion group, as in the plagioclase feldspars which may be expressed as:

$$NaAlSi_3O_8 + Ca^{2+} + Al^{3+} \rightarrow CaAl_2Si_2O_8 + Na^+ + Si^{4+}.$$

It is clear that other combinations are possible and also that the exchange of two elements may be constrained by the availability and properties of an additional element (or even elements). Hence, element partition between melt and crystal, and involving altervalent substitution, is sensitive to changes in the composition of the melt. The activity coefficients of species of the same valency tend to change in parallel manner with change in melt composition, whereas changes in those of different valency often diverge.

Specific properties of the chemical bonding can also affect element substitution but before these are examined a more detailed treatment of the effects of ionic size and charge must be given.

6.5 Ionic radius and charge

The recognition of the importance of ionic radius and charge in determining element distribution was one of the major advances in theoretical geochemistry. In 1937 Goldschmidt (in a summary of some of his earlier work) enunciated the principles of this control. The ideas embodied in his statements have since become known as Goldschmidt's Rules and they may be expressed as follows:

1. Ions of similar radii and the same charge will enter into a crystal in amounts proportional to their concentration in the liquid.
2. An ion of a smaller radius but with the same charge as another, will be incorporated preferentially into a growing crystal.
3. An ion of the same radius but with a higher charge than another will be incorporated preferentially into a growing crystal.

The rules offer a simple explanation of element distribution as observed in minerals but it was soon realized that there are some cases where the rules are invalid. Fundamental criticisms of the rules were made in an important paper by Shaw (1953). One of the objections raised by him was that the rules fail to explain element distribution in a binary phase system with a maximum or minimum melting point. Consider a solid solution system AX–BX in which the ionic radius of A^+ is smaller than B^+, and where the phase relationship is of the kind in Fig. 6.9. A melt of composition M_1, on cooling, will crystallize a compound

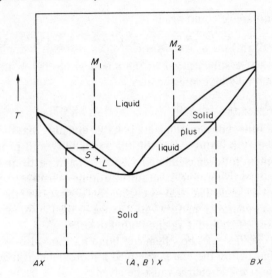

Fig. 6.9. Binary phase diagram with minimum melting point. For discussion, see text.

enriched in A relative to the melt. Similarly in the case of M_2 the solid phase will be enriched in B relative to the melt. Goldschmidt's rule 2 is in accord with the observed distribution for a melt M_1 but not for melt of composition M_2.

Despite this and other criticisms of Goldschmidt's rules (Burns and Fyfe, 1967) it is clear that ionic size and charge play an important part in determining element distribution, but they must not be considered in isolation—attention must be paid also to the bonding characteristics. However, to be able to predict or understand the geochemical distribution of an element it is necessary to have an accurate set of ionic radii values and to know the existent valency state or states.

Two approaches have been used in attempts to establish a set of ionic radii. One is to derive radius values from theoretical models of element bonding, and bond length. This may be extended so as to use the relationships between the size of an ion in an ionic crystal and certain physical properties (e.g. molar refractivity). The other approach is the geometrical one, based on the packing of spheres and the use of interatomic distances from X-ray structural analysis. Here the well-known 'radius ratio' concept has been used.

An important compilation of ionic radii was made by Ahrens (1952). He noted that there are regularities between radius, ionization potential and charge, and he revised earlier ionic radii values (principally those derived by Pauling) with the aid of these regularities and the currently available data on interionic distances. This set of radii, for six-fold co-ordination, has been much used and quoted in the geochemical literature (see Appendix V). The serious disadvantage of this set, as well as earlier ones, is that it is applicable only to ions in six-fold co-ordination and it did not take account of different oxidation states of the ion. Shannon and Prewitt (1969, 1970) compiled an extensive list of interionic distance data from X-ray structural analyses of oxides and fluorides and used these to derive a set of radii for different oxidation states, co-ordination numbers and spin state (where appropriate, see Section 6.6) of the ions. In deriving the radii the approximately linear relationship between ionic volume and unit cell volume of over sixty isotypic series of oxides and fluorides was used. Their effective ionic radii values, however, still require an assumed value for the radius of either the oxygen or fluorine anion—a value of 1.40 Å for O^{2-}, six-fold co-ordinated was taken (the same value as used by Ahrens).* The compilation by Shannon and Prewitt has been updated and revised by Shannon (1976); the revised values are given in Table 6.7 where the trends of decreasing radius (for a particular co-ordination number) across a period as well as down a group within the table can be seen.

Whittaker and Muntus (1970) used the data provided by Shannon and Prewitt to produce a set of radii which are consistent with the radius ratio criterion. They found that an ionic radius for the O^{2-} (VI) ion of 1.32 Å is the best value for giving radii values to other ions which then comply with co-ordination requirements as determined by the radius ratio concept. Because the oxide ion is smaller by 0.08 Å compared with the value used by Shannon and Prewitt, and as the bond lengths are fixed, so the cation radii are all 0.08 Å larger in the Whittaker and Muntus compilation than in the set by Shannon and Prewitt (1969, 1970). There is, in the author's view, no clear case for preferring either set of radii except that the Shannon (1976) revision used more data and is more recent. The values given by Shannon are used in this text (both sets are listed in Appendix V). Readers of geochemical publications should be aware that confusion could arise through the existence of two recent compilations. There is, for example, no consistent usage of one set of radii in the *Handbook of Geochemistry*, Part II (editor, Wedepohl, 1969–1978).

Although the Shannon and Prewitt radii are for cations co-ordinated to oxide or fluoride ions, this obviously raises little restriction in their use in geochemistry. However, it is inappropriate to apply ionic radii values to covalently bonded structures. In general the covalent bond distance is different from that expected if the bond were ionic. For example, the sum of the ionic radii of Fe^{2+} (VI) and S^{2-} (IV) (based on values by Shannon and Prewitt, and

---

* Shannon and Prewitt also list values derived according to the description by Fumi and Tosi (1964) of 'ionic' radii in the alkali halides with the NaCl structure. Here the F^- radius was taken as 1.19 Å and so the cation radii are all larger by 0.14 Å. These values are rarely quoted in the literature.

TABLE 6.7. Ionic radii (Å) for different co-ordination numbers (roman numerals), oxidation states (arabic numerals) and, where appropriate, spin states (H = high, L = low); py = pyramidal; sq = square.
(Values from Shannon, 1976.)
* Indicates that data for additional oxidation states may be found in the list given by Shannon (1976).

Li		Be	
1 IV	0.590	2 III	0.16
VI	0.76	IV	0.27
VIII	0.92	VI	0.45

Na		Mg	
1 IV	0.99	2 IV	0.57
V	1.00	V	0.66
VI	1.02	VI	0.720
VII	1.12	VIII	0.89
VIII	1.18		
IX	1.24		
XII	1.39		

K		Ca		Sc		Ti		V		Cr	
1 IV	1.37	2 VI	1.00	3 VI	0.745	2 VI	0.86	2 VI	0.79	2 VI	L 0.73
VI	1.38	VII	1.06	VIII	0.870	3 VI	0.670	3 VI	0.640		H 0.80
VII	1.46	VIII	1.12			4 IV	0.42	4 V	0.53	3 VI	0.615
VIII	1.51	IX	1.18			V	0.51	VI	0.58	4 IV	0.41
IX	1.55	X	1.23			VI	0.605	VIII	0.72	VI	0.55
X	1.59	XII	1.34			VIII	0.74	5 IV	0.355	5 IV	0.345
XII	1.64							V	0.46	VI	0.49
								VI	0.54	VIII	0.57
										6 IV	0.26
										VI	0.44

Rb		Sr		Y		Zr		Nb		Mo	
1 VI	1.52	2 VI	1.18	3 VI	0.900	4 IV	0.59	3 VI	0.72	3 VI	0.69
VII	1.56	VII	1.21	VII	0.96	V	0.66	4 VI	0.68	4 VI	0.650
VIII	1.61	VIII	1.26	VIII	1.019	VI	0.72	VIII	0.79	5 IV	0.46
IX	1.63	IX	1.31	IX	1.075	VII	0.78	5 IV	0.48	VI	0.61
X	1.66	X	1.36			VIII	0.84	VI	0.64	6 IV	0.41
XI	1.69	XII	1.44			IX	0.89	VII	0.69	V	0.50
XII	1.72							VIII	0.74	VI	0.59
XIV	1.83									VII	0.73

Cs		Ba		La-Lu		Hf		Ta		W	
1 VI	1.67	2 VI	1.35			4 IV	0.58	3 VI	0.72	4 VI	0.66
VIII	1.74	VII	1.38			VI	0.71	4 VI	0.68	5 VI	0.62
IX	1.78	VIII	1.42	See page		VII	0.76	5 VI	0.64	6 IV	0.42
X	1.81	IX	1.47	131		VIII	0.83	VII	0.69	V	0.51
XI	1.85	X	1.52					VIII	0.74	VI	0.60
XII	1.88	XI	1.57								
		XII	1.61								

Fr		Ra		Ac-Lw	
		2 VIII	1.48	See page	
		XII	1.70	131	

Mn*		Fe*		Co		Ni		Cu		Zn	
2 IV	H 0.66	2 IV	H 0.63	2 IV	H 0.58	2 IV	0.55	1 II	0.46	2 IV	0.60
V	H 0.75	IV sq	H 0.64	V	0.67	IV sq	0.49	IV	0.60	V	0.68
VI	L 0.67	VI	L 0.61	VI	L 0.65	V	0.63	VI	0.77	VI	0.740
	H 0.830		H 0.780		H 0.745	VI	0.690	2 IV	0.57	VIII	0.90
VII	H 0.90	VIII	H 0.92	VIII	0.90	3 VI	L 0.56	IV sq	0.57		
VIII	0.96	3 IV	H 0.49	3 VI	L 0.545		H 0.60	V	0.65		
3 V	0.58	V	0.58		H 0.61	4 VI	L 0.48	VI	0.73		
VI	L 0.58	VI	L 0.55	4 IV	0.40			3 VI	L 0.54		
	H 0.645		H 0.645	VI	H 0.53						
4 IV	0.39	VIII	H 0.78								
VI	0.530	4 VI	0.585								

Tc		Ru		Rh		Pd		Ag*		Cd	
4 VI	0.645	3 VI	0.68	3 VI	0.665	1 II	0.59	1 II	0.67	2 IV	0.78
5 VI	0.60	4 VI	0.620	4 VI	0.60	2 IV sq	0.64	IV	1.00	V	0.87
7 IV	0.37	5 VI	0.565	5 VI	0.55	VI	0.86	IV sq	1.02	VI	0.95
		7 IV	0.38			3 VI	0.76	V	1.09	VII	1.03
		8 IV	0.36			4 VI	0.615	VI	1.15	VIII	1.10
								VII	1.22	XII	1.31
								VIII	1.28		
								2 IV sq	0.79		
								VI	0.94		

Re		Os		Ir		Pt		Au		Hg	
4 VI	0.63	4 VI	0.630	3 VI	0.68	2 IV sq	0.60	3 IV sq	0.68	1 III	0.97
5 VI	0.58	5 VI	0.575	4 VI	0.625	4 VI	0.625			VI	1.19
6 VI	0.55	6 V	0.49	5 VI	0.57	5 VI	0.57	5 VI	0.57	2 II	0.69
7 IV	0.38	VI	0.545							IV	0.96
VI	0.53	7 VI	0.525							VI	1.02
		8 IV	0.39							VIII	1.14

Continued

Table 6.7 (*Continued*)

B		C*		N*		O		F	
3 III	0.01	—		—		−2 II	1.35	−1 II	1.285
IV	0.11					III	1.36	III	1.30
VI	0.27					IV	1.38	IV	1.31
						VI	1.40	VI	1.33
						VIII	1.42		

Al		Si		P		S		Cl	
3 IV	0.39	4 IV	0.26	5 IV	0.17	6 IV	0.12	5 III py	0.12
V	0.48	VI	0.400	V	0.29	VI	0.29	7 IV	0.08
VI	0.535			VI	0.38				

Ga		Ge		As		Se		Br	
3 IV	0.47	4 IV	0.390	5 IV	0.335	6 IV	0.28	3 IV sq	0.59
V	0.55	VI	0.530	VI	0.46	VI	0.42	5 III py	0.31
VI	0.620							7 IV	0.25

In		Sn		Sb		Te		I	
3 IV	0.62	4 IV	0.55	3 IV py	0.76	4 III	0.52	5 III py	0.44
VI	0.800	V	0.62	V	0.80	IV	0.66	VI	0.95
VIII	0.92	VI	0.690	5 VI	0.60	VI	0.97	7 IV	0.42
		VII	0.75			6 IV	0.43	VI	0.53
		VIII	0.81			VI	0.56		

Tl		Pb		Bi		Po		At	
1 VI	1.50	2 IV py	0.98	3 V	0.96	4 VI	0.94		
VIII	1.59	VI	1.19	VI	1.03	VIII	1.08		
XII	1.70	VII	1.23	VIII	1.17				
3 IV	0.75	VIII	1.29	5 VI	0.76				
VI	0.885	IX	1.35						
VIII	0.98	X	1.40						
		XI	1.45						
		XII	1.49						
		4 IV	0.65						
		V	0.73						
		VI	0.775						
		VIII	0.94						

La		Ce		Pr		Nd		Pm	
3 VI	1.032	3 VI	1.01	3 VI	0.99	2 VIII	1.29	3 VI	0.97
VII	1.10	VII	1.07	VIII	1.126	IX	1.35	VIII	1.093
VIII	1.160	VIII	1.143	IX	1.179	3 VI	0.983	IX	1.144
IX	1.216	IX	1.196	4 VI	0.85	VIII	1.109		
X	1.27	X	1.25	VIII	0.96	IX	1.163		
XII	1.36	XII	1.34			XII	1.27		
		4 VI	0.87						
		VIII	0.97						
		X	1.07						
		XII	1.14						

Sm		Eu		Gd		Tb		Dy	
2 VII	1.22	2 VI	1.17	3 VI	0.938	3 VI	0.923	2 VI	1.07
VIII	1.27	VII	1.20	VII	1.00	VII	0.98	VII	1.13
IX	1.32	VIII	1.25	VIII	1.053	VIII	1.040	VIII	1.19
3 VI	0.958	IX	1.30	IX	1.107	IX	1.095	3 VI	0.912
VII	1.02	X	1.35			4 VI	0.76	VII	0.97
VIII	1.079	3 VI	0.947			VIII	0.88	VIII	1.027
IX	1.132	VII	1.01					IX	1.083
XII	1.24	VIII	1.066						
		IX	1.120						

Ho		Er		Tm		Yb		Lu	
3 VI	0.901	3 VI	0.890	2 VI	1.03	2 VI	1.02	3 VI	0.861
VIII	1.015	VII	0.945	VII	1.09	VII	1.08	VIII	0.977
IX	1.072	VIII	1.004	3 VI	0.880	VIII	1.14	IX	1.032
X	1.12	IX	1.062	VIII	0.994	3 VI	0.868		
				IX	1.052	VII	0.925		
						VIII	0.985		
						IX	1.042		

Ac		Th		Pa		U		Np	
3 VI	1.12	4 VI	0.94	3 VI	1.04	3 VI	1.025	2 VI	1.10
		VIII	1.05	4 VI	0.90	4 VI	0.89	3 VI	1.01
		IX	1.09	VIII	1.15	VII	0.95	4 VI	0.87
		X	1.13	5 VI	0.78	VIII	1.00	VIII	0.98
		XI	1.18	VIII	0.91	IX	1.05	5 VI	0.75
		XII	1.21	IX	0.95	XII	1.17	6 VI	0.72
						5 VI	0.76		
						VII	0.84		
						6 II	0.45		
						IV	0.52		
						VI	0.73		
						VII	0.81		
						VIII	0.86		

Pu		Am		Cm		Bk		Cf	
3 VI	1.00	2 VII	1.21	3 VI	0.97	3 VI	0.96	3 VI	0.95
4 VI	0.86	VIII	1.26	4 VI	0.85	4 VI	0.83	4 VI	0.821
VIII	0.96	IX	1.31	VIII	0.95	VIII	0.93	VIII	0.92
5 VI	0.74	3 VI	0.975						
6 VI	0.71	VIII	1.09						
		4 VI	0.85						
		VIII	0.95						

Pauling respectively) lead to a Fe—S bond length of 2.62 Å. In pyrite this bond length is 2.26 Å (Section 6.2.2). Covalent radii are briefly discussed in Section 6.7.3.

The influence of ionic radius and charge on element distribution has been well demonstrated by the work of Onuma *et al.* (1968) and Jensen (1973). A simple partition coefficient, k, for an element, M, distributed between melt and coexisting crystal is defined by:

$$k = \frac{[M]_{\text{crystal}}}{[M]_{\text{melt}}}. \tag{6.15}$$

The k-values were determined by analysing phenocrysts and the groundmass matrix of a number of igneous extrusives. A graphical plot of k-values versus ionic radius for different ions gives relatively smooth curves for ions of the same charge or valency (Fig. 6.10(a) and (b)).

Consider the substitution of a trace component, Mn^{2+}, into orthopyroxene, where the exchange reaction may be written

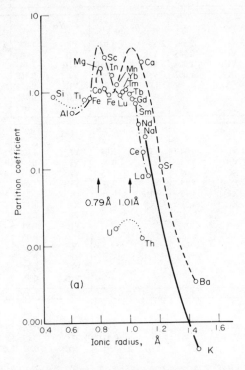

FIG. 6.10a. Partition coefficient versus ionic radius for augite-matrix pair. Note that ionic radius values are those by Whittaker and Muntus (1970). (After Jensen, 1973)

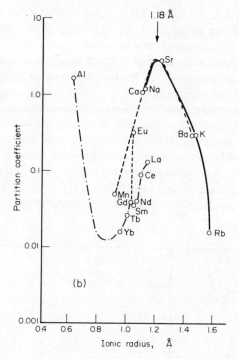

FIG. 6.10b. As Fig. 6.10a but for plagioclase-matrix pair. Note that ionic radius values are those by Whittaker and Muntus (1970). (After Jensen, 1973)

$$Mg_2Si_2O_6 + 2Mn^{2+}_{melt} \rightleftharpoons Mn_2Si_2O_6 + 2Mg^{2+}_{melt}.$$

The partition of Mn^{2+} between melt and enstatite is bound up inextricably with the partition of magnesium. However, the form of the plots of k versus ionic radius would not be changed in essence if a partition coefficient defined as

$$K_D^{opx} = \frac{[Mn]_{crystal}[Mg]_{melt}}{[Mn]_{melt}[Mg]_{crystal}} \tag{6.16}$$

or

$$K_D^{opx} = \frac{k_{Mn}}{k_{Mg}} \tag{6.17}$$

was used, as this would only shift the values on the ordinate by a constant amount.

Although the data points on the graphs are few, so that the curve fitting is subject to variation, nonetheless the resultant patterns show clearly the controlling effects that ionic radius and structure have on partition. The presence of two peaks (in Fig. 6.10(a)) at two distinct ionic radii positions is accounted for by the presence of the M1 and M2 co-ordination sites, of different sizes, in the clinopyroxene structure. The peak position itself gives the ionic radius of an ion that may not actually exist but which would give the greatest relative stabilization to the system by exchanging with a major element in the crystal structure.

Certain elements often show deviations from the curve defined by the other ions. These deviations are commonly shown by certain members of the transition elements, and can be illustrated by the case of chromium partition between matrix and clinopyroxene (Fig. 6.11). Many transition metal ions are non-spherical and are subject to crystal field effects, described and discussed in the next section. Thus, although ionic radius and charge are important factors in determining element distribution, there are other energy factors that need to be considered.

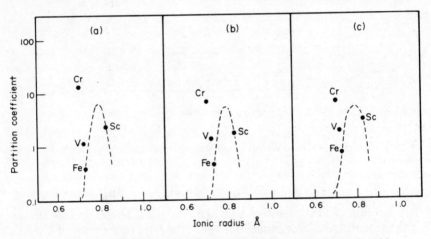

FIG. 6.11 Partition coefficient versus ionic radius for the ions Sc^{3+}, V^{3+}, Cr^{3+}, Fe^{3+} in three clinopyroxene–matrix pairs from three basalts (a, b and c). The dashed curve is drawn for reference purposes; its peak position and form are based on curves constructed by Jensen (1973) for augite–matrix pairs. The radii by Whittaker and Muntus (1970) are used to facilitate comparison with Fig. 6.10(a). (After Henderson, 1979)

The construction of the partition coefficient versus ionic radii plots involve a number of simplifying assumptions. Usually we do not know the co-ordination number or radius of an ion in the melt (or matrix). The ionic radius value that is assigned to the ion is the one relevant to its position in the crystal structure, despite the fact that we are studying element exchange between crystal and melt. Also, the ions of a particular element may be located in more than one type of site in the crystal—each with its appropriate radius. However, provided due caution is exercised the curves provide a predictive function. The partition coefficients of ions that have not been determined may be established by interpolation at the correct ionic radius value on the appropriate valency curve. This clearly is possible only where the curves are well defined by a relatively large number of points (e.g. that part of the curve determined by the rare-earth element ions in Fig. 6.10(a)) and where other parameters such as crystal field effects do not operate. Similarly, it is possible to obtain an approximate value of the proportions of two oxidation states of an element where they exist, provided the partition coefficients for the two states are very different. Approximate values of the proportion of Eu^{2+} and Eu^{3+} in a magma may be determined from the observed partition of Eu into plagioclase, compared with partition coefficients, at the appropriate radii, if the europium was all $2+$ and $3+$ respectively.

To be able to define the curves of k versus ionic radius it has been necessary to assume or know the valency or oxidation state of each ion. In most cases the valency is well known and Table 6.8 lists the commoner valency states of elements in igneous and metamorphic rock systems. In a number of cases (e.g. P, S, As, Se, the platinum group metals) the oxidation states are formal because the bonding exhibited by these elements is commonly covalent in minerals.

The oxidation state of an element within a magmatic or metamorphic system depends upon the temperature, composition and, most important, on the redox potential in the system. Under oxidizing conditions the right-hand side of the reaction $2FeO_{magma} + \frac{1}{2}O_2 \rightarrow Fe_2O_{3\,magma}$ will tend to be favoured. This in turn determines the fractionation sequence of a magma—e.g. oxide minerals will form earlier during fractional crystallization of a magma that is more oxidized compared with another of similar composition. However, it can be seen from Table 6.8 that the majority of elements exhibit only one oxidation state in terrestrial igneous or metamorphic rocks. The presence of two oxidation states can usually be explained on the basis of the relative stability of the electronic configurations shown by the different ions; e.g. the Eu^{2+} ion has the configuration $[Xe]4f^7$, which has an enhanced stability as a result of the relatively low energy of the half-filled f shell (see Appendix III).

In Chapter 5 it was shown that a number of elements have very low crystal/basaltic magma partition coefficients. Their ionic radii and/or oxidation states preclude their entry (to any significant degree) into the rock-forming minerals, and they are concentrated in the final products of fractional crystallization, or in the first liquid phase on partial melting. Large ions such as Rb^+, Cs^+, Ba^{2+} and highly charged ions such as Zr^{4+} and Hf^{4+} are included among this group (although the crystallization of minor mineral phases (e.g. zircon) can affect this general trend). Elements which show this behaviour have

136   Inorganic Geochemistry

TABLE 6.8. Valency states shown by the elements in terrestrial minerals*

No.	Elem.	VALENCY STATE					
		1	2	3	4	5	6
3	Li	+					
4	Be		+				
5	B			+			
6	C		−		+		
8	O		−				
9	F	−					
11	Na	+					
12	Mg		+				
13	Al			+			
14	Si				+		
15	P					+	
16	S		−				+
17	Cl	−					
19	K	+					
20	Ca		+				
21	Sc			+			
22	Ti			+	+		
23	V			+	+	+	
24	Cr		(+)	+			
25	Mn		+	+	+		
26	Fe		+	+			
27	Co		+	+			
28	Ni		+				
29	Cu	+	+				
30	Zn		+				
31	Ga			+			
32	Ge				+		
33	As			+		+	
34	Se		−		+		
35	Br	−					
37	Rb	+					
38	Sr		+				
39	Y			+			
40	Zr				+		
41	Nb					+	
42	Mo				+		+
44	Ru				+		
45	Rh			+			

No.	Elem.	VALENCY STATE					
		1	2	3	4	5	6
46	Pd		+				
47	Ag	+					
48	Cd		+				
49	In			+			
50	Sn		+		+		
51	Sb			+			
52	Te		−		+		(+)
53	I	−					
55	Cs	+					
56	Ba		+				
57	La			+			
58	Ce			+	+		
59	Pr			+	(+)		
60	Nd			+			
62	Sm			+			
63	Eu		+	+			
64	Gd			+			
65	Tb			+	+		
66	Dy			+			
67	Ho			+			
68	Er			+			
69	Tm			+			
70	Yb			+			
71	Lu			+			
72	Hf				+		
73	Ta					+	
74	W				+		+
75	Re				+	+	
76	Os				+		+
77	Ir			+	+		
78	Pt		+		+		
79	Au	+		+			
80	Hg	+	+				
81	Tl	+		+			
82	Pb		+		+		
83	Bi			+		+	
90	Th				+		
92	U				+		+

* Some elements have other oxidation states in non-terrestrial minerals (e.g. Cr^{2+} may exist in lunar minerals). The values are not necessarily applicable to aqueous and sedimentary systems. The oxidation states are formal in the cases of those elements which are covalently bonded in many minerals. Some elements occur also in the 'native' form.

been called the *incompatible elements*. Ions such as Ba^{2+}, K^+, Sr^{2+}, Rb^+ and those of the lanthanides have also been called *large ion lithophiles* (LIL elements). However, the usage of both these terms is somewhat variable and sometimes ions that are not particularly large (e.g. Zr^{4+} VI, 0.72 Å) have been included in the LIL group. This means that in current usage the term LIL is almost synonymous with the term 'incompatible element', and unless the term LIL is applied strictly to *large* ions it would be better if it were no longer used.

Small differences in ionic radius of a group of elements that show similar chemical properties may be used in solving petrological problems. In the rare-earth elements (i.e. lanthanides, La to Lu), the ionic radius for the $3+$ oxidation state changes steadily from 1.03 Å in La^{3+} (six-fold co-ordination) to 0.86 Å in Lu^{3+}. In most igneous systems all the rare-earth elements (REE) exist in the $3+$ state except for europium, which may occur as Eu^{2+} and Eu^{3+}, the proportion of each state depending upon such factors as the oxygen fugacity and composition of the parent magma. In Chapter 5, data show that the partition coefficients, k, for the individual REE sometimes change systematically with atomic number and, therefore, with ionic radius. Garnets discriminate against the larger (or 'light') REE ions (La to Sm), so that a chondrite-normalized, abundance pattern for the mineral often shows a marked enrichment in the 'heavy' REE (see Fig. 5.1). Plagioclase feldspar, on the other hand, discriminates against the smaller REE ions (Gd to Lu), resulting in the typical normalized pattern shown in Fig. 5.1. The frequent existence in plagioclase of a 'europium anomaly' has also been mentioned in Chapter 5, and can be seen as a deviation from the curve defined by the other REE ions in Fig. 6.10(b).

As an example of the use of these aspects of REE distribution, the generation of basaltic melts by the partial melting of a plagioclase, clinopyroxene, olivine assemblage may be considered. The effect of successive melting of the mineral phases on the normalized abundance pattern for the melt, excluding any effect on the europium anomaly, will be such that if the plagioclase were to start melting first, then the initial liquid will have a pattern showing a relative enrichment in the light REE over the heavier REE (Fig. 6.12(a)). This type of pattern will be associated with relatively high concentrations of other incompatible elements such as Ba, Rb and K. The solid residue will be depleted in these incompatible elements and the light REE. If each successive melt were tapped off and isolated from the solid then the pattern of a late stage melt would show light REE depletion, Fig. 6.12(c). An intermediate stage melt would have a pattern similar to that shown in Fig. 6.12(b). This hypothetical example shows how REE patterns of successive partial melt fractions depend upon the sequence of mineral melting. The precise abundance levels of these elements and the precise nature of the patterns for the melts, will depend upon:

(a) the modal proportions of the minerals in the source material;

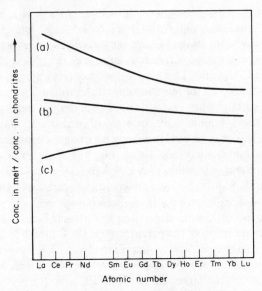

FIG. 6.12 Schematic plot of chondrite-normalized REE concentrations in melts at successive stages of fractional fusion of an olivine–clinopyroxene–plagioclase source. Patterns are for: (a) initial melt fraction, (b) and (c) subsequent melt fractions, see text.

(b) the concentration of the REE in each mineral phase;

(c) the sequence of melting;

(d) the extent of segregation of the successive melts and residual solids.

These factors are considered further in Chapter 8, where mathematical expressions are given that enable some quantification of the extent of these processes to be made.

The differential uptake by plagioclase of europium over the other REE ions, especially Sm^{3+} and Gd^{3+}, may also be used as an interpretive aid in the elucidation of rock geneses. The preferential uptake results from the existence of both Eu^{2+} and Eu^{3+} oxidation states in the magma, whereas the other REE ions are usually present only in the 3+ state. Thus, if there is a significant amount of plagioclase feldspar involved in fractional crystallization of a magma then the accumulated solids will have a positive europium anomaly, while the residual liquids have a negative one. This type of reasoning has been used in elucidating the genesis of some lunar rocks (see Chapter 3).

Considerations of ionic size and charge may also be helpful in interpreting the detailed structure of a mineral. The sizes of the constituent atoms as well as their charges determine the molar volume of the crystal. If a larger or smaller ion is substituted into the structure then the lattice energy will change with the change in the equilibrium separation distance of the ions. Thus, the calculation of lattice energies of particular structures can enable the determination of the most stable configuration of ion distribution in a mineral. This has been demonstrated with some success for cation distribution over available sites (M1, M2, M3, M4, T1 and T2) in amphiboles. Whittaker (1971) has calculated the lattice energy for about 120 different ionic arrangements in amphiboles (based on the cummingtonite and glaucophane structures) in order to establish which configuration has the lowest energy. The results agree qualitatively and semi-quantitatively with the observed preferences of trivalent ions in the M2 sites, and for other distributions in specific minerals of the amphibole group. The calculations do not take account of covalency or crystal field effects so caution must be exercised in their interpretation.

6.6 Crystal field effects

Crystal field theory is concerned with the chemical bonding characteristics of those atoms having varying numbers of d (or f) electrons—namely the transition metals (or the lanthanides). It treats the interaction between a transition metal ion and its surrounding anions or *ligands* as being purely electrostatic in nature, i.e. the ligands are represented as point charges. The theory has been used with particular success in the qualitative and quantitative interpretation of the geochemical behaviour of the first transition series, from scandium to copper, and so the account of the application of the theory is confined here to these nine elements.

Functions which describe the spatial distribution of electron density around the nucleus are termed *orbitals*, and each one is defined uniquely by its quantum numbers. The spatial distribution may be spherical or it may be localized and, therefore, of a directional form. In ions of the transition elements Sc to Cu, the five $3d$ orbitals are only partially filled with electrons. Because of the directional nature of these orbitals, Fig. 6.13, the transition metal ions are, in many cases, non-spherical. Thus, the interaction energy between a transition element cation and the anions (treated as point charges) will vary as functions of: the type of co-ordination; the distance between the ions; and the strength of the point charges.

Octahedral co-ordination

When the negatively-charged ligands are arranged in octahedral co-ordination around the transition element ion as shown in Fig. 6.14, it can be seen that the interaction energy must be higher for the $d_{x^2-y^2}$ and d_{z^2} orbitals

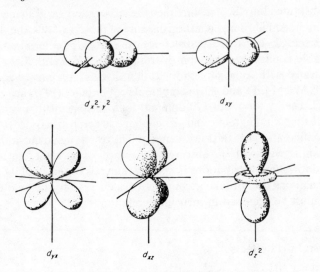

FIG. 6.13. Boundary surfaces of the five *d*-orbitals.

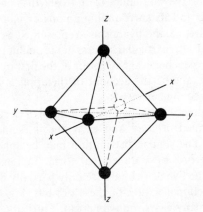

FIG. 6.14. Diagrammatic sketch of the position of ligands in octahedral co-ordination.

than for the d_{xy}, d_{yz} and d_{xz} orbitals, as the former point towards the ligands. The three d orbitals d_{xy}, d_{yz} and d_{xz} have their maximum electron density localized between groups of four coplanar ligands (i.e. in each of the planes xy, yz and xz) and each one is stabilized, or lowered in energy, by the same amount in relation to that energy level produced by the same total ligand charge at the same distance but when hypothetically and uniformly distributed as a sphere around the central ion (i.e. *a spherical field*).

Thus, the five d orbitals are split by the electrostatic field into two groups: one, the 't_{2g}' orbital group, is stabilized and the other, the 'e_g' orbital group, is destabilized relative to the mean d orbital energy. The energy separation between the two is called the *crystal field splitting* (denoted by Δ_0); it is shown in the energy level diagram of Fig. 6.15. The e_g orbital is raised in energy by $\frac{3}{5}.\Delta_0$ while the t_{2g} orbital is lowered by $\frac{2}{5}.\Delta_0$ relative to the mean energy level.

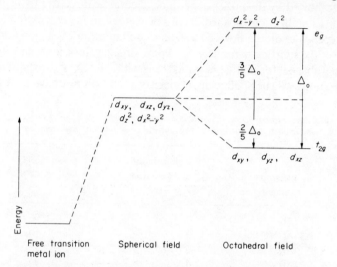

FIG. 6.15. Relative energy levels of d-orbitals of a transition metal ion when 'free' and in spherical and octahedral co-ordination fields.

The filling of the orbitals with electrons takes place according to Hund's Rule of electronic configuration so that each of the first three electrons in the $3d$ orbitals will occupy singly each of the lower energy orbitals of the t_{2g} group and their spins will be parallel. If four or five electrons are present in the $3d$ orbitals, then two possible configurations are available and which one is adopted depends upon the value of Δ_0. If Δ_0 is small, it is energetically more favourable for the fourth and fifth electrons to occupy singly each of the two orbitals of the e_g group. If Δ_0 is sufficiently large, it will be energetically more stable for the fourth and fifth electrons to occupy orbitals already containing one electron. In this situation the extra energy involved in pairing two electrons in one orbital is less than the energy Δ_0. The former configuration is called the *high-spin* state; it results from a relatively low electrostatic field. The latter configuration is the *low-spin* state created by a high field.

Once an ion contains more than five electrons in the $3d$ orbitals, spin pairing must occur; electrons being paired off successively, first in the t_{2g} group and then in the e_g group until each of the five $3d$ orbitals contains two electrons (i.e.

a maximum of ten $3d$ electrons). Both high-spin and low-spin states are possible in ions with $3d^6$ or $3d^7$ configurations. However, it seems that in all minerals, at least at pressures existing in the Earth's crust, Δ_0 is never sufficiently large to produce the low-spin state for all $3d$ configurations.

The electronic configuration of transition metal ions are given in Table 6.9. Ions in the high-spin state and with configurations other than $3d^0$, $3d^5$ or $3d^{10}$ will be stabilized by the presence of an octahedral field relative to a spherical field. This stabilization is called *crystal field stabilization energy* (CFSE). Its value depends upon the precise electronic configuration of the ion and upon the size of the splitting parameter Δ_0. Approximate values, in terms of Δ_0, are given in Table 6.9 for high-spin states. Ions with $3d^0$, $3d^5$ (high-spin) or $3d^{10}$ configurations will be spherically symmetrical and do not have a CFSE.

TABLE 6.9. *Electronic configurations and crystal field stabilization energies for some transition element ions in octahedral co-ordination, high-spin state*

Ion	Electronic configuration	Configuration of 3d electrons t_{2g}			e_g		CFSE
Sc^{3+}; Ti^{4+}	(Ar)						0
Ti^{3+}	(Ar)$3d^1$	↑					$\frac{2}{5}\Delta_0$
V^{3+}	(Ar)$3d^2$	↑	↑				$\frac{4}{5}\Delta_0$
Cr^{3+}	(Ar)$3d^3$	↑	↑	↑			$\frac{6}{5}\Delta_0$
Mn^{2+}; Fe^{3+}	(Ar)$3d^5$	↑	↑	↑	↑	↑	0
Fe^{2+}	(Ar)$3d^6$	↑↓	↑	↑	↑	↑	$\frac{2}{5}\Delta_0$
Co^{2+}	(Ar)$3d^7$	↑↓	↑↓	↑	↑	↑	$\frac{4}{5}\Delta_0$
Ni^{2+}	(Ar)$3d^8$	↑↓	↑↓	↑↓	↑	↑	$\frac{6}{5}\Delta_0$
Cu^{2+}	(Ar)$3d^9$	↑↓	↑↓	↑↓	↑↓	↑	$\frac{3}{5}\Delta_0$
Zn^{2+}	(Ar)$3d^{10}$	↑↓	↑↓	↑↓	↑↓	↑↓	0

(Ar) = configuration of argon: $1s^2 2s^2 2p^6 3s^2 3p^6$

Tetrahedral co-ordination

In tetrahedral co-ordination, Fig. 6.16, it is the d_{xy}, d_{xz} and d_{yz} orbitals that are destabilized and the $d_{x^2-y^2}$ and d_{z^2} orbitals that are stabilized relative to the mean d orbital energy. These groups are designated t_2 and e respectively. The crystal field splitting parameter is denoted by Δ_t (and is approximately equal to $\frac{4}{9}.\Delta_0$ for the same ligands at the same interionic distance). Therefore, the t_2 orbital is destabilized by $\frac{2}{5}.\Delta_t$ and the e orbital is stabilized by $\frac{3}{5}.\Delta_t$ (see Fig. 6.17). As in octahedral co-ordination, high-spin and low-spin states are possible but in minerals in the Earth's crust the low-spin state is absent.

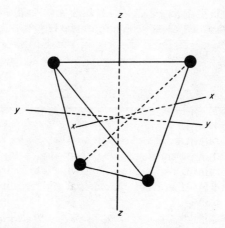

Fig. 6.16. Diagrammatic sketch of the position of ligands in tetrahedral co-ordination.

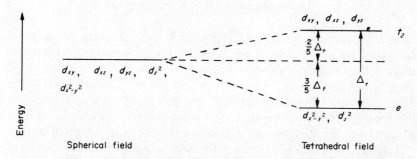

Fig. 6.17. Relative energy levels of d-orbitals of a transition metal ion in spherical and tetrahedral co-ordination fields.

Approximate CFSE values, in terms of Δ_t, are given in Table 6.10 for the high-spin states.

Co-ordination polyhedra besides regular octahedra or tetrahedra also occur in minerals and probably occur in silicate melts. Burns (1970) has discussed the nature of crystal field splitting in polyhedra of tetragonal, trigonal and monoclinic symmetry. The lower symmetry leads to a further energy-splitting of the e and t orbital groups, compared with that under cubic symmetry. For example, the M2 site in orthopyroxene is a distorted six-co-ordinate polyhedron (see Section 6.2.2). The low symmetry leads to splitting of the e_g and t_{2g} orbitals such that Fe^{2+} ions are more stable in the distorted M2 site than in the more regular M1 site. The observed cation ordering in orthopyroxene can, therefore, be explained on the basis of crystal field theory.

TABLE 6.10. *Electronic configurations and crystal field stabilization energies for some transition element ions in tetrahedral co-ordination, high-spin state*

Ion	Electronic configuration	Configuration of 3d electrons					CFSE
		e			t_2		
Sc^{3+}; Ti^{3+}	(Ar)						0
Ti^{3+}	$(Ar)3d^1$	↑					$\frac{3}{5}\Delta t$
V^3	$(Ar)3d^2$	↑	↑				$\frac{6}{5}\Delta t$
Cr^{3+}	$(Ar)3d^3$	↑	↑	↑			$\frac{4}{5}\Delta t$
Mn^{2+}; Fe^{3+}	$(Ar)3d^5$	↑	↑	↑	↑	↑	0
Fe^{2+}	$(Ar)3d^6$	↑↓	↑	↑	↑	↑	$\frac{3}{5}\Delta t$
Co^{2+}	$(Ar)3d^7$	↑↓	↑↓	↑	↑	↑	$\frac{6}{5}\Delta t$
Ni^{2+}	$(Ar)3d^8$	↑↓	↑↓	↑↓	↑	↑	$\frac{4}{5}\Delta t$
Cu^{2+}	$(Ar)3d^9$	↑↓	↑↓	↑↓	↑↓	↑	$\frac{2}{5}\Delta t$
Zn^{2+}	$(Ar)3d^{10}$	↑↓	↑↓	↑↓	↑↓	↑↓	0

(Ar) = configuration of argon: $1s^2 2s^2 2p^6 3s^2 3p^6$

With certain electronic configurations it can be energetically more favourable for a co-ordination polyhedron to distort from a regular arrangement as there may be a gain in CFSE as a result of further splitting of the energies associated with the t- and e-orbital groups. This is known as the *Jahn–Teller effect* and can be shown by ions with d^4, d^6 and d^9 configurations in a high-spin state for octahedral co-ordination, and ions with d^3, d^4, d^8 and d^9 configurations in a high-spin state in tetrahedral co-ordination. In igneous and metamorphic rock systems the presence of only certain oxidation states of the transition element ions restricts the operation of the Jahn–Teller effect to Cu^{2+} (octahedral), Fe^{2+} (octahedral), Cr^{3+} (tetrahedral) and Ni^{2+} (tetrahedral).

It is to be noted that the account of crystal field theory has been necessarily brief. The interested reader may obtain fuller details of the theory from Cotton and Wilkinson (1966), Phillips and Williams (1966) or Burns (1970).

Determination of splitting parameters Δ_0 and Δ_t

The values of the splitting parameter for octahedral co-ordination, Δ_0, have been determined from optical absorption spectroscopy* of aqueous solutions of the transition element ions and of glasses doped with these elements. Dunitz

* The frequency v of the radiation that is absorbed is related to the splitting parameter by $v = \Delta/h$, where h is Planck's constant.

and Orgel (1957) and McClure (1957) give values of Δ_0 for oxide structures and from these they calculated the octahedral CFSE for a number of transition element ions (Table 6.11).

TABLE 6.11. *Crystal field stabilization energies (CFSE) and octahedral site preference energies (OSPE) for transition metal ions in spinels*

Number of 3d electrons	Ion	Octahedral CFSE (kJ mol^{-1})	Tetrahedral CFSE (kJ mol^{-1})	OSPE (kJ mol^{-1})
0	Sc^{3+}, Ti^{4+}	0	0	0
1	Ti^{3+}	87.5	58.6	28.9
2	V^{3+}	160.2	106.7	53.5
3	Cr^{3+}	224.7	66.9	157.8
4	Cr^{2+}	100.4	29.3	71.1
5	Mn^{2+}, Fe^{3+}	0	0	0
6	Fe^{2+}	49.8	33.1	16.7
7	Co^{2+}	92.9	61.9	31.0
8	Ni^{2+}	122.2	36.0	86.2
9	Cu^{2+}	90.4	26.8	63.6
10	Zn^{2+}	0	0	0

There were very few reliable data for transition element ions in tetrahedral co-ordination, so that in determining Δ_t, Dunitz and Orgel (1957) and McClure (1957) were forced to rely on the theoretical ratio of 4/9 between Δ for ions in tetrahedral and octahedral sites. This means that the Δ_t values may be as much as 25% in error. Table 6.11 gives the calculated, tetrahedral CFSE values as determined by Dunitz and Orgel.

The octahedral and tetrahedral CFSE values have been used extensively in interpretations of the geochemical behaviour of transition element ions, but it needs to be remembered that although the splitting parameter for an ion may be similar in value from one oxide structure to the next, this is not always the case. For example, the Δ_0 splitting is smaller for Cr^{3+} in the oxide Cr$_2$O$_3$ than it is in the ruby spinel or ruby. Burns (1970) has reviewed the experimental data on absorption spectra of minerals and the derived CFSE for a number of transition element ions. Table 6.12 gives octahedral (or slightly distorted octahedral) CFSE for a few minerals; these values may be compared with those in Table 6.11, and it can be seen that the differences are not large.

The difference between the octahedral CFSE and tetrahedral CFSE of an ion is called the octahedral site preference energy (OSPE); values are given in Table 6.11. Ions which have high OSPE values will gain considerable stability by occupying any available octahedral sites than by occupying available tetrahedral sites. The approximate measure of the distribution pattern of ions

TABLE 6.12. *Some octahedral CFSE in mineral structures*

Ion	Mineral and site	CFSE (kJ mol^{-1})	Reference
Fe^{2+}	Fayalite M1	57.7	1,2
	Fayalite M2	53.6	1,2
	Forsterite M1	61.7	1,2
	Forsterite M2	54.8	1,2
	Bronzite M1	48.1	1
	Bronzite M2	49.0	1
	Orthoferrosilite M1	46.0	1
	Orthoferrosilite M2 (Fs$_{86.4}$)	47.3	1
Ni^{2+}	Olivine, synthetic M1	114.2	3
	(Mg$_{0.9}$Ni$_{0.1}$)$_2$SiO$_4$ M2	107.5	3
Cr^{3+}	Uvarovite	236.4	1

1. Burns, R. G. (1970) *Mineralogical Applications of Crystal Field Theory*. Cambridge University Press.
2. Walsh, D., Donnay, G. and Donnay, J. D. H. (1976) *Can. Mineral.* **14**, 149–150.
3. Wood, B. J. (1974) *Am. Mineral.* **59**, 244–248.

over available sites may be obtained from the relationship:

$$\frac{n_{oct}}{n_{tet}} = e^{-(OSPE)/RT}. \tag{6.18}$$

In practice, the actual site occupancies depend upon the proportions and sizes of the available tetrahedral and octahedral sites, as well as the effect of other ions which are 'competing' for the same sites.

The influence of OSPE on the distribution of transition metal ions is illustrated in Fig. 6.18, where the use of the van't Hoff relationship between the partition coefficient, $k = [M]_{olivine}/[M]_{melt}$, and the exchange enthalpy

$$\ln k = -\Delta H/RT + B \tag{6.19}$$

has been used. In this expression the enthalpy or heat of exchange for one mole of an ion between melt (or magma) and co-existing olivine crystals could be dominated by the octahedral site preference energy. This suggestion is reasonable in that a significant proportion of transition element ions may occupy tetrahedral co-ordination positions in a magma, but not in olivine where they are octahedrally co-ordinated. If this assumption regarding the enthalpy term is correct then there should be a linear relationship between $\ln k$ and OSPE for different ions in the same system. The experimentally determined partition coefficients of some transition metal ions: Mn^{2+}, Fe^{2+}, Co^{2+} and Ni^{2+}, between olivine phenocrysts and groundmass of some basalts follows the expected trend, Fig. 6.18. (The figure also shows that similar results are found for clinopyroxenes.)

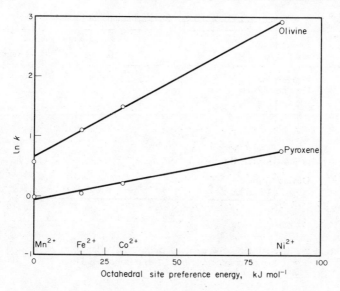

FIG. 6.18. Relationship between the crystal/matrix partition coefficient of four transition metal ions for olivine and clinopyroxene from a picrite. (After Dale and Henderson, 1972)

The slope of the line in Fig. 6.18 is a function of temperature as indicated by equation (6.19) and of the relative availability of different cation sites in the melt. It is also clear, from Section 6.5, that ionic radius is an important factor affecting element partition. The ions Mn^{2+}, Fe^{2+}, Co^{2+}, and Ni^{2+} have similar but not identical radii (Table 6.7), and the differences will help also to determine the form of the plots shown in Fig. 6.18. Since there is a smooth variation in ionic radius with OSPE of these four ions (Fig. 6.19) it is not in practice possible to determine for this system the precise contribution to partitioning made by crystal field and ionic radii effects. Furthermore, the bond length between a transition metal ion and an anion, and hence the cation radius, is determined partly by crystal field effects.

However, in the case of the ions of $3+$ oxidation state, there is not a smooth variation between radius and OSPE, Fig. 6.19; some ions (e.g. Cr^{3+} and Fe^{3+}) have similar radii but very different OSPE values. Figure 6.11 shows a plot of $\ln k$ against ionic radius for the partition of $3+$ ions between a clinopyroxene and the co-existing basaltic groundmass. The curve is drawn in for reference and is based on the curves presented in the work of Jensen (1973). It is seen that Cr^{3+} departs significantly from the curve by having a much higher partition coefficient than Fe^{3+}, despite the similarity in ionic radius. V^{3+} also shows a slight departure from the reference curve. These observations suggest that

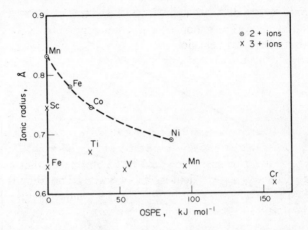

FIG. 6.19. Relationship between ionic radius and octahedral site preference energy of some 2+ and 3+ transition metal ions.

crystal field phenomena, as distinct from ionic radius effects, play an important role in determining element distribution (Henderson, 1979).

The spinel structure, with its available tetrahedral and octahedral coordination sites for cations (Section 6.2.2), also shows this role. In normal spinels the 2 + ions are located in the tetrahedral sites. Where, however, the 2 + cations have a greater octahedral site preference energy than the 3 + ions, there arises the possibility of the spinel being inverse or partially inverse. Table 6.2 lists the normal–inverse relationships of some natural spinels and it can be seen that where the spinel contains a 2 + ion with a high crystal field octahedral site preference energy that the spinel is inverse. For example, $NiAl_2O_4$ is inverse whereas $MgAl_2O_4$ is normal, Mg^{2+} and Al^{3+} having no associated crystal field stabilization energy. However, as discussed in Section 6.2.2, other factors besides crystal field energy determine the distribution of cations over the available sites in a spinel.

We are now in a position to see the importance of these structural parameters in element distribution as, for example, during the formation of the Skaergaard layered intrusion (see Chapter 5). We would expect on the basis of crystal field octahedral site preference energies that Ni^{2+} would be more rapidly depleted from the magma during crystallization of olivine and pyroxene than would Co^{2+}. This is observed to be so in the Skaergaard intrusion. The orders of decreasing octahedral site preference energy for the 2 + and 3 + transition metal ions, viz.

$$M^{2+} : Ni > Co > Mn$$
$$M^{3+} : Cr > V > Sc$$

are the observed orders of uptake of these trace elements from the fractionating magma. The major element, iron, departs from the theoretical sequence in its observed behaviour.

6.7 Covalency effects

Significant advances can be made in the description of the chemical properties of minerals by use of the ionic model of chemical bonding. There are some cases, particularly sulphide minerals, where the model is no longer applicable and attention has to be paid to the covalence that exists in the bonding of elements. The covalent model of chemical bonding is concerned with the overlap of the electron orbitals of the bonded atoms and can be treated either from the standpoint of the localization of electron pairs between adjoining nuclei, as in the valence bond theory, or by consideration of the build-up of orbitals through the addition of electrons to a multi-centred set of nuclei, as in the molecular orbital theory. This last theory treats the orbitals in a molecule in much the same way as those in a single atom.

It is not within the scope of this book to discuss these models of chemical bonding. Accounts of valence bond theory incorporating the concept of hybridization of orbitals, and of molecular orbital theory may be found in many standard chemical texts. In this section a brief discussion is given of the application of molecular orbital theory to the bonding of elements in mineral sulphides. This is followed by short accounts of electronegativity and covalent radii.

6.7.1 Application of molecular orbital theory

Any overlap of electronic orbitals in a chemical bond is ignored in crystal field theory. There the ligands are treated as point charges and the interaction of the ligands with the transition element ion is considered to be purely electrostatic. Molecular orbital theory deals with both the orbital overlap and electrostatic factors in bond formation. It involves the creation of molecular orbitals through the combination of atomic orbitals.

These aspects may be illustrated by considering the bonding of six sulphur atoms, in octahedral co-ordination, to a central, transition element ion. Two of the five $3d$, one $4s$ and three $4p$ orbitals of the central ion can hybridize (d^2sp^3) to form six orbitals directed towards the ligands. If these orbitals contain no electrons they will overlap sufficiently with the s and p orbitals of the sulphur ligands to form two sets of molecular orbitals—a bonding orbital (symbol σ) set with a high electron-density between the atoms, and an anti-bonding orbital (σ^*) set with an associated low electron density. The bonding set is of lower energy, and the anti-bonding set of higher energy, than the uncombined atomic orbitals of the transition element and sulphur atoms. The filling of the molecular orbitals with electrons follows the Pauli exclusion principle, there being a maximum of two electrons in each.

The same approach may now be followed in describing the bonding in the mineral pyrite FeS_2 (for the description of the structure see Section 6.2.2). The two sets of hybridized orbitals, the sp^3 of the sulphur atoms and the d^2sp^3 of the ion atoms, overlap and give rise to bonding and anti-bonding molecular orbitals. Only two of the five $3d$ orbitals of the iron atom are used in the σ bonding (these are the e_g group). The three remaining $3d$ orbitals (t_{2g}) are *non-bonding* (Bither *et al.*, 1968). A schematic representation of the energy levels of the molecular orbitals in pyrite is shown in Fig. 6.20. The energy splitting between the t_{2g} and e_g* groups of the iron $3d$ orbitals is designated Δ_{cov}.

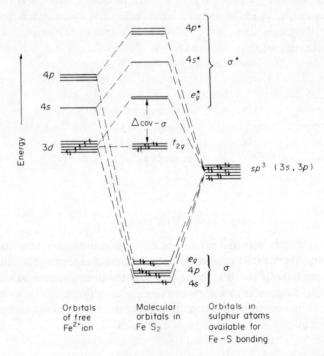

FIG. 6.20. Schematic energy level diagram for σ molecular orbitals in pyrite, FeS_2.

Of the twenty electrons involved in the bonding (six from each sulphur atom and eight from the iron atom in FeS_2), two occupy the one σ S—S bond and twelve the six metal—sulphur σ bonds. The remaining six occupy the t_{2g} group of the Fe $3d$ orbitals. The filling of the molecular orbitals follows the *aufbau* principle in that the lower energy molecular orbitals are filled first. In the case of FeS_2 there are none occupying the σ^* orbitals. This means that it is energetically more favourable for the electron to pair off in the σ orbitals than to occupy singly the σ and σ^* orbitals. Fe II, therefore, is in the low-spin state and considerable stabilization is achieved.

The pairing of all the electron spins in pyrite accounts for the mineral being a diamagnetic semiconductor. This is in contrast to CoS_2 (pyrite structure), where there is an extra unpaired electron in the e_g^* orbital, which leads to the mineral (cattierite) being a ferromagnetic metallic conductor.

Burns and Vaughan (1970) have suggested that instead of the Fe t_{2g} orbitals being non-bonding they overlap with the empty $3d$ orbitals of the sulphur ligands to form relatively delocalized π-orbitals. This increases the energy splitting between the t_{2g} and e_g^* groups and further stabilizes the t_{2g} group. It may be that the larger Δ value created by π-bonding leads to the observed spin-pairing in pyrite. Burns and Vaughan also relate the variations in reflectivity of a number of sulphide minerals to a molecular orbital bonding model involving π-bond formation.

Molecular orbital theory has been applied successfully in the qualitative and semi-quantitative interpretation of a number of physical and chemical properties of sulphides. These properties include conducting and magnetic behaviour and solid-solution relationships, as well as variations in cell dimensions, all of which depend upon the electronic configuration in the molecular orbitals. A good example of the use of the theory in this way may be found in the work of Vaughan et al. (1971) on thiospinels.

The application to properties of silicate minerals has been less successful and less extensive than for sulphides. O'Nions and Smith (1973), in a study of orthopyroxene, used a bonding model, involving both σ- and π-bonding between metal and oxygen, in which there is qualitative agreement with the absorption spectra assignments for the M2 site, but in general the complexity of most silicate minerals precludes the quantitative application of the theory. More success has been obtained in applying the theory quantitatively to the physical properties of certain simple oxides, such as rutile and hematite, through the development of new spectroscopic and computing methods (e.g. Tossell et al., 1974).

6.7.2 Electronegativity

In predicting the degree of ionic character in a bond that is largely covalent, the concept of electronegativity has been used. Electronegativity may be defined as the power of an atom in a molecule to attract electrons to itself, and is given the units of electron volts. For the concept to be useful it is necessary to quantify the electronegativities of elements in specific valency states. There have been many approaches to this problem, some of which are reviewed by Cotton and Wilkinson (1966) and Pauling (1960). Two sets of electronegativity values by Pauling and by Mulliken are those more commonly used in geochemistry.

Pauling's derivation of electronegativity values rests on the assumption that the bond energy ($E_{covalent}$) of a purely covalent bond between atoms of A and B (symbol, $D(A - B)$) is the same as the arithmetic mean of the covalent bond energies $D(A - A)$ and $D(B - B)$. If, however, the electronegativities of A and B are not the same as each other then the bond will have some ionic character and an associated bond energy, $D(A - B)$, different from that of the purely covalent bond such that:

$$D(A - B) = E_{covalent} + \Delta(A - B) \tag{6.20}$$

It has been shown empirically that this excess energy, Δ, has the following relationship to the electronegativity difference $(x_A - x_B)$ of the atoms A and B:

$$0.208\Delta^{\frac{1}{2}} = |x_A - x_B|. \tag{6.21}$$

The assignment of an electronegativity value to a particular 'reference' atom allows the values for other elements to be established (see Appendix VI); these values are relative electronegativities which depend in many cases on the valency state of the element. For example, the electronegativity of Fe(II) is about 1.8 but for Fe(III) it is about 1.9 eV. The value also depends upon the nature of the hybridization of the orbitals.

Mulliken showed theoretically that the electronegativity of an element is proportional to the average of the electron affinity (E.A.) and ionization potential (I.P.) from the ground state of an element. If the Mulliken electronegativity of an element, A, is x_A^M then:

$$x_A^M = \tfrac{1}{2}(\text{E.A.}_A + \text{I.P.}_A) \tag{6.22}$$

and it has been shown that if E.A.$_A$ and I.P.$_A$ are in units of electron volts, then:

$$x_A^{\text{Pauling}} \approx x_A^M/3.15. \tag{6.23}$$

Appendix VI lists the Mulliken values normalized by the factor 3.15.

While differences in electronegativities may be used to give an assessment of the ionic contribution to a predominantly covalent bond between two elements, their usefulness is otherwise very restricted. Attempts to use electronegativities in predicting the distribution of elements between a crystal and co-existing melt have proved inadequate and theoretically unsound (e.g. see discussion by Burns and Fyfe, 1967, and by Whittaker, 1967). However, if in general the bond strength increases for metal–oxygen bonds as the electronegativity of the metal atom increases it has been suggested by

Whittaker (1967) that electronegativities will then provide a qualitative assessment of differences in partitioning behaviour of atoms between a crystalline solid and liquid. Consider the case where atom X has a higher electronegativity than atom Y. The shorter bond lengths that are present in liquids (compared to the crystalline solid of the same composition) will lead to increased overlap of the electron orbitals. This will be greater for X than for Y and so X will be energetically more stable in the liquid than will Y, *all other things being equal.*

The concept of electronegativity has not yet proved to be helpful in geochemistry. As mentioned above, attempts at applying it to problems in element distribution have led to confusion and misunderstanding, but when applied to the problem of the nature of bonding between elements, electronegativity values are useful semi-empirical data.

6.7.3 Covalent radii

The bond length between atoms that are covalently bonded is usually different from the sum of their respective ionic radii, as for example in the Fe—S bond distance in pyrite. The *covalent radius* of an atom can differ substantially in value from that of its ionic radius, and it is helpful therefore to know covalent radii values for elements which commonly form covalent bonds in minerals.

The covalent radius may be determined by halving the internuclear distance observed in appropriate monatomic molecules. Thus, the covalent radius for carbon ($0.77\,\text{Å}$) is obtained from the C—C distance ($1.54\,\text{Å}$) measured in diamond. The carbon covalent radius may then be used to establish the radius of other atoms bonded to it, but it is then necessary to correct for any difference in electronegativities between the two elements. This may be done by the Schomaker and Stevenson relationship (Pauling 1960, 1970)

$$\text{I.D.}_{A-B} = r_A + r_B - C|x_A - x_B| \qquad (6.24)$$

where I.D._{A-B} is the internuclear distance between atoms A and B, r is the covalent radius of the atom and C is a constant which has the value $0.08\,\text{Å}$ for all bonds involving one atom of the first row of the periodic table, or $0.06\,\text{Å}$ for bonds between Si, P, or S and a more electronegative atom not of the first row. Unfortunately the correction does not work well in a number of cases, including the bonding of silicon, e.g. it predicts a longer bond distance in Si—O than is observed.

As with the ionic radius, so it is that a covalent radius is not an invariant property of an atom but depends upon orbital hybridization and bond order. In general, the higher the bond order, the shorter the radius because there is greater overlap of the elements' orbitals. Variation in other parameters (such as the electronegativity of sulphur with change in its assigned formal charge, Pauling, 1970) and their effects on covalent radii (e.g. equation (6.24)) tend to make covalent radii values only of qualitative use in geochemistry. Some single bond radii are listed in Table 6.13.

TABLE 6.13. *Single bond covalent radii (Å)*
(from Pauling, 1960 and 1970)

C	0.77	Ge	1.22
N	0.70	As	1.21
O	0.66	Se	1.17
Si	1.17	Sn	1.40
P	1.10	Sb	1.41
S	1.03	Te	1.37

Suggested further reading

BURNHAM, C. W. (1979) Magmas and hydrothermal fluids. In *Geochemistry of Hydrothermal Ore Deposits* (ed. H. C. BARNES), pp. 71–136. Wiley.

BURNS, R. G. (1970) *Mineralogical Applications of Crystal Field Theory*, Cambridge. 224 pp.

GREENWOOD, N. N. (1968) *Ionic Crystals, Lattice Defects and Nonstoichiometry*, Butterworths. 194 pp.

HESS, P. C. (1980) Polymerization model for silicate melts. In *Physics of Magmatic Processes* (ed. R. B. HARGRAVES), pp. 3–48. Princeton Univ. Press.

7

Thermodynamic Controls of Element Distribution

7.1 Introduction

The quantitative use of thermodynamics in petrology has gained significant impetus in recent years from the growth of available experimental data on the thermodynamic properties of minerals. Relatively big advances have been made in the use of thermodynamic principles and data to establish the temperatures and pressures of formation of both igneous and metamorphic assemblages, and we now have a better understanding of the relationship between activity and composition in silicate systems. Although a prerequisite for the many applications of the principles is that the system under study should be at chemical equilibrium, this fortunately does not appear to be very restricting as many petrological systems have attained equilibrium either on a large or local scale. Even where the 'equilibrium condition' may be confined to the rims of adjacent minerals in a multimineralic assemblage, modern analytical techniques (such as the electron-microprobe analyser) allow the nature of the equilibrium state to be studied and used in thermodynamically based interpretations. This is not to suggest, however, that the recognition of equilibrium is rarely a difficult problem, for sometimes there is no alternative but to assume that the assemblage under study is an equilibrium one.

It is not the purpose of this chapter to give an account of thermodynamic principles. These may be found in a number of excellent books on chemical thermodynamics (e.g. Denbigh, 1971) and there are now many publications that deal with thermodynamics in geology or petrology (Wood and Fraser, 1976; Greenwood, 1977; Froese, 1976; Powell, 1978a). The purpose is to show, by examples, how thermodynamics may be applied so as to obtain quantitative information relating to element distribution. A start is made, however, by giving some of the basic equations and relationships for ease of reference and to introduce the symbolism and terminology used in the chapter.

7.2 Free energy and equilibrium

7.2.1 Gibbs function and chemical potential

The Gibbs function, or Gibbs free energy, G, is one of the most useful thermodynamic functions in petrology. It is referred to here simply as free energy, and is related to the enthalpy or heat content (H) and entropy (S) of a system by equation (7.1)

$$G = H - TS \tag{7.1}$$

where T is thermodynamic temperature (i.e. kelvins).

For any chemical reaction, generalized as in (7.2),

$$bB + cC = dD + eE \tag{7.2}$$

the change in the free energy (ΔG) between the products and reactants is

$$\Delta G = \Delta H - T\Delta S \tag{7.3}$$

where ΔH and ΔS are the enthalpy and entropy changes respectively. By convention a negative ΔG for a reaction indicates that the reaction can proceed spontaneously. (The free energy is an additive, or extensive, function. This means that the free energy of the system is dependent upon the amount of material present. Thus the free-energy values for each of the components B, C, D and E in equation (7.2) must be multiplied by the appropriate coefficients b, c, d and e in determining ΔG at any particular temperature and pressure.) The free-energy change is the driving force of a reaction but it does not indicate anything about the rate at which the reaction proceeds.

Free-energy values for many compounds may be found in tables (e.g. Robie *et al.*, 1978). Usually the values are for the *standard free energy of formation, $G°$*, i.e. the free energy of reaction to form one mole of the compound in a defined standard state from pure elements also under standard conditions. There are various standard states which are used, the most convenient one being chosen. Most tables give standard free-energy values at 298.15 K (25°C) and 1 atm (1.013 bars or 101.325 kPa) pressure, but many experimental petrologists prefer to work with a standard state of the temperature of interest and 1 bar (10^5 Pa) pressure. By convention, the pure elements in their standard state, i.e. 1 bar pressure and at the temperature of interest, also have enthalpy of formation values of zero. Free-energy and enthalpy values are now given in units of $kJ\,mol^{-1}$, and entropy values in units of $J\,K^{-1}\,mol^{-1}$.

At equilibrium the free energy, G, of a closed system is at a minimum, and ΔG (which can be considered as the derivative of G) is equal to zero. This is the criterion of equilibrium but it may be expressed in another way with the use of the chemical potential, μ. For a system of one component, i, the chemical potential, μ_i, is the free energy per mole. For a multi-component system:

$$G_{syst} = \sum_j \mu_j N_j \tag{7.4}$$

where N_j are the number of moles of the jth component.

Chemical potential is an intensive property—it does not depend upon the mass of the system. In a system at equilibrium the chemical potential of any component is the same in all compounds or phases within the system, e.g. if an olivine $(Mg, Fe)_2SiO_4$ and an orthopyroxene $(Mg, Fe)SiO_3$ coexist at equilibrium then

$$\mu_{Fe}^{oliv} = \mu_{Fe}^{opx}. \tag{7.5}$$

This holds for all other components however defined. In non-equilibrium systems chemical potential gradients or differences will exist and these will be the driving forces for reaction or moving of components so as to bring the system towards the equilibrium state.

7.2.2 Variation in H, S and G with temperature and pressure

Enthalpy, entropy and, therefore, free energy of a substance are temperature- and pressure-dependent. The change in molar enthalpy (dH) with change in temperature (dT), at constant pressure, is given by

$$dH = Cp\,dT \tag{7.6}$$

where Cp is a constant of proportionality, called the molar heat capacity, which is characteristic of the substance. Equation (7.6) needs to take account of the variation of Cp with temperature, which can be represented by the following empirical equation:

$$Cp = a + bT + c/T^2 \tag{7.7}$$

where a, b and c are experimentally determined constants for the substance and T is temperature in kelvins. The equation may provide a good fit to the observed variation in molar heat capacity over only a restricted temperature range.

The substitution of equation (7.7) into (7.6), followed by integration within temperature limits, gives the molar enthalpy at any required temperature (H_T) and at constant pressure

$$H_T = H° + [aT + bT^2/2 - c/T]_{298}^T. \tag{7.8}$$

Similarly, the entropy increase with temperature change can readily be shown to be:

$$S_T = S^\circ + [a \ln T + bT - c/2T^2]_{298}^T. \tag{7.9}$$

Hence, the variation of free energy with temperature, at constant P, may be obtained by the simple substitution of equations (7.8) and (7.9) into equations (7.1) or (7.3). In some solid–solid reactions little error is introduced if the assumption is made that $\Delta Cp = 0$ (Wood and Fraser, 1976) but the steady improvement in the precision of thermochemical data makes it worthwhile in some situations to take account of the variation in ΔCp (Carmichael, 1977).

For a closed system the variation of free energy with temperature and pressure is given by

$$dG = V\,dP - S\,dT, \tag{7.10}$$

so the isothermal variation of G with pressure is

$$\left(\frac{\partial G}{\partial P}\right)_T = V. \tag{7.11}$$

It is usually justifiable in most geological situations relating to the crust to assume in thermochemical calculations that the volume change for solids in a closed-system reaction is independent of pressure and temperature.

Equation (7.10) may be used to derive an equation that gives the slope of a reaction boundary between two states, A and B, of a system. Substitution of the molar free energy, molar volume and molar entropy values of each state into equation (7.10), followed by subtraction gives

$$dG_B - dG_A = (V_B - V_A)\,dP - (S_B - S_A)\,dT. \tag{7.12}$$

At the reaction boundary $dG_B - dG_A$ can be considered as being equal to zero. Therefore,

$$\frac{dP}{dT} = \frac{S_B - S_A}{V_B - V_A} = \frac{\Delta S}{\Delta V}. \tag{7.13}$$

This is known as the Clausius–Clapeyron equation.

7.2.3 *Activity and fugacity*

An ideal solution may be defined simply as one that follows the relationship

$$\mu_i = \mu_i^\circ + RT \ln x_i \tag{7.14}$$

for every component of mole fraction x_i and where μ_i° is a function only of temperature and pressure. μ_i° is the chemical potential of component i in the solution under conditions of the chosen standard state (e.g. 1 bar pressure and temperature T).

Non-ideal solutions do not follow equation (7.14) but, in order to maintain the form of the relationship, the concept of activity is introduced, so that for a non-ideal solution

$$\mu_i = \mu_i^{\circ} + RT \ln a_i, \tag{7.15}$$

where a_i is the activity of the ith species, and it is related to concentration by:

$$a_i = \gamma_i x_i \tag{7.16}$$

where γ_i is called the activity coefficient, and is such that as $x_i \to 1$, $\gamma_i \to 1$.

The activity coefficient may be seen as a measure of the extent of departure from ideality. Its value may be greater (positive deviation) or less (negative deviation) than 1. Elements or compounds at a standard state have, by definition, $\gamma = 1$. The coefficient may be P- and T-dependent, and cannot be considered a constant even at any given pressure and temperature as it may vary with the concentration of the component.

The investigation of activity–composition relationships is, therefore, an important part of the study of non-ideal solutions. However, when a solution is dilute, the activity of a solute component is considered proportional to its mole fraction:

$$a_i = k_h x_i \quad \text{as} \quad x_i \to 0, \text{ at constant } T \text{ and } P, \tag{7.17}$$

where k_h is a proportionality constant. This relationship (7.17) is *Henry's law* and k_h is referred to as Henry's law constant, its value being determined from experiment. The concentration range over which Henry's law operates for a particular solute may be very restricted (Fig. 7.1); however, the behaviour of trace elements in magmatic systems often appears to follow this law, thereby helping to make the study of this branch of geochemistry relatively simple (but see Section 7.5.2 below).

A perfect gas may be defined by an equation (7.18) of the same type as for an ideal solution (7.14), and where μ_i° is a function of T only,

$$\mu_i = \mu_i^{\circ} + RT \ln p_i, \tag{7.18}$$

where p_i is the pressure of the pure gas i. If the gas is not ideal, the concept of fugacity (f) is introduced:

$$\mu_i = \mu_i^{\circ} + RT \ln f_i, \tag{7.19}$$

such that $f_i/p_i \to 1$ as $P \to 0$, and μ_i° is a function of T only. As with other thermochemical parameters it is convenient to define a standard state to which

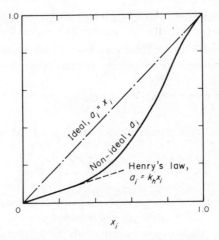

FIG. 7.1. Possible activity–composition relationship of a component, i, showing non-ideal solution behaviour but obeying Henry's law over part of its concentration range. Ideal behaviour is shown for comparison. Ordinate is activity.

the fugacity of a gas in a non-standard state is referred, and so we write (7.20):

$$\mu_i = \mu_i^\circ + RT \ln \frac{f_i}{f_i^\circ}. \qquad (7.20)$$

The relative fugacity, f_i/f_i°, is the same as activity, a_i. The standard state of a gas is often taken to be at unit fugacity at the temperature of interest and at 1 bar pressure. Hence, if $f_i^\circ = 1$, $f_i = a_i$ for a gas.

It is only for a gas that fugacity is equal to activity, as for solids and liquids the activity is given by the ratio of fugacities where the denominator (f°) is generally not equal to 1. This is readily seen in the case of a pure liquid compound at a standard state, which by convention has activity equal to 1. The vapour pressure may be significantly less than the pressure of the system at the standard state. For example, if the system is at a standard state pressure of 1 bar and the measured vapour pressure is found to be (say) 0.3 bar, then the fugacity of the vapour $f/f^\circ = 0.3$ as $f^\circ = 1$.

This may be clarified further by considering a hypothetical mixture of two miscible liquid compounds A and B. At 1 bar pressure, pure liquid A has a vapour pressure of 0.4 bar, and pure liquid B of 0.3 bar. If there are equal molar proportions of A and B in the mixture, the resultant vapour pressure of the mixture, if ideal, would be 0.35 bar. The mixture is, however, non-ideal and shows a negative deviation such that the observed vapour pressure is only 0.28 bar. The vapour pressure of B in the mixture is measured to be 0.12 bar, so the fugacity of B is $0.12/1 = 0.12$ bar, for as the pressure is low, the vapour pressure is equal to the fugacity. The *activity* of $B(a_B)$ in the liquid mixture is:

$$a_B = \frac{f_{B\,\text{liq}}}{f^{\circ}_{B\,\text{liq}}} = \frac{0.12}{(0.30/1)} = 0.40 \text{ bar.}$$

If the liquid mixture was ideal then clearly a_B would be 0.5 bar as compound B constitutes half the mixture.

7.2.4 Equilibrium constant

For the generalized reaction at equilibrium:

$$bB + cC = dD + eE \tag{7.21}$$

where b, c, d and e are the number of moles of the components, the standard free-energy change ΔG° is given by

$$\Delta G^{\circ}_{P,T} = -RT \ln \left(\frac{a^d_D \cdot a^e_E}{a^b_B \cdot a^c_C} \right) \tag{7.22}$$

$$= \Delta G^{\circ}_{1,T} + \int^P_1 \Delta V^{\circ} \, dP \tag{7.23}$$

for the standard state of pure components at 1 bar pressure and temperature of interest. Equation (7.22) may be written

$$\Delta G^{\circ} = -RT \ln K, \tag{7.24}$$

where
$$K = \left(\frac{a^d_D \cdot a^e_E}{a^b_B \cdot a^c_C} \right) \tag{7.25}$$

and is called the equilibrium constant of the reaction (7.21).

Expression (7.24) shows the temperature dependence of the free-energy change for the reaction but it must be remembered that the equilibrium constant, K, is also temperature-dependent, the dependence being given by van't Hoff's equation (7.26), at constant pressure:

$$\frac{d \ln K}{dT} = \frac{\Delta H^{\circ}}{RT^2}. \tag{7.26}$$

If ΔH° is constant for a reaction over a considerable temperature range, the above equation can be readily integrated to give

$$\ln K = -(\Delta H/RT) + B, \tag{7.27}$$

where B is a constant of integration which can be evaluated from a knowledge of the values of K and ΔH at a standard state.

The uses of the equilibrium constant are discussed in some of the sections that follow.

7.3 Activity–composition relations

To be able to use thermodynamic relationships in petrology it is necessary to have the appropriate data such as standard Gibbs free energies of reactions, and the concentrations of components in the various phases of the system. As many silicate systems are non-ideal, the relationships between composition and activity of the components will also need to be known. Robie *et al.* (1978) give a list of the thermodynamic properties of minerals, which allows computation of free energies, etc., of reaction, while a number of research and review papers have dealt with activity–composition relations in individual minerals or in particular mineral reactions (e.g. Saxena and Ribbe, 1972; Powell, 1974, 1977; Wood, 1977; Carmichael *et al.*, 1970; Nicolls *et al.*, 1971).

One of the ways in which the effect of change in composition of a multi-component system on activity of one component can be measured is to conduct experiments in which one phase only is a solid solution of two components while the other mineral phases are pure. For example, the reaction

$$\underset{\text{plagioclase}}{CaAl_2Si_2O_8} = \underset{}{CaAl_2SiO_6} + \underset{\text{quartz}}{SiO_2} \qquad (7.28)$$

could be used to determine the activity of the $CaAl_2SiO_6$ component in a clinopyroxene containing other components (e.g. $CaMgSi_2O_6$), if the other existing phases were pure anorthite and quartz as then the equilibrium constant for reaction (7.28) would be given by:

$$K = a^{cpx}_{CaAl_2SiO_6},$$

the activities of pure anorthite and pure quartz both being 1. Hence from equation (7.24) we have for reaction (7.28) at equilibrium:

$$\Delta G^\circ = -RT \ln a^{cpx}_{CaAl_2SiO_6}. \qquad (7.29)$$

The calculated activity values may then be related to experimentally determined compositions.

A more general way of using the equilibrium constant is to study an exchange equilibrium in which the activity–composition relations of one phase are already known and allow the determination of the relations for the second phase. This approach makes use of the Gibbs–Duhem equation* which shows that activity coefficients of components in a binary system are not independent

---

* For a discussion of the Gibbs–Duhem equation and its application to activity coefficient see Denbigh (1971), sections 2.10 (d), 9.7 and 9.8.

of each other. One way that the Gibbs–Duhem equation may be expressed is as (7.30) for a two-component system in which the mole fractions are x_a and x_b:

$$\ln \gamma'_b = -\int_0^{x'_b} \left(\frac{1-x_b}{x_b}\right) d\ln \gamma_a \tag{7.30}$$

where γ'_b is the activity coefficient value of b at the particular mole fraction of b, which is the upper limit of the integration. We will show the usefulness of this expression (7.30) in determining the activity–composition relations in olivine solid solutions using the equilibrium reaction (7.31) as studied by Nafziger and Muan (1967):

$$\underset{\text{oxide}}{\text{FeO}} + \underset{\text{olivine}}{\text{MgSi}_{0.5}\text{O}_2} = \underset{\text{oxide}}{\text{MgO}} + \underset{\text{olivine}}{\text{FeSi}_{0.5}\text{O}_2} \tag{7.31}$$

The equilibrium constant for reaction (7.31) may be written:

$$K = \frac{a^{\text{ox}}_{\text{MgO}} \cdot x^{\text{ol}}_{\text{FeSi}_{0.5}\text{O}_2} \cdot \gamma^{\text{ol}}_{\text{FeSi}_{0.5}\text{O}_2}}{a^{\text{ox}}_{\text{FeO}} \cdot x^{\text{ol}}_{\text{MgSi}_{0.5}\text{O}_2} \cdot \gamma^{\text{ol}}_{\text{MgSi}_{0.5}\text{O}_2}} \quad \text{at constant } T \text{ and } P. \tag{7.32}$$

The activity–composition relations of the oxides were known, and the mole fractions of the components in the olivine were established from the experiments. Therefore, the quotient C' is known:

$$C' = \frac{a^{\text{ox}}_{\text{MgO}} \cdot x^{\text{ol}}_{\text{FeSi}_{0.5}\text{O}_2}}{a^{\text{ox}}_{\text{FeO}} \cdot x^{\text{ol}}_{\text{MgSi}_{0.5}\text{O}_2}} \tag{7.33}$$

and the equilibrium constant can be expressed as:

$$\ln K = \ln C' + \ln \gamma^{\text{ol}}_{\text{FeSi}_{0.5}\text{O}_2} - \ln \gamma^{\text{ol}}_{\text{MgSi}_{0.5}\text{O}_2}. \tag{7.34}$$

Wood (1977) has shown that the activity coefficient terms in equation (7.34) may be determined from a slightly modified form of the Gibbs–Duhem equation (7.30) as follows—(7.35) and (7.36):

$$\ln \gamma^{\text{ol}}_{\text{MgSi}_{0.5}\text{O}_2} = \int_0^{x^{\text{ol}}_{\text{FeSi}_{0.5}\text{O}_2}} x_{\text{FeSi}_{0.5}\text{O}_2} \, d\ln C', \tag{7.35}$$

$$\ln \gamma^{\text{ol}}_{\text{FeSi}_{0.5}\text{O}_2} = \int_0^{x^{\text{ol}}_{\text{MgSi}_{0.5}\text{O}_2}} x_{\text{MgSi}_{0.5}\text{O}_2} \, d\ln C'. \tag{7.36}$$

As the mole fraction and C' values are known the appropriate activity coefficients and hence activities may be calculated. The results from the work of Nafziger and Muan are shown in Fig. 7.2, where it can be seen that the olivine solid solution displays, at 1200°C, slight positive deviation from ideality.

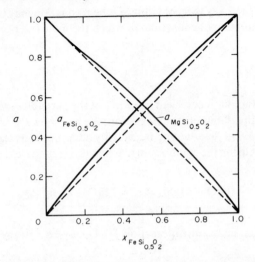

FIG. 7.2. Activity–composition relationships along the olivine join of the system MgO–FeO–SiO$_2$ at 1200° C and 1 atmosphere pressure. (After Nafziger and Muan, 1967)

A study of this kind gives us a further quantitative understanding of the distribution behaviour of elements between mineral phases but it is important to remember that the activity–composition relation is a function of temperature (and in some instances of pressure also). The next two sections extend the treatment into that of cation ordering and Section 7.4.2 discusses the effect of temperature on Fe and Mg distribution in orthopyroxenes and the use of the regular solution model.

The determination of activities of components or species in a melt that coexists with mineral phases can be done using some of the principles outlined above. Carmichael *et al.* (1970) have taken a number of mineral–liquid reactions of relevance to common igneous rocks in order to establish the relationship between silica activity in a melt and temperature. For example, the equilibrium between forsterite and enstatite (7.37) may be used to establish silica activity in the melt at any particular temperature and pressure.

$$Mg_2SiO_4 + SiO_2 = 2MgSiO_3.$$ (7.37)

olivine   liquid   pyroxene

The equilibrium constant for this reaction is

$$K = \frac{(a^{pyx}_{MgSiO_3})^2}{a^{ol}_{Mg_2SiO_4} \cdot a^{l}_{SiO_2}}$$ (7.38)

but since the pyroxene and olivine are pure phases their activities are each one, so (7.38) reduces to:

$$K = \frac{1}{a^l_{SiO_2}}. \tag{7.39}$$

The use of equation (7.24) gives:

$$\Delta G^\circ = RT \ln a^l_{SiO_2} \tag{7.40}$$

and from available free energy data (Robie *et al.*, 1978) for equilibrium (7.37), the silica activity at any given temperature may be calculated. Carmichael *et al.* (1974) have used this approach in a novel system for the classification of extrusive igneous rocks.

7.4 Order–disorder in minerals

In Chapter 6 (Section 6.2.2) it was shown that orthopyroxenes and spinels have more than one type of site available to cations such as Fe^{2+}, Mg^{2+}, Al^{3+}, etc., and that the observed preference for a particular site by an occupying cation, through a gain in chemical stability, leads to a non-random distribution, or 'mixing', of cations over the available sites. The spinels, along with other oxide structures, have been used to determine quantitatively the site preferences of particular cations, using optical absorption spectroscopy (Section 6.6). The empirically determined distribution of different cations may also be used in a thermodynamic treatment to establish the same thing, provided allowance is made for the temperature dependence of the distribution. These aspects are discussed in the following two sections with specific reference to spinels and orthopyroxenes but the implications extend to many rock-forming minerals that show solid solution and where the mixing of the cations is non-random over the available sites.

7.4.1 Site preferences

These may be obtained from the observed distribution of cations over the available sites in a mineral if that distribution is at equilibrium. This method may be demonstrated using a simplified thermodynamic model which describes the cation interchange in spinels,* by Navrotsky and Kleppa (1967).

The interchange reaction for inversion of an initially normal spinel AB_2O_4 may be represented as:

$$A_{tet} + B_{oct} = A_{oct} + B_{tet}. \tag{7.41}$$

* For details of the spinel structure see Section 6.2.2.

The degree of inversion of a spinel has been denoted by λ (Section 6.2.2). When $\lambda = 0$, the spinel is normal, when $\lambda = 1$, it is inverse. For a particular degree of inversion, λ, the formula of a spinel will be $A_{1-\lambda}B_\lambda(A_\lambda B_{2-\lambda})O_4$, where the cations in brackets are located in the octahedral sites. Hence the equilibrium constant for reaction (7.41) may be written

$$K = \frac{(A_{\text{oct}})(B_{\text{tet}})}{(B_{\text{oct}})(A_{\text{tet}})} = \frac{\lambda^2}{(2-\lambda)(1-\lambda)}. \tag{7.42}$$

In the simple model of cation interchange, full ideality of the system is assumed. This means that:

(a) The activity of each ion on a particular site is equal to its ionic fraction.
(b) The entropy change as a result of the cation interchange is only configurational. (Experimental evidence suggests that this is a reasonable assumption for a number of spinels, Navrotsky and Kleppa 1967.)
(c) The total enthalpy change for the exchange reaction $= \lambda \Delta H_{\text{interchange}}$. The configurational entropy difference between a spinel of formula $A_{1-\lambda}B_\lambda(A_\lambda B_{2-\lambda})_2O_4$ and the normal spinel of the same composition is obtainable from the relation:

$$\Delta S_{\text{config}} = -R\Sigma \lambda_i \ln \lambda_i. \tag{7.43}$$

Since there are two filled octahedral sites for every filled tetrahedral site we have:

$$\Delta S_{\text{config}} = -R[\lambda \ln \lambda + (1-\lambda)\ln(1-\lambda)] - 2R\left[\frac{\lambda}{2}\ln\frac{\lambda}{2} + \left(1-\frac{\lambda}{2}\right)\ln\left(1-\frac{\lambda}{2}\right)\right]. \tag{7.44}$$

The free-energy change for the interchange, at constant P, is given by $\Delta G = \Delta H - T\Delta S$, and so for the simplified model

$$\Delta G = \lambda \Delta H_{\text{int}} - RT\left[\lambda \ln \lambda + (1-\lambda)\ln(1-\lambda) + \lambda \ln\frac{\lambda}{2} + (2-\lambda)\ln\left(1-\frac{\lambda}{2}\right)\right]. \tag{7.45}$$

At equilibrium $(\mathrm{d}\Delta G/\mathrm{d}\lambda) = 0$. This condition together with the differentiation and re-arrangement of equation (7.45) leads to the equilibrium relation:

$$\frac{-\Delta H_{\text{int}}}{RT} = \ln\frac{\lambda^2}{(2-\lambda)(1-\lambda)}. \tag{7.46}$$

This expression may now be used to obtain interchange enthalpies for cations in spinels for which the degree of inversion has been measured at known temperatures. The simple model is clearly inapplicable to normal ($\lambda = 0$) or completely inverse ($\lambda = 1$) spinels and it is not satisfactory when λ is close to 0 or 1.

As the interchange involves two different cations it is necessary to have an ion of known octahedral (or tetrahedral) site preference energy to use as a standard against which the site preferences of the other ions may be measured. Navrotsky and Kleppa (1967) chose Al^{3+} as the reference ion and concluded, from independent experimental work on enthalpy changes associated with structural transformations of alumina, that the ion has an octahedral site preference energy of 4.17 kJ/mol in spinels. With this value and interchange energies calculated by equation (7.46) for spinels of known degree of inversion at a particular temperature, site preferences have been calculated. The actual values are unlikely to be very accurate because the λ-values of some spinels were measured on crystals quenched from high temperatures and without the rate of cation re-equilibration being known. However, the relative positions of cations in their preference for particular sites are probably near to the true situation. They are:

Large tetrahedral preference: Zn^{2+}, Cd^{2+}, In^{2+}
Tetrahedral preference: $Mn^{2+} \geq Fe^{3+} \geq Ga^{3+} > Co^{2+}$
Octahedral preference: $Cr^{3+} \gg Ni^{2+} > Al^{3+} > Cu^{2+} > Fe^{2+}$

The sequence, except for the position of Co^{2+}, is the same as that determined by absorption spectra for transition element ions in oxides but the 'thermodynamic' values are generally lower by a few kJ/mol (e.g. Cu^{2+} thermodynamically derived octahedral site preference energy: ~ 42 kJ/mol; spectral determination; 63 kJ/mol). This difference probably results from a combination of a number of factors including the error involved in the use of $\Delta_t = \frac{4}{9}\Delta_0$ for determining crystal-field tetrahedral stabilization energies (see Chapter 6) and in the assumption of full ideality (see above).

7.4.2 Temperature dependence

The approach used to determine site preferences of ions may be used also to determine how cation ordering in a particular mineral varies with temperature. In an ideal solid solution in which the interchange of cations A and B on different structural sites occurs as:

$$A \text{ (site 1)} + B \text{ (site 2)} = B \text{ (site 1)} + A \text{ (site 2)} \qquad (7.47)$$

the equilibrium constant is:

$$K = \frac{(1 - X_A^{\text{site 1}}) X_A^{\text{site 2}}}{X_A^{\text{site 1}} (1 - X_A^{\text{site 2}})}, \qquad (7.48)$$

where $X_A^{\text{site 1}}$ is the ionic fraction $A/(A + B)$ on site 1 and so $X_B^{\text{site 1}} = 1 - X_A^{\text{site 1}}$.

If the assumptions used in the simple model for the spinels hold also for this interchange and provided the molar enthalpy of interchange for each temperature is known, then the relation (see equation (7.46))

$$-\frac{\Delta H_{\text{int}}}{RT} = \ln K$$

may be used to calculate the degree of order at any required temperature. This approach could be used to relate cation ordering in orthopyroxenes to temperature, provided that temperatures are greater than about 1000°C. It is shown below that the ordering in orthopyroxene solid solution approximates to being ideal at high temperatures but is discernibly non-ideal at temperatures below $\sim 900°$C. At these lower temperatures any thermodynamic analysis of the system must allow for the departure from ideality.

Saxena and Ghose (1971) have studied, by Mössbauer spectroscopy, the order–disorder relationship in some natural orthopyroxenes* as a function of temperature. Their results are shown in Fig. 7.3. The interchange may be

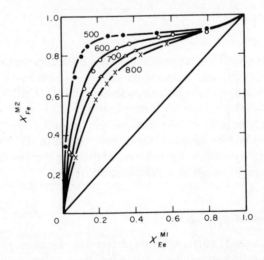

FIG. 7.3. Distribution, as ionic fraction, of Fe^{2+} between M1 and M2 sites in orthopyroxene heated at 500, 600, 700 and 800° C. (After Saxena and Ghose, 1971)

expressed by:

$$Fe^{2+}(M2) + Mg^{2+}(M1) \rightleftharpoons Fe^{2+}(M1) + Mg^{2+}(M2), \qquad (7.49)$$

---

* A description of the structure of orthopyroxene is given in Section 6.2.2.

for which an apparent equilibrium constant (or distribution coefficient) K_D may be written

$$K_D = \frac{X_{Fe}^{M1}\,(1 - X_{Fe}^{M2})}{X_{Fe}^{M2}\,(1 - X_{Fe}^{M1})}, \tag{7.50}$$

where X_{Fe}^{M1} stands for ionic fraction of Fe^{2+} in site M1, etc. However, since the behaviour of cation ordering may be non-ideal we need to consider the activity of each ionic species in each site. We will designate the partial activity coefficient for an ion (e.g. Fe^{2+}) in a particular site (M1) by $\overline{\gamma}_{Fe}^{M1}$. (This partial activity coefficient is to be distinguished from the activity coefficient γ_i for species i in pyroxene, i.e. that $a_i^{pyx} = \gamma_i^{pyx} X_i^{pyx}$.) Therefore, the equilibrium constant for the interchange K_{int} may be written:

$$K_{int} = \frac{X_{Fe}^{M1}\,(1 - X_{Fe}^{M2})}{(1 - X_{Fe}^{M1})X_{Fe}^{M2}} \cdot \frac{\overline{\gamma}_{Fe}^{M1} \cdot \overline{\gamma}_{Mg}^{M2}}{\overline{\gamma}_{Mg}^{M1} \cdot \overline{\gamma}_{Fe}^{M2}} = K_D \frac{\overline{\gamma}_{Fe}^{M1} \cdot \overline{\gamma}_{Mg}^{M2}}{\overline{\gamma}_{Mg}^{M1} \cdot \overline{\gamma}_{Fe}^{M2}}. \tag{7.51}$$

The partial activity coefficients are indeterminate by direct means and so the free energy for the interchange ΔG_{int}° (where $\Delta G_{int}^{\circ} = -RT \ln K_{int}$) is also indeterminate as a function of temperature unless some suitable thermodynamic model can be shown to fit the ion-exchange characteristics. One such simple model is that of 'regular solutions' as first defined by Hildebrand in 1929. These solutions are characterized by an ideal entropy of mixing but a non-zero enthalpy of mixing.* The partial molar heat of solution, $\Delta \overline{H}_{soln}$, of a component is directly proportional to the square of its mole fraction but is independent of temperature. For solutions of this type it can be readily shown (e.g. Guggenheim, 1952) that the relative activity coefficients of components A and B in a two-component system are given by:

$$\ln \gamma_A = (1 - x_A)^2 W/RT, \tag{7.52}$$
$$\ln \gamma_B = (x_A)^2 W/RT, \tag{7.53}$$

where W is a constant independent of composition and temperature and with energy units (e.g. J/mol). (W is a measure of the difference between the interaction energy of unlike atoms and that of like atoms in the solution. If W is negative in sign then the solution has greater stability than that of the ideal solution and vice-versa if the sign is positive.)

---

* The strict definition, by Hildebrand, of a regular solution is 'one involving no entropy change when a small amount of one of its components is transferred to it from an ideal solution of the same composition, the total volume remaining unchanged'. Guggenheim (1952) defined 'strictly regular solutions' by the expression $\ln \gamma = Wx^2/RT$ and where W is a constant independent of composition and temperature. A number of chemical solutions have properties amenable to theoretical treatment according to the more restricted definition by Guggenheim, and it is the definition often used in mineralogical studies. Theoretical discussions of regular solutions are beyond the scope of this book. The interested reader is referred to texts by Guggenheim (1952), Rock (1969), Hildebrand and Scott (1962).

In their study of order–disorder relationships in orthopyroxenes, Saxena and Ghose (1971) adopted the regular solution model and defined W such that:

$$\ln \gamma_{Fe}^{M1} = \frac{W^{M1}}{RT} (1 - X_{Fe}^{M1})^2 \tag{7.54}$$

but with W not necessarily being constant with temperature. The substitution of four expressions of type (7.54) above, for $\overline{\gamma}_{Fe}^{M1}$, $\overline{\gamma}_{Mg}^{M2}$, $\overline{\gamma}_{Mg}^{M1}$ and $\overline{\gamma}_{Fe}^{M2}$ into equation (7.51) and with rearranging gives:

$$\ln K_D - \ln K_{int} = \frac{W^{M2}}{RT} (1 - 2X_{Fe}^{M2}) - \frac{W^{M1}}{RT} (1 - 2X_{Fe}^{M1}). \tag{7.55}$$

A curve of the form of (7.55), with the best fit to the experimentally determined points on the X_{Fe}^{M1} versus X_{Fe}^{M2} plot (Fig. 7.3) is then found, and this gives the values for: $\ln K_{int}$; W^{M2}/R and W^{M1}/R at a particular temperature. The fitting of appropriate curves (see Fig. 7.3) to the orthopyroxene order–disorder relationships give the results in Table 7.1. As W is a measure of the departure of the solution from ideality it can be seen that as the temperature increases the degree of ideality increases.

TABLE 7.1. *Values of the factor W for sites M1 and M2 in orthopyroxene at different temperatures*

$T °C$	K_{int}	W^{M1}	W^{M2}
		(kJ/mol)	
500	0.175	10.9	4.91
600	0.277	10.0	6.60
700	0.273	8.02	5.09
800	0.289	6.87	4.42

Below about $500 °C$ the ordering in orthopyroxenes appears to be frozen such that there is virtually no change with decreasing temperature. For this reason and the fact that ordering above $600 °C$ probably proceeds at a relatively fast rate, order–disorder relationships in orthopyroxenes are generally unsuitable as a geothermometer. However, it is worthy of note that Virgo and Hafner (1970) recorded higher degrees of ordering in orthopyroxenes from some granulites, corresponding to temperatures of about $480 °C$, while orthopyroxenes from some volcanic rocks had lower degrees of order corresponding to a temperature of $500–600 °C$. These differences may result from the different rates of cooling of the granulite and volcanic rocks.

Equation (7.55) shows the nature of the cation ordering in orthopyroxenes as a function of temperature. Cation ordering in some other silicate and non-silicate solid solution mineral series may follow the regular solution model while other minerals may follow other solution models. (Some of these other models are listed by Grover, 1977.)

7.5 Element distribution between phases

7.5.1 Major element distribution

The distribution of elements between co-existing minerals follows the same general principles as discussed for element distribution over available sites within a mineral but the situation may be complicated by the relatively large number of different types of sites in the system. Many studies of element distribution have been directed towards the practical application of geothermometry and geobarometry. Some of these are described in the next section. A few studies have shown that element distribution in certain mineral pairs is insensitive to temperature (or pressure) effects. Such is the case with the distribution of Fe and Mg between olivine and orthopyroxene (Fig. 7.4), which stays remarkably constant from 900°C to 1300°C. The observed partitioning can be matched by a model of distribution involving an ideal single-site phase (olivine, with M1 treated as being equivalent to M2) and a double-site phase (orthopyroxene) in which there is ideal solution behaviour between the sites (Medaris, 1969). This ideality in cation mixing in the orthopyroxene is consistent with the findings described in Section 7.3.2, where the application of the regular solution model is for temperatures of mixing below 900°C.

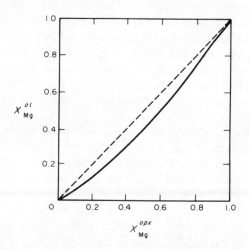

FIG. 7.4. Distribution of magnesium between olivine and orthopyroxene. The continuous line represents the theoretical partitioning at 900°C, which closely matches the experimentally determined partitioning (not shown). (After Medaris, 1969)

In co-existing mineral systems where experimental work has shown that element distribution is insensitive to temperature and pressure, it is possible to use the partitioning as an indicator of the attainment of chemical equilibrium between the two minerals, as the equilibrium partition is a known constant.

7.5.2 *Trace-element distribution*

In the absence of a more rigorous thermodynamic definition of a trace element, we will adopt the common usage by placing into this category those elements below about 0.1 wt% in a rock system. Usually these elements are distributed at trace levels amongst the various phases of the system and are rarely significant constituents of any mineral.

When a solution is sufficiently dilute with respect to a particular solute component, the component's behaviour will follow Henry's law:

$$a_i = k_h x_i \text{ as } x_i \to 0 \text{ at constant } T \text{ and } P, \tag{7.17}$$

where k_h is Henry's law constant (Section 7.2.3). In recent years there have been many experimental studies on the degree of adherence to Henry's law shown by trace elements in silicate systems but unfortunately some of the data are contradictory and have created a controversy which is not yet resolved (e.g. Mysen, 1976b; Drake and Holloway, 1978; Mysen, 1979; Drake and Holloway, 1981). For example, experimental results by Mysen (1979) suggest that adherence to Henry's law in the distribution of Ni between olivine and silicate melt (Fig. 7.5) occurs only when the Ni concentration in the olivine is below about 1000 ppm. Drake and Holloway (1981) were unable to replicate Mysen's findings, and have shown instead that the law is followed at all concentrations studied in the range 10 to 60,000 ppm Ni in olivine (Fig. 7.5). The importance of careful experimental procedure has been stressed (Drake and Holloway, 1978). It is possible that there will be deviations from Henry's law when the trace element is at extremely low concentration because of the likely existence of defect sites, such as those caused by structural dislocations or intergrowths, which cause the 'anomalous' behaviour in element distribution. Defect sites are relatively few in number and so soon become saturated with occupying trace elements (Navrotsky, 1978). The distribution behaviour of samarium between synthetic garnet and a liquid (Fig. 7.6) has been interpreted in this way (Wood, 1976; Harrison and Wood, 1980).

The distribution of trace elements between minerals or phases may be described by a partition coefficient. The exact form of the coefficient which is used depends on the requirements of the investigator. A simple partition coefficient such as defined in equation (7.56) and used in Chapter 5

$$k = \frac{[M]_{\text{phase } a}}{[M]_{\text{phase } b}} \tag{7.56}$$

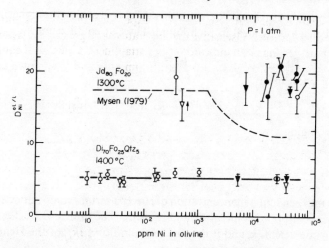

FIG. 7.5. Plot of the Ni partition coefficient (oliving/synthetic melt, $D_{Ni}^{ol/l}$) as a function of Ni concentration in the olivine, determined experimentally by Drake and Holloway (1981) for two synthetic systems. In both systems the partitioning is independent of the concentration. The dashed line shows the variation in partition coefficient as measured by Mysen (1979); see text for discussion. (After Drake and Holloway, 1981)

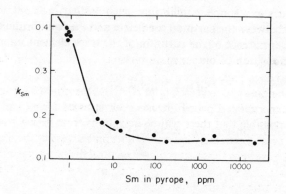

FIG. 7.6. Variation in the partition coefficient Sm (pyrope garnet/liquid, k_{Sm}) as a function of the concentration of Sm in the garnet, at 1300° C and 30 kbar (3 GPa) pressure. (After Harrison and Wood, 1980)

is a perfectly valid coefficient but its usefulness and applicability is, at least in theory, very restricted.

The coefficient may be put into a thermodynamic context by considering two partially miscible liquids ξ and ψ (i.e. non-ideal system) in which a third component, j, is distributed between them and is at such a low concentration that its behaviour is ideal in each liquid. Thus, at equilibrium

$$\mu_j^\xi = \mu_j^\psi$$

therefore, from (7.14)

$$\mu_j^{0,\xi} + RT \ln x_j^\xi = \mu_j^{0,\psi} + RT \ln x_j^\psi \quad \text{at constant } P,$$

and rearranging gives

$$\ln \frac{x_j^\xi}{x_j^\psi} = \frac{\mu_j^{0,\psi} - \mu_j^{0,\xi}}{RT} \tag{7.57}$$

which is independent of concentration of j (over a certain compositional range) but dependent upon temperature and pressure. The ratio x_j^ξ/x_j^ψ is the Nernst partition coefficient, k, and it is equal to the ratio k_{hj}^ξ/k_{hj}^ψ of the Henry's law constants in the two media ξ and ψ.

The Nernst partition coefficient has been used extensively in studies of the distribution of trace elements, but because of its sensitivity to changes in major element concentrations it is of little use in geothermometry and geobarometry (see below). It is, however, a particularly useful coefficient when applied to relatively simple rock-generating processes such as those discussed in Chapter 8.

We may use any simple equilibrium, such as that in (7.58), to show the relationship between the partition coefficient and the equilibrium constant, as well as the dependence of the partition of the trace element upon the major element composition of one or more phases:

$$SiO_2 + 2NiO = Ni_2SiO_4. \tag{7.58}$$

liquid   liquid   olivine cryst

The equilibrium constant for reaction (7.58) is:

$$K = \frac{a_{Ni_2SiO_4}^{ol}}{a_{SiO_2}^l \cdot (a_{NiO}^l)^2} \tag{7.59}$$

which may be written

$$K = \frac{k_h^{ol} x_{Ni_2SiO_4}^{ol}}{a_{SiO_2}^l \cdot (k_h^l x_{NiO}^l)^2}, \tag{7.60}$$

where k_h signifies the appropriate Henry's law constant. Therefore,

$$\ln K = \ln x_{Ni_2SiO_4}^{ol} - 2 \ln x_{NiO}^l - \ln a_{SiO_2}^l + \ln k_h^{ol} - 2 \ln k_{NiO}^l. \tag{7.61}$$

The first two terms on the right-hand side of (7.61) are clearly related to the partition coefficient, k, and are seen to be dependent upon silica activity in the liquid as well as on the equilibrium constant, K.

Equation (7.58) is, however, for a particular simple reaction of only a little relevance to natural rock systems, since the presence of magnesium (and/or iron) in the latter requires consideration of the following type of exchange reaction (7.62):

$$Mg^{2+} + Ni^{2+} = Ni^{2+} + Mg^{2+}. \qquad (7.62)$$

olivine　　liquid　　olivine　　liquid

For the equilibrium (7.62) an alternative partition coefficient, K_D, may be defined:

$$K_D = \frac{[Ni]_{ol}[Mg]_{liq}}{[Ni]_{liq}[Mg]_{ol}}. \qquad (7.63)$$

This partition coefficient, provided it represents the actual exchange process, is far less sensitive to changes in composition than is the Nernst one, since it allows for the fact that the partition of any element (e.g. Ni) must involve the exchange of at least one other (e.g. Mg).

In the partition of trace amounts of an element into a mineral (e.g. Ni into olivine), the major element for which it substitutes (e.g. Mg), is named the 'carrier' element, and K_D is sometimes called the Henderson and Kracek (1927) partition coefficient after the authors who introduced it (McIntire, 1963).

There have been a number of studies of the distribution of elements between a magma, or synthetic silicate melt, and crystalline phase, and some of the data from these have been given in Chapter 5. One study by Takahashi (1978) relates directly to the examples given above. He showed a marked compositional dependence of the Nernst partition coefficient for a number of transition element ions, including Ni^{2+}, in olivine basalt melt systems but an almost constant K_D (involving Mg) for many of the elements over a wide compositional range.

The identification of the carrier element in defining the partition coefficient K_D for nickel between olivine and melt is straightforward. In other cases the carrier element may be one of a small number of possibilities and where altervalent substitutions occur it will be necessary to consider additional charge-compensating substitutions (see Section 6.4).

7.6 Geothermometry and geobarometry

There are many (at least fifty) geothermometers and geobarometers available for application, based on a wide range of different principles including stable isotope distribution, mineral stability, liquid–vapour homogenization points (in fluid inclusion studies), as well as element distribution between mineral phases. This section is concerned with the last of these, i.e. the temperature and pressure dependence of element distribution, and the establishment of reliable geothermometers and geobarometers, using this approach. It should be noted that while the main purpose in using any geothermometer and geobarometer is to obtain the temperature and pressure at which the rock's mineral assemblage formed, subsequent re-equilibration processes during rock cooling or pressure changes may, in some cases, preclude this from being realized.

Preceding sections have demonstrated the relationships between the equilibrium constant, K, and temperature and also between these parameters and activity–composition relationships, cation ordering and element partitioning. If the appropriate thermodynamic data are available then, in suitable cases, these relationships may be used to determine temperatures and/or pressures of mineral formation by application of the basic thermodynamic equation (see Section 7.2.4):

$$\Delta G^\circ = - RT \ln K \quad \text{at constant } P \tag{7.24}$$

and where

$$\Delta G^\circ = \Delta H^\circ - T\Delta S^\circ + \int_1^P \Delta V^\circ \, dP,$$

where T and P are pressure and temperature of interest and ΔH° and ΔS° are at temperature of interest and 1 bar P, for any relevant equilibrium.

Where K for a particular reaction is dependent principally on temperature (i.e. there is a large enthalpy but little volume change between reactants and products) then that reaction could be a good potential geothermometer. Similarly, reactions involving a large volume change but a small enthalpy change offer a potential geobarometer. For example, the reaction (7.64):

$$\underset{\text{plagioclase}}{CaAl_2Si_2O_8} = \underset{\text{'clinopyroxene'}}{CaAl_2SiO_6} + \underset{\text{quartz}}{SiO_2} \tag{7.64}$$

has shallow lines of constant K-value on a pressure versus temperature plot (Fig. 7.7) and so is a possible geobarometer. By contrast, the reaction

$$\underset{\text{clinopyroxene}}{CaFeSi_2O_6} + \underset{\text{garnet}}{\tfrac{1}{3}Mg_3Al_2Si_3O_{12}} = \underset{\text{garnet}}{\tfrac{1}{3}Fe_3Al_2Si_3O_{12}} + \underset{\text{clinopyroxene}}{CaMgSi_2O_6} \tag{7.65}$$

has steep lines of constant K (Fig. 7.8) and so is a good potential geothermometer (Wood, 1977). However, the potential use of equilibria (7.64) and (7.65) in rocks can be realized only when the relationship between the activities of the components of reaction (7.64) or (7.65) and composition of the natural multi-component system is known (see Section 7.3).

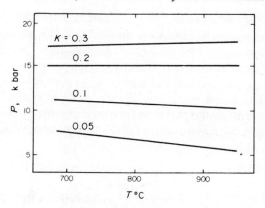

FIG. 7.7. Lines of constant K for the equilibrium:

$$CaAl_2Si_2O_8 = CaAl_2SiO_6 + SiO_2.$$
plagioclase clinopyroxene quartz

(Standard states are the pure phases at P and T of interest.) (After Wood, 1977, with permission.)

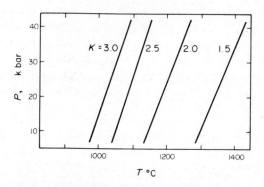

FIG. 7.8. Approximate K lines for the equilibrium:

$$CaFeSi_2O_6 + \tfrac{1}{3}Mg_3Al_2Si_3O_{12} = \tfrac{1}{3}Fe_3Al_2Si_3O_{12} + CaMgSi_2O_6.$$
clinopyroxene garnet garnet clinopyroxene
(After Wood, 1977)

As it is necessary to establish the activity–composition relationships of the major components before an equilibrium can be used as a geothermometer (or geobarometer), it might be asked if the use of trace-element partitioning could be an easier approach, provided that the trace-element solution obeyed Henry's law. In fact, the Nernst distribution law has formed the basis of many attempts at establishing geothermometers as, for example, in that by Hakli and Wright (1967), who determined the distribution of nickel between coexisting olivine, pyroxene and glass of some lava samples quenched from different known temperatures. The results are shown in Fig. 7.9. It has already been shown in the discussion of the equilibrium given in equation (7.58) that the Ni partition coefficient between crystal and liquid depends upon the activities of major components in the liquid, so crystal–liquid partition is unlikely to be a reliable geothermometer except over a very restricted compositional range. Similarly, the partition of nickel between olivine and pyroxene depends upon the activities of some major components. The equilibrium can be written in terms of the components in (7.66)

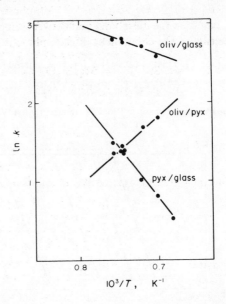

Fig. 7.9. Partition of Ni between olivine-glass, pyroxene-glass, and olivine-pyroxene as function of temperature (K^{-1}). (After Hakli and Wright, 1967)

$$CaMgSi_2O_6 \; + \; \tfrac{1}{2}Ni_2SiO_4 \; = \; CaNiSi_2O_6 \; + \; \tfrac{1}{2}Mg_2SiO_4 \quad (7.66)$$

$$\text{clinopyroxene} \qquad \text{Ni-olivine} \qquad \text{Ni-clinopyroxene} \qquad \text{olivine}$$

for which

$$K = \frac{a^{cpx}_{CaNiSi_2O_6} \cdot (a^{ol}_{Mg_2SiO_4})^{\frac{1}{2}}}{a^{cpx}_{CaMgSi_2O_6} \cdot (a^{ol}_{Ni_2SiO^4})^{\frac{1}{2}}}. \qquad (7.67)$$

Using a treatment similar to that for reaction (7.58) it is seen that the partition coefficient, k, is dependent upon the activities of $CaMgSi_2O_6$ and $CaNiSi_2O_6$. The activity coefficients of these components are unlikely to have the value of one as other elements, especially Al, are common constituents of natural clinopyroxenes. The use of the Nernst partition law in extending geothermometers outside the compositions for which they were calibrated requires that the system is ideal which, in general, it is not. Inevitably, the application of the nickel partition Ni geothermometer has had limited success.

In certain cases the use of simple, but not unreasonable assumptions concerning activity–compositions relationships may lead to the establishment of sufficiently reliable geothermometers. Wood and Banno (1973) used the assumption that both orthopyroxene and clinopyroxene phases behave as ideal two-site solutions of $CaMgSi_2O_6$ and $Mg_2Si_2O_6$ components, to formulate a geothermometer based on the miscibility gap between diopside and enstatite. The reaction in terms of the component in each phase is (7.68)

$$Mg_2Si_2O_6 \quad = \quad Mg_2Si_2O_6$$
$$\text{orthopyroxene} \qquad \text{clinopyroxene} \tag{7.68}$$

for which, at equilibrium

$$\Delta G^\circ = -RT \ln \left(\frac{a^{cpx}_{Mg_2Si_2O_6}}{a^{opx}_{Mg_2Si_2O_6}} \right) = -RT \ln K, \quad \text{at constant } P. \tag{7.69}$$

The assumption of ideality leads to (see Wood and Banno, 1973):

$$a^{cpx}_{Mg_2Si_2O_6} = (X^{M1}_{Mg} \cdot X^{M2}_{Mg})_{cpx},$$
$$a^{opx}_{Mg_2Si_2O_6} = (X^{M1}_{Mg} \cdot X^{M2}_{Mg})_{opx}. \tag{7.70}$$

The ratio of activities (i.e. the equilibrium constant, K) can be determined from use of equation (7.70) and by assuming that large cations (e.g. Ca^{2+}, Na^+, Mn^{2+}) occupy the M2 site while the smaller ones (e.g. Al^{3+}, Cr^{3+}, Ti^{4+} Fe^{3+}) occupy the M1 site, and with Fe^{2+} and Mg^{2+} randomly distributed over the remaining sites of both types after allowing for the site occupancies of the other cations above. So the activity of $Mg_2Si_2O_6$ in each pyroxene phase is given by:

$$a_{Mg_2Si_2O_6} = \left(\frac{Mg^{2+}}{Ca^{2+} + Mg^{2+} + Fe^{2+} + Mn^{2+} + Na^+} \right) M2 \cdot$$
$$\left(\frac{Mg^{2+}}{Fe^{3+} + Fe^{2+} + Al^{3+} + Ti^{4+} + Cr^{3+} + Mg^{2+}} \right) M1. \tag{7.71}$$

This treatment was applied to the distribution of $Mg_2Si_2O_6$ between coexisting orthopyroxene and clinopyroxene phases crystallized at known temperatures in a number of published experimental studies, and the following expression for temperature was obtained:

$$T = \frac{-10,202}{\ln K - 7.65\,X_{Fe}^{opx} + 3.88\,(X_{Fe}^{opx})^2 - 4.6}, \tag{7.72}$$

$$\text{where } X_{Fe}^{opx} = \left(\frac{Fe^{2+}}{Fe^{2+} + Mg^{2+}}\right)_{opx}.$$

This geothermometer contains the implicit assumption that pressure effects are small or negligible. The apparently successful application to a number of rock types with co-existing orthopyroxene and clinopyroxene indicates that the assumptions are reasonable for this system (cf. Powell, 1978b). In a more recent consideration of the available experimental data, Wells (1977) has confirmed that for most there is a linear relationship between $\ln K$ and T^{-1} using an ideal two-site mixing model for reaction (7.68) and that the equilibrium is insensitive to a pressure range of 1 kbar (10^8 Pa) to 40 kbar (4×10^9 Pa). For a best fit to the available experimental data, he obtained a revised relationship (7.73)

$$T = \frac{-7341}{\ln K - 2.44\,X_{Fe}^{opx} + 3.355}. \tag{7.73}$$

Another geothermometer based on similar principles as in the two-pyroxene case is one which uses the distribution of the component, albite ($NaAlSi_3O_8$), between coexisting plagioclase and alkali-feldspar at equilibrium (Barth, 1951). In the earlier formulation of the *two-feldspar geothermometer* the distribution of the albite component (abbreviated: ab) was considered to follow the Nernst partition law so that for reaction (7.74), equation (7.75) was applied

$$(NaAlSi_3O_8)_{\text{plagioclase}} = (NaAlSi_3O_8)_{\text{alkali-feldspar}} \tag{7.74}$$

$$\Delta G^\circ = RT \ln \frac{x_{ab}^{\text{alk feld}}}{x_{ab}^{\text{plag}}}. \tag{7.75}$$

However, the solution of albite in alkali-feldspar deviates significantly from Henry's law in the range of composition of geological interest, so strictly equation (7.75) is inapplicable. Furthermore, Barth (1951) ignored the effect of pressure.

In a subsequent and more accurate formulation Stormer (1975) considered the activities of ab in the two phases. The solution behaviour of ab in plagioclase is nearly ideal, and in alkali-feldspar can be matched using a subregular solution model. Pressure effects also were considered. More recently, Powell and Powell (1977) have modified the geothermometric relationship to allow for the possible presence of Ca in the alkali-feldspar. The two-feldspar geothermometer is a particularly useful one because of the common co-existence of these minerals in igneous rocks.

Geothermometers and geobarometers have been developed also for sulphide and for oxide mineral assemblages. The compositional dependence of pyrrhotine (approximately FeS) on the temperature and sulphur fugacity at the time of formation has been used to construct an experimentally calibrated geothermometer (the *pyrrhotine-pyrite geothermometer*) of especial use in the study of sulphide ore deposits (Toulmin and Barton, 1964). The effect of pressure on the pyrrhotine composition is small, unlike that on the composition of sphalerite in equilibrium with other sulphides. The FeS content of sphalerite is sufficiently sensitive to pressure to make it a useful geobarometer (the *sphalerite geobarometer*) in the temperature range 265–600°C, when the mineral is in equilibrium with pyrite and hexagonal pyrrhotine. Scott and Barnes (1971) and Scott (1976) have calculated curves for these conditions, showing the relationship between sphalerite composition and temperature at various isobars. The steepness of the curves, shown in Fig. 7.10, demonstrates the insensitivity to temperature variation. The sphalerite geobarometer has been applied to a number of different geological situations including deposits in regionally metamorphosed terrains (Scott, 1976). However, one of the principal problems is the occasional absence of complete equilibration among

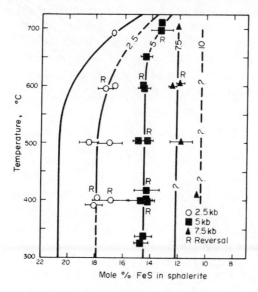

FIG. 7.10. Relationship between sphalerite composition, temperature and pressure. The curves are isobars at 0, 2.5, 5, 7.5 and 10 kbar. Sphalerite and pyrrhotine coexist to the left of each isobar while sphalerite and pyrite coexist to the right. (After Scott, 1976)

the coexisting sulphides – a prerequisite for the successful application of the geobarometer.

The *iron–titanium oxide geothermometer* (Buddington and Lindsley, 1964) is one of the most important because not only may it be applied to a relatively wide range of igneous and metamorphic rocks, but also it provides a measure of the oxygen fugacity at the time of mineral equilibration. Its basis is the compositional dependence of coexisting solid solutions of mag-netite–ulvöspinel and hematite–ilmenite on temperature and f_{O_2}. The equi-librium may be depicted by equation (7.76)

$$y\mathrm{Fe_2TiO_4} + (1-y)\mathrm{Fe_3O_4} + \tfrac{1}{4}\mathrm{O_2} = y\mathrm{FeTiO_3} + (3/2-y)\mathrm{Fe_2O_3}. \quad (7.76)$$
ulvospinel-magnetite   ilmenite-hematite
solid solution   solid solution

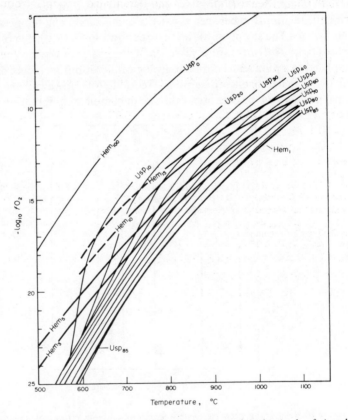

FIG. 7.11. Relationship between temperature, oxygen fugacity (as $\log f_{O_2}$) and composition (mole percent) of coexisting ilmenite–hematite and mag-netite–ulvospinel solid solutions. (After Buddington and Lindlsey, 1964)

From experimental data Buddington and Lindsley (1964) constructed a calibration figure shown in Fig. 7.11. Temperatures are probably accurate to $\pm 50°C$, and f_{O_2} to ± 1 log unit. Effects of pressure (at least up to 10 kbar $(10^9$ Pa)) and of trace amounts of additional components are probably small. Because reaction (7.76) involves a gas phase (O_2) it might be expected to be very sensitive to pressure, but the quantity of oxygen evolved is so small that total, inert pressure has a negligible effect on the equilibria.

The geothermometer has been applied to many rock suites. One interesting application (Duchesne, 1972) is to the petrogenesis of the Bjerkrem–Sogndal massif, an igneous, layered lopolith of the anorthosite-mangerite suite in south-west Norway. The evidence shows that the magma became more reducing during its differentiation and that the sudden changes in f_{O_2} of the magma are related to the rhythms of the layering. Duchesne (1972) favours the idea that intermittent supplies of undifferentiated magma produced the observed rhythms.

The iron–titanium oxide geothermometer has the added interest in that the mineral pair usually undergoes post-crystallization re-equilibration during slow cooling from magmatic temperatures. Application of the geothermometer to such cases could help to elucidate the processes of phase re-equilibration in plutonic rocks, especially by indicating the change in oxygen fugacity during cooling.

Suggested further reading

FRASER, D. G. (editor) (1977) *Thermodynamics in Geology.* D. Reidel Publ. Co. 410 pp.

GREENWOOD, H. J. (editor) (1977) *Application of Thermodynamics to Petrology and Ore Deposits.* Mineralogical Association of Canada, Short Course Handbook, volume 2.

NAVROTSKY, A. (1976) Silicates and related minerals: solid state chemistry and thermodynamics applied to geothermometry and geobarometry. *Progress in Solid State Chemistry,* **11,** 203–264.

WOOD, B. J. (1975) The application of thermodynamics to some subsolidus equilibria involving solid solutions. *Fortschritte der Mineralogie,* **52,** 21–45.

WOOD, B. J. and FRASER, D. G. (1977) *Elementary Thermodynamics for Geologists.* Oxford University Press. 303 pp.

8

Kinetic Controls of Element Distribution

8.1 Introduction

The generation of igneous and metamorphic rocks entails the motion of chemical entities on a scale ranging from the atomic, as in diffusion and crystal growth, to the massive, as can be involved in magma emplacement. Over the whole of this range the nature, rate and duration of any rock-generating process all help to determine the final chemical make-up of the product. There are many possible processes, some acting in places inaccessible to direct observation, so that the unravelling of a particular petrological history can be a very complicated task. It is not surprising, therefore, that it has been found necessary to assume certain constraints when investigating a rock's origin—for example, the assumption that rocks act as a closed chemical system during regional metamorphism.

A list of some of the major variables involved in igneous rock formation will serve to illustrate the complexity inherent in this field of study. The variables can be put under three headings—those affecting:

    (a) Magma generation.
    (b) Magma transport and accumulation.
    (c) Magma consolidation.

  (a) Magma generation.
      (i) Nature and composition of source material.
     (ii) Rate and extent of melting of source material.
    (iii) Rate and extent of separation of liquid from solid phases during melting.
    (iv) Physico-chemical conditions of melting: T, P, P_{H_2O}, etc.
  (b) Magma transport and accumulation.
      (i) Interaction of magma with wall-rock.
     (ii) Crystal fractionation or separation during transport.
    (iii) Mixing of different magmas.
    (iv) Mode of accumulation in magma reservoir.

(c) Magma consolidation.
   (i) Physico-chemical conditions of crystallization, P, T, etc.
  (ii) Rates of cooling and of crystallization.
 (iii) Extent of fractional crystallization.
 (iv) Openness of the magma system. (This may include loss of volatiles: interaction of magma with ground water, etc.).

To this list could be added a section dealing with post-consolidation changes.

A number of the aspects in the list form the contents of this chapter. After a discussion of diffusion, there follows a review of crystal nucleation and growth from a melt, and also of those transformation mechanisms in the solid state that can affect igneous and metamorphic rocks and minerals. Finally, there are sections on the theoretical modelling of magma compositions as governed by crystallization as well as by fusion processes.

8.2 Diffusion

The rates at which some processes in geology operate can be dependent upon the type and rate of diffusion of the chemical components in the system. Processes of rock alteration or mineral equilibration are controlled, in part, by this diffusion of chemicals over the surfaces (*surface diffusion*) or along the grain boundaries (*grain-boundary diffusion*), as well as through the body of the mineral grains (*volume diffusion*), Fig. 8.1. Rates of crystal growth can be dependent upon diffusion rates in the melt or magma. Hence, in order to obtain a proper understanding of these processes it is necessary first to establish the nature and rates of diffusion.

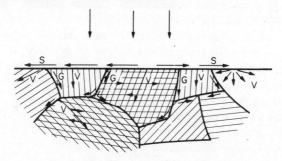

Fig. 8.1. Schematic representations of paths for surface diffusion (S), grain boundary diffusion (G), and volume diffusion (V).

8.2.1 Fick's laws and their solutions

The presence of a concentration gradient in an otherwise uniform melt or solid will, in general, create a flux of the appropriate element or chemical species down that gradient. When at a steady state, the net flux (J) across a plane is proportional to the concentration ($\partial C/\partial x$) normal to that plane:

$$J = -D\left(\frac{\partial C}{\partial x}\right)_t. \tag{8.1}$$

Equation (8.1) is Fick's first law of diffusion. D is the diffusion coefficient which generally is not independent of ($\partial C/\partial x$)$_t$; the negative sign arises because the diffusion is in the direction of decreasing concentration. The flux, J, is in units of mass per unit area, unit time; ($\partial C/\partial x$)$_t$ is in units of concentration per unit length, therefore, D is in units of area per unit time (usually $cm^2 s^{-1}$).

Equation (8.1) is for a steady-state flow but such a situation rarely occurs in practice; the concentration gradient and, therefore, the flux change with time. Fick's second law, equation (8.2), is appropriate for this situation (for simple proof, see Crank, 1975):

$$\frac{\partial C}{\partial t} = \frac{\partial}{\partial x}\left[D\frac{\partial C}{\partial x}\right]. \tag{8.2}$$

If D is a constant (i.e. independent of concentration) then:

$$\frac{\partial C}{\partial t} = D\frac{\partial^2 C}{\partial x^2}. \tag{8.3}$$

Fick's laws may be applied to a number of situations where different diffusion mechanisms are active. The type of diffusion must be known if the determined diffusion coefficient is to be a meaningful value. Four types of diffusion coefficient are described qualitatively below, although it is possible to define them in strict mathematical terms (e.g. see Darken, 1948; Wei and Wuensch, 1976; Crank, 1975).

(a) A *chemical diffusion coefficient* appertains when diffusion occurs down a chemical potential gradient in a system. (If, however, the component producing the gradient is present at a sufficiently low level, so that the solution can be treated as being very dilute, then the coefficient closely approximates to a tracer diffusion coefficient.)

(b) A *tracer diffusion coefficient* is applicable when the system contains only isotopic gradients. Experiments in which radioactive tracers are used to measure the rate of diffusion (e.g. the diffusion of radioactive ^{55}Fe in pure fayalite, incorporating a $^{55}Fe_2SiO_4-^{56}Fe_2SiO_4$ couple), yield tracer diffusion coefficients. (In most situations these coefficients are indistinguishable in value from intrinsic diffusion coefficients.)

(c) *Intrinsic diffusion coefficients* are obtained when there is no bulk-flow across the section of interest. Such a situation applies, for example, in the diffusion of Fe in pure fayalite, Fe_2SiO_4, *but* in the absence of any chemical or isotopic gradients (cf. (b) above).

Tracer and intrinsic diffusion coefficients are sometimes referred to as 'self-diffusion coefficients'.

(d) An *interdiffusion coefficient* arises when the diffusion of one chemical species is dependent on the opposing diffusion of another species in order to maintain a constant matrix volume and/or electrical neutrality. The diffusion in olivine of Mg^{2+} in one direction and the compensatory diffusion of Fe^{2+} in the opposite direction is one example.

It will be seen below that most experiments relating to geological systems have yielded tracer and interdiffusion coefficients.

A number of solutions to the diffusion equation (8.3) are possible, each one for particular initial and boundary conditions. Two types of solution are of especial relevance to the experimental work done on silicate systems; one is for the diffusion of a substance deposited (amount M per unit area) as a plane source (at $x = 0$) and at time $t = 0$ (equation (8.4)):

$$C_{(x,\,t)} = \frac{M}{(\pi\,Dt)^{1/2}} \exp\left(-x^2/4Dt\right). \qquad (8.4)$$

Equation (8.4) is for diffusion in one direction, results for two values of Dt being shown in Fig. 8.2. (If diffusion is in two directions, i.e. negative and positive values of x, then the right-hand side of equation (8.4) is multiplied by 0.5.) Many diffusion coefficients in silicates have been determined experimentally using an initially solid charge of the material in a capillary tube, with an element tracer deposited as a very thin layer (plane source) at one end (e.g. Lowry *et al.*, 1981). The expression in (8.4) is applicable to this type of experiment and yields the diffusion coefficient, D, at the temperature of the experimental run.

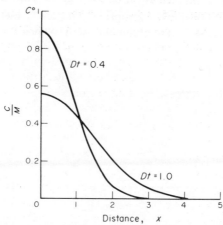

Fig. 8.2. Concentration–distance curves for the condition of a plane source, for two different Dt values.

The second type of solution is for a diffusion couple, or for an initial distribution of the diffusing substance in a finite volume (i.e. not in a plane), equation (8.5).

$$C_{(x,\,t)} = \tfrac{1}{2}C_0 \; \text{erfc} \; x/2 \; (Dt)^{1/2} \tag{8.5}$$

where

$$\text{erfc} \; x = 2/(\pi)^{1/2} \int_x^\infty \exp(-z^2)\,dz. \tag{8.6}$$

C_0 is the concentration at time, t, $= 0$, and z in equation (8.6) is an independent variable. The *error function* is a mathematical function defined by equation (8.7)

$$\text{erf} \; x = 2/(\pi)^{1/2} \int_0^x \exp(-z^2)\,dz. \tag{8.7}$$

From equations (8.6) and (8.7) it can be shown that the *complementary error function* (erfc x) is:

$$\text{erfc} \; x = 1 - \text{erf} \, x. \tag{8.8}$$

Values of the error function for different values of x can be found directly in some sets of mathematical tables (e.g. Milne-Thomson and Comrie, 1944) or by using available algorithms (e.g. Samford, 1981).

Figure 8.3 shows the concentration distribution as given by equation (8.5), when $t = 0$ and at $Dt = 0.5$ for a diffusion couple. Equation (8.5) has been used, for example, in the study of the diffusion of Ni in olivine crystals (Clark and Long, 1971).

This section has shown that appropriate solutions to Fick's Law have been obtained which may be used to obtain diffusion coefficients of species in some

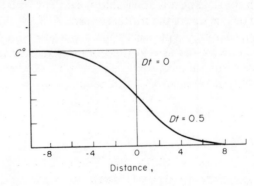

FIG. 8.3. Concentration–distance relations for a diffusion couple when $t = 0$ and at Dt value of 0.5.

geochemical systems. The derivation and discussion of these and related solutions may be found in the book by Crank (1975).

8.2.2 *Temperature dependence and compensation*

Rates of diffusion generally increase with increase in temperature, the relationship being expressed by equation (8.9) of Arrhenius type:

$$D = D_0 \exp\left(-Q/RT\right) \qquad (8.9)$$

where D_0 is called the frequency factor; it is usually constant for a diffusing species in a particular medium (units of $cm^2\,s^{-1}$);

Q is the activation energy (units of $kJ\,mol^{-1}$);

T is absolute temperature.

The temperature dependence is readily shown, therefore, on an Arrhenius plot of log D versus reciprocal temperature (Fig. 8.4 (a)).

The activation energy, Q, gives a quantitative indication of the energy which is required to initiate the movement of the chemical component involved in the diffusion process. Its value can be similar for a particular species and matrix composition, whether the matrix is molten, glassy or crystalline, or it can be very different.

Of interest is the fact that the so-called *compensation effect* is observed in the diffusion of components in many silicates (Winchell, 1969); it is simply that Q correlates with D_0 (Fig. 8.4(b)). This means that for a particular matrix and a *single* diffusion mechanism, the diffusion rates of the different species are identical at a particular temperature (called the critical temperature), Fig. 8.4(a). Thus, it is meaningless to make general statements about the relative diffusivities of different species in a matrix, without stating the temperature range for which the statement is applicable; the order of relative diffusivities will be reversed on one side of the critical temperature from that on the other. In practice, compensation is not 'exact', since the relationship between Q and D_0 can rarely be expressed as a perfect straight line for more than two experimentally determined data points. This aspect of compensation and its relevance to diffusion mechanisms is discussed below.

8.2.3 *Diffusion in melts*

Recent years have seen a considerable increase in the body of data on diffusion in silicate melts of petrological relevance; much work has been done on the diffusion of the alkalis, the alkaline earths and some transition elements. Table 8.1 gives some examples of typical diffusion coefficients and

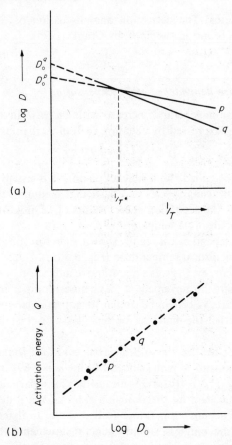

FIG. 8.4. Schematic diagrams showing: (a) Arrhenius plot of $\log D$ versus reciprocal temperature (K^{-1}) for two species, p and q, in the same matrix. The temperature (T^*) represented by the point of intersection of the two lines is called the critical temperature (see text). (b) A compensation plot showing data points, including those for the species p and q, which define a 'compensation line' (dashed), see also Fig. 8.6.

activation energies for some cations in basaltic and andesitic melts. Data are shown on an Arrhenius plot in Fig. 8.5.

Sufficient data are now available to be able to observe some general relationships:

(a) Many cations, but not all, diffuse at faster rates in a basaltic melt than in a more silicic one.

TABLE 8.1. *Examples of tracer diffusion data for cations in two types of magmas (at atmospheric pressure)*

Cation	Temp. °C	Melt type	D ($cm^2 s^{-1}$)	D_0 ($cm^2 s^{-1}$)	Q ($kJ mol^{-1}$)
Na^+	1300	Basalt	3.5×10^{-6}	0.96	163
	1400	"	6.5×10^{-6}		
	1300	Andesite	4.3×10^{-6}	0.0093	100
	1400	"	6.6×10^{-6}		
Cs^+	1300	Basalt	1.0×10^{-7}	110	272
	1400	"	3.5×10^{-7}		
	1300	Andesite	0.2×10^{-7}	4.9	251
	1400	"	0.7×10^{-7}		
Sr^{2+}	1300	Basalt	2.9×10^{-7}	3.5	213
	1400	"	7.1×10^{-7}		
	1300	Andesite	1.0×10^{-7}	0.0045	138
	1400	"	2.2×10^{-7}		
Ba^{2+}	1300	Basalt	2.0×10^{-7}	0.1	172
	1400	"	4.4×10^{-7}		
	1300	Andesite	0.5×10^{-7}	0.44	209
	1400	"	1.2×10^{-7}		
Co^{2+}	1300	Basalt	3.7×10^{-7}	1.5	201
	1400	"	8.2×10^{-7}		
	1300	Andesite	0.7×10^{-7}	160	280
	1400	"	2.2×10^{-7}		
Sc^{3+}	1300	Basalt	0.9×10^{-7}	0.28	197
	1400	"	2.3×10^{-7}		
	1300	Andesite	1.2×10^{-8}	0.018	184
	1400	"	3.4×10^{-8}		

Based on data in Lowry *et al.* (1982).

(b) The activation energies for diffusion in a basaltic melt of the different cations are neither consistently greater than or less than the activation energies for cations in a more silicic melt.

(c) There tends to be a direct relationship between activation energy and ionic radius for cations of a particular oxidation state. For example, in a basaltic melt the activation energy increases with increase in ionic radius of M^+ ions but decreases with radius of some M^{3+} ions.

(d) Addition of water to a granitic melt can increase the diffusion coefficients of cations by many orders of magnitude (Jambon *et al.*, 1978).

Attempts are being made to relate these observations to the nature of silicate melt structure and to the diffusion mechanisms of the different ions. The recognition that different diffusion mechanisms exist is aided by use of 'compensation plot', which shows the relationship between the activation energy, Q, and the frequency factor, D_0, for a particular matrix (Fig. 8.4(b)).

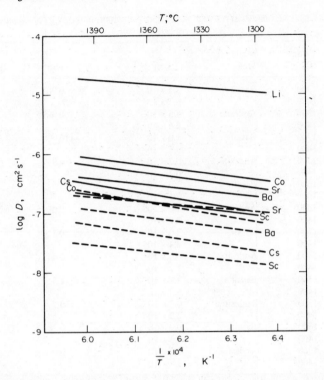

FIG. 8.5. Arrhenius plot for tracer diffusion of various cations in a melt of basaltic composition (continuous lines) and in one of andesitic composition (dashed lines). (Based on data in Lowry *et al.*, 1982)

Where more than one diffusion mechanism exists, the different ions will tend to lie on two lines (or more, according to the number of mechanisms) on the plot. Each line is said to define a *compensation law*. An example of a compensation plot showing more than one compensation law is given in Fig. 8.6; in that example, it is very probable that the diffusion of Cs^+ involves a mechanism different in some way from that for Na^+ but it is not yet possible to identify this difference.

The use of diffusion data in helping to define the properties of a melt seems to be promising but is still in its early stages. The application of such data to more general petrological problems is, however, very restricted for the simple reason that melt turbulence and convection, rather than diffusion, are likely to be the more active processes in controlling element movement in a magma. In a few cases, e.g. kinetics of crystal growth, diffusion data *sometimes* can be very relevant (see below).

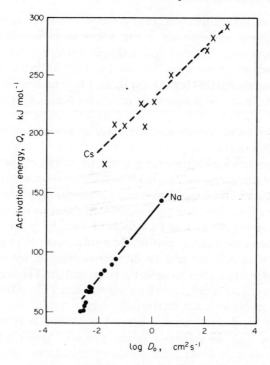

FIG. 8.6. Compensation plots for tracer diffusion of sodium and caesium in silicate matrices (melts and glasses) of various compositions. (After Lowry *et al.*, 1982)

8.2.4 *Diffusion in solids*

Experimental data on diffusion of cations in silicate glasses show that the activation energies, Q, for diffusion of species are not necessarily much different from those for species in melts of similar composition, but that the values of the frequency factor, D_0, tend to be lower for the case of a glass than for those of a melt.

The actual rates of diffusion of cations in glasses are likely to be of only restricted use to the petrologist but the data on activation energies might be more useful to the geochemist. These data suggest, for example, that the diffusion mechanisms for some cations are similar in both glasses and melts of the same composition. Such data could further our understanding of silicate melt and glass structures.

Diffusion data for crystalline silicates are likely to be of occasional use in petrology, especially in helping to establish the rates of metamorphic reactions or of re-equilibration processes in igneous rocks. The diffusion rates of most ions in minerals are very slow but, in the context of geological time, can be significant. Some diffusion data are given in Table 8.2; they follow the Arrhenius relationship and show that the rates are direction-dependent in anisotropic minerals and that both Q and D_0 vary with cation size. Hart (1981) has shown that the compensation effect occurs for diffusion in feldspars and olivines.

Volume diffusion in crystals can take place by any one of four principal mechanisms (Fig. 8.7):

(a) Exchange mechanism involving the interchange of the position of adjacent atoms or ions. This mechanism has a large activation energy.

(b) Interstitial mechanism in which ions (or atoms) of relatively small size can move from one interstitial position in the structure to another. A low activation energy may be associated with this process.

(c) Interstitialcy mechanism involves the displacement of an atom (or ion) from a structural site into an interstitial position, by an atom of a similar size which then occupies the structural site. (The region centred on an interstitial atom is called an interstitialcy.) The activation energy for this mechanism can be relatively high.

(d) Vacancy mechanism usually has a low activation energy and is a common type of volume-diffusion process. Atoms adjacent to a vacancy or point defect in the crystal structure, move into the vacancy and so create a vacancy behind them which, in turn, can be occupied by another diffusing atom and so on.

Diffusion data can give an indication of the diffusion mechanism. For example, the diffusivities of the alkalis in orthoclase decrease, and activation energies increase with increasing ionic radius, but the large difference between the diffusivity of Na and of K and Rb indicate the operation of different mechanisms. Foland (1974) tentatively suggests that Na is transported by an interstitial mechanism, while K and Rb move by a vacancy or perhaps an interstitialcy mechanism. It has been suggested also that cation diffusion in olivines is predominantly by a vacancy mechanism, whereas in sulphides an interstitial mechanism is probably common.

It can be seen that, in general, the diffusion rates are too slow for reactions involving only crystalline phases to be important in determining element distribution in the Earth's crust. Even though intergranular transport may be rapid, it need not be the rate-determining process since the diffusion of species within a crystal may still be required for reaction to proceed. Element redistribution will tend to occur more rapidly when a fluid phase is present. A discussion of some of the possible effects of diffusion on reaction rates may be found in a study by Loomis (1977) on a garnet granulite reaction involving the development of a corona around garnet crystals.

TABLE 8.2. *Examples of diffusion data for a glass and minerals (experimentally determined at atmospheric pressure or less)*

Matrix	Species	Diffusion type	Temperature range, °C	D_0 (cm²s⁻¹)	Q (kJ mol⁻¹)	D at 1000°C*	Reference
Basaltic glass	Na	T	745–985	5×10^{-6}	42	9×10^{-8}	1
	Cs	T	770–968	1.1	230	4.6×10^{-10}	1
	Sr	T	770–995	0.01	163	2×10^{-9}	1
	Ba	T	780–988	79	255	2.6×10^{-9}	1
MINERALS							
Olivine (Fo 93) c-axis	Fe–Mg	I	1200–1400	2×10^{-2}	274	1.1×10^{-13}	2
Olivine (Fo 93) b-axis	Fe–Mg	I	1200–1400	0.44	331	1.2×10^{-14}	2
Olivine (Fo 94) c-axis	Ni	C	1149–1234	1.1×10^{-5}	189	2.4×10^{-13}	3
Olivine (Fo 94) b-axis	Ni	C	1149–1234	0.34	355	1.4×10^{-15}	3
Orthoclase (Or 94) bulk	Na	T	500–800	4.97^+	221^+	4.3×10^{-9}	4
Orthoclase (Or 94) bulk	K	S	600–800	7.19^+	285^+	1.4×10^{-11}	4
Almandine garnet (10 wt % Mn)	Mn–Fe	I	900–1000	1.1×10^{-9}	94	1.5×10^{-13}	5
Spinel (Mg, Fe) Al₂O₄ (5 wt % Fe)	Mg–Fe	I	800–1034	4615	334	9.1×10^{-11}	6
Spinel (25 wt % Fe)	Mg–Fe	I	800–1034	0.011	221	9.4×10^{-12}	6

Diffusion types: I = Interdiffusion; C = Chemical; S = 'Self', T = Tracer.

* Calculated from D_0 and Q-values, for ease of comparison.

⁺ Values at 2 kbar pressure and assuming a spherical model for the diffusion.

References: 1: Jambon and Carron (1978); 2: Misener (1974); 3: Clark and Long (1971); 4: Foland (1974); 5: Freer (1979); 6: Freer and O'Reilly (1980).

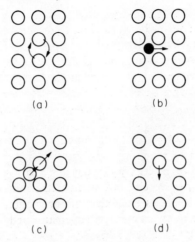

FIG. 8.7. The four main types of volume diffusion mechanisms: (a) Exchange mechanism. (b) Interstitial mechanism. (c) Interstitialcy mechanism. (d) Vacancy mechanism.

The rates of volume diffusion are important, however, to our understanding of the process of exsolution in minerals; e.g. the presence or absence of an exsolved phase in some minerals should be guide to the rate of cooling of a rock mass. This aspect is discussed in Section 8.4 on transformation processes, but before that a brief review is given on crystal nucleation and growth from a melt.

8.3 Nucleation and crystal growth from the melt

8.3.1 Nucleation

When a totally liquid magma is cooled it will pass through a particular temperature at which the magma is saturated with respect to a dissolved mineral phase. However, crystallization of this mineral does not occur until the magma has cooled even further to a level of supersaturation and a temperature where small, stable nuclei of crystals begin to form and grow— i.e. the process of nucleation. As the crystals nucleate at a temperature below that at which they would melt in the magma, the magma is said to be 'supercooled' and before the onset of crystallization is in a metastable state. The existence of the metastable, supercooled liquid arises because at low levels of supersaturation any crystal nucleus that may form is soon disrupted by molecular motion before it can grow to a stable size. Supersaturation is the driving force for crystal growth and, within certain limits, the greater the degree of supersaturation, then the faster is the growth rate. Hence there is a level of supersaturation at which there is a greater probability of crystal nuclei growing than of them being disrupted over a period of time. The extent of the metastable region is dependent upon a number of factors including the rate of cooling, and the movement of the magma. A magma may also become saturated with a particular mineral phase by processes other than simple cooling, or a combination of these. For example, the early fractionation stages of a basic magma do not involve the crystallization of a phosphate mineral and with increasing fractionation the amount of phosphorus progressively rises in the residual magma until a concentration and temperature are reached where apatite nucleates and grows.

In complicated silicate melts and magmas the minerals which crystallize have different compositions from that of the magma. Crystallization can continue until the magma is no longer supersaturated with a mineral phase. The rate of nucleation, the rate of crystal growth, and the order in which different minerals crystallize from a cooling magma are important factors in the production of varied textures in igneous rocks.

Nucleation of crystals may commence if by the production of crystals the free energy, G, of the system is reduced. However, the initial step of forming a crystal nucleus requires an increase in free energy relative to the value at equilibrium. This increase is the surface free energy of the solid liquid interface. Hence crystallization will proceed if the reduction in the free energy of the whole system is greater than the excess free energy created by the production of the solid liquid interface and such a condition exists only when the melt or magma is supersaturated with respect to one (or more) mineral phases. The free energy change, ΔG, of the system for the formation of a crystal of the same composition as the melt can be given (for a spherical nucleus) by:

$$\Delta G = 4\pi r^2 \sigma + \tfrac{4}{3}\pi r^3 \Delta G_v \qquad (8.10)$$

where r is the radius of the crystal particle,

σ is the surface free energy per unit area,

ΔG_v is the free energy change of transformation per unit volume; it is negative in a supersaturated solution.

The equation shows, for a supersaturated solution (ΔG_v negative) and for small values of r, that the free energy change increases with increase in r until some critical radius (r_{crit}) after which the free energy decreases with further increase in r. The variation in free energy with crystal radius is shown in Fig. 8.8.

Theoretical considerations show that the number of new nuclei produced per unit time per unit volume (the so-called 'rate of nucleation') is governed by three main variables: temperature of nucleation; surface free energy; and degree of supersaturation. This last factor is the dominant one and, in theory,

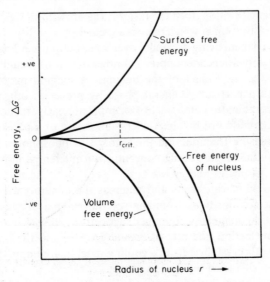

FIG. 8.8 The relationship between free energy and radius of a crystal nucleus. r_{crit} is critical radius.

there is a very rapid rise in nucleation rate with increase in supersaturation. Experimental results, however, do not agree with the predicted nucleation rates in that there is usually a maximum followed by a decline with increasing supersaturation (Figs. 8.9 and 8.10). This may be explained by the higher viscosities of melts with large supersaturations which in turn rapidly increase the activation energy for molecular motion across the incipient nucleus–liquid interface.

The process of nucleation described above is one that occurs spontaneously at some degree of supersaturation and is called *homogeneous nucleation*. Nucleation may also be initiated by other means—such as by the addition of 'seed' nuclei, or by the supersaturated magma being subjected to some shock or sudden movement. Magmatic stoping and convection currents carrying earlier formed crystals may both contribute to the nucleation of mineral phases at a lower degree of supersaturation than in homogeneous nucleation. The induction of nucleation by processes involving other factors as well as supersaturation (e.g. presence of some 'seed' nuclei) is referred to as *heterogeneous nucleation*.

In their discussion of the origin of textures in the rocks of the Skaergaard intrusion, East Greenland, Wager and Brown (1968) have used the varying rate of nucleation with degree of supersaturation of a mineral phase. A good example is shown by the chilled marginal gabbro of this intrusion, which has

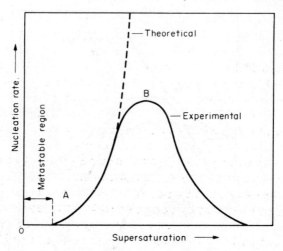

FIG. 8.9. Nucleation rate as a function of supersaturation. Regions A and B are referred to in the text.

large numbers of small crystals of plagioclase feldspar and olivine, poikilitically enclosed by relatively few crystals of pyroxene and ilmenite. The intruding magma was rapidly chilled against the country rocks and became supersaturated with olivine and plagioclase such that many nuclei of these minerals formed and then grew (region B of Fig. 8.9). The magma did not become saturated with respect to pyroxene and ilmenite until further cooling had taken place and by then the country rock had warmed up and the temperature was falling more slowly. These conditions give rise to a low nucleation rate such as in region A of Fig. 8.9, and the slow cooling allowed the few nuclei of pyroxene and ilmenite to develop into large poikilitic crystals.

The process of adcumulus growth of crystals, in which settled crystals at the base of a magma chamber continue to grow by addition of material of the same composition, can be considered as a special case of heterogeneous nucleation. Supercooled magma, within the metastable region for homogeneous nucleation, may be carried by convection currents over the floor of the chamber, where the earlier formed crystals will grow by diffusion of the required elements from the supercooled magma.

In a set of interesting experiments on nucleation and crystal growth of alkali feldspars from melts in the system $NaAlSi_3O_8$–$KAlSi_3O_8$–H_2O, it has been shown that the water content of the liquid has a dramatic effect (Fenn, 1977). Figure 8.10 depicts the nucleation density and growth rate for the feldspars from two melts of differing water content. At higher water contents, the nucleation density reaches a maximum at smaller degrees of supercooling.

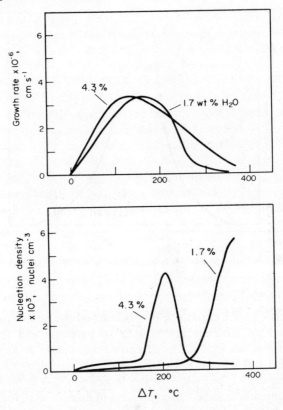

FIG. 8.10. Plots of nucleation density and growth rate of alkali feldspar $(Na_{0.7}K_{0.3}AlSi_3O_8)$ as function of supercooling (ΔT) in melts with 1.7 and 4.3 wt% water. (After Fenn, 1977)

Figure 8.10 is useful also in showing the typical forms of the nucleation-density and growth-rate curves. Kirkpatrick (1977) has estimated the rates of nucleation and growth of plagioclase feldspar in the lava lakes of Kilauea Volcano, Hawaii. Nucleation rates vary from 6.8×10^{-3} to $2.0\,cm^{-3}\,s^{-1}$, and growth rates (perpendicular to (010)) from 1.7 to $11.0 \times 10^{-10}\,cm\,s^{-1}$. In these lakes the feldspar nucleation occurs heterogeneously, and the growth rate is interface-controlled (see Section 8.3.3 below).

8.3.2 Crystal growth—latent heat of crystallization

The latent heat of crystallization produces a higher temperature at the crystal surface than in the body of the melt. The temperature profile is shown schematically in Fig. 8.11 where the degree of supercooling is given by $(T^* - T)$ and the driving force for crystallization is proportional to $(T_i - T)$. The crystal growth rate (in terms of mass per unit time) is related to the temperature

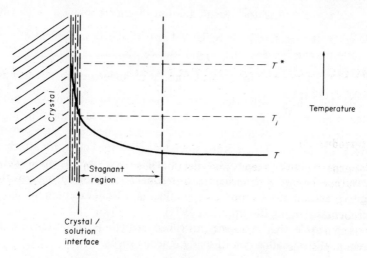

FIG. 8.11. Temperature gradient in the liquid adjacent to a growing crystal; see text for discussion.

difference $(T^* - T = \Delta T)$ by the expression:

$$\frac{dm}{dt} = \theta A \Delta T^n \qquad (8.11)$$

where A is the crystal surface area,

θ is the overall mass transfer coefficient for growth,

n is a constant for the system, generally with a value above 1 and below 3.

The thickness of the stagnant region in Fig. 8.11 depends upon the rate of flow of latent heat away from the crystal interface, which partially depends upon the rate of mixing or stirring of the melt. Thus, the rate of crystal growth is dependent upon the rate of conduction of heat away from the crystal. However, this does not appear to be an important effect in determining the crystallization rate of igneous rock melts (e.g. see Bottinga *et al.*, 1966).

8.3.3 Interface-controlled growth

The actual mechanism of growth also determines the growth rate provided that diffusion of the required ions or elements is not the rate-controlling step.

For example, if the mechanism involves continuous growth (i.e. uniform development of the crystal surface) the linear growth rate (i.e. rate of advance of a cyrstal surface), Y is related to ΔT simply by:

$$Y = \theta \Delta T \text{ for a small amount of undercooling.} \tag{8.12}$$

If, however, the mechanism is by lateral growth (i.e. development of a step, one-molecule thick, across the crystal surface) the rate is given by:

$$Y = \theta \Delta T^2 \text{ for a small amount of undercooling.} \tag{8.13}$$

In establishing the growth mechanism it is necessary to determine ΔT, but this is often impracticable. If the growth rate is not determined by the diffusing rate of the attaching species then it will not be time dependent. Such growth is *interface controlled* and there is an associated activation energy for attachment of the atom or molecular species on the crystal surface. The activation energy is a measure of the ease of molecular rearrangement occurring during growth but the interpretation of activation-energy values is difficult except in the simplest experimental systems (Kirkpatrick, 1975).

It is probable that interface-controlled growth is the rate-controlling process in the crystallization of many igneous rocks.

8.3.4 Diffusion-controlled growth

Mineral growth from natural silicate melts may in some cases be diffusion controlled. The theory for diffusion-controlled growth is not so well developed as for interface-controlled growth but in practice the study of the former in igneous systems is made easier by the existence of measurable concentration gradients.

Under conditions of a negligible growth rate or of interface-controlled growth, the concentrations of an element in the coexisting crystal and melt will be shown in Fig. 8.12(a), for an element (M) with partition coefficient C_s/C_l < 1. Where growth rate is relatively fast the concentration of M rises above C_l at the crystal–melt interface, Fig. 8.12(b). If the partition coefficient of M remains constant so that

$$C'_s/C^i_l = C_s/C_l,$$

the concentration of M in the solid increases with increasing growth rate until, in the limit, $C''_s = C_l$ so that the effective partition coefficient, k_{ef}, = 1 (see Fig. 12(c)).

At any particular growth rate a steady state in the concentration profile will develop, as shown in Fig. 8.12(b). The actual form of the profile is determined by: the diffusion rate of M in the melt; the growth rate Y, and also by the degree

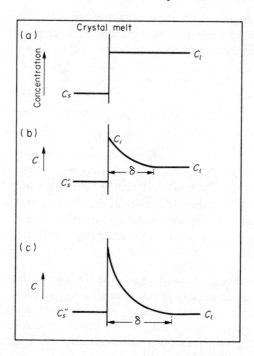

FIG. 8.12. Concentration profile of a trace element at a crystal-liquid interface. (a) Equilibrium distribution at infinitely slow crystal growth rate. (b) Typical distribution during finite crystal growth. (c) Steady-state distribution at very fast crystal-growth rate. δ is the thickness of the diffusion boundary layer. (After Henderson and Williams, 1979)

of motion or stirring of the melt. The amount of any element or component diffusing into an area of distance x from the interface is

$$D \frac{dC(x)}{dx},$$

and the amount diffusing out of an area at distance $x + dx$ is

$$D \frac{dC}{dx}(x + dx)$$

where D is the diffusion coefficient. The net flow is, therefore,

$$D \frac{d^2C}{dx^2}$$

per unit volume (defined by the distances $x + dx$). If the interface is considered to be stationary and the crystal growth treated as the flow of liquid with velocity Y towards the interface, then the flow of the element in the liquid as a result of this movement will be

$$Y\frac{dc}{dx}$$

for the same volume. Under steady-state conditions:

$$D\frac{d^2C(x)}{dx^2} = -Y\frac{dC(x)}{dx}. \tag{8.14}$$

From this may be derived the relationship between the concentration of M in the liquid as a function of distance x' away from the interface (Tiller *et al.*, 1953):

$$(C_l)_{x'} = C_l^* \exp\left(-\frac{Yx'}{D}\right) + C_l \tag{8.15}$$

where C_l^* is the excess solute concentration (i.e. $C_l^i - C_l$, Fig. 8.12(b)) at the interface. In theory this equation may be used to determine the growth rate of a crystal if a suitable concentration profile for an element could be determined, and the diffusion coefficient at the appropriate temperature for the element in that melt is known, but in practice a concentration profile (if present) cannot be measured.

An alternative means of determining the growth rate is to measure the effective partition coefficient of an element between the crystal and melt, and then to apply the equation derived by Burton *et al.* (1953). They solved a one-dimensional steady-state diffusion equation expressing conservation of the solute across the crystal–melt interface coupled to a boundary condition that in a liquid layer immediately around the crystal, diffusion is the sole mass transfer process while beyond this layer the element concentration in the bulk liquid is maintained at a uniform level by convective mixing. The derived equation is:

$$k_{ef} = \left[1 + \left(\frac{1}{k_0} - 1\right)\exp\left(-Y\frac{\delta}{D}\right)\right]^{-1} \tag{8.16}$$

where k_{ef} is the effective or measured partition coefficient (C_s/C_l, Fig. 8.12(b)),

k_0 is the equilibrium partition coefficient (C_s/C_l, Fig. 8.12(a)),

δ is the effective thickness of the diffusion boundary layer; beyond, in the liquid, fluid motion is the dominant transporting mechanism.

The thickness, δ of the diffusion boundary depends upon the extent of movement (or 'stirring') of the melt, its value being smaller for greater fluid motion. Since δ depends also on the diffusion coefficient, its value may vary from one element to the next, even for the same crystal–liquid interface, and it may vary, to a small extent, as a function of the crystal growth rate. The few experimental results (e.g. Bottinga *et al.*, 1966) on magmatic systems show that δ is of the order of $10\,\mu m$ for elements with partition coefficient values other than near to, or equal to, 1. Hence, the value of δ/D will be about $10^5\,cm^{-1}\,s$ if D is taken to be $10^{-8}\,cm^2\,s^{-1}$. Curves showing this variation in the partition coefficient, k_{ef}, with growth rate are given in Figs. 8.13 and 8.14; which were constructed using a δ/D of $10^5\,cm^{-1}\,s$. The diagrams show the rapid change in effective partition coefficient and its approach to unity with increase in growth rate. It should be noted that elements with very high or very low equilibrium partition coefficients will be the most sensitive to variations in growth rate. Thus, the concentrations of certain trace elements in mineral phases could be determined partly by the growth rate of that mineral. For the chosen δ/D value, growth rates of between 10^{-7} and $10^{-5}\,cm\,s^{-1}$ (see Figs. 8.13 and 8.14) could be distinguished by this method. These rates are of the same order of magnitudes as must have existed during the formation of some extrusive or intrusive rocks (see, for example, comments by Donaldson (1975) on the rate of crystal growth during the formation of the Skaergaard layered intrusion).

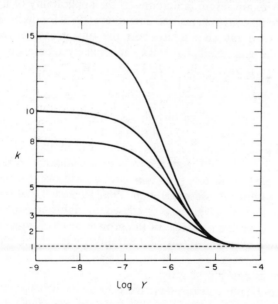

FIG. 8.13. Variation in partition coefficient, k (with $k \geqslant 1$), as a function of crystal-growth rate (Y) as calculated from the model by Burton *et al.* (1953) for a δ/D value of $10^5\,cm^{-1}\,s$. (After Henderson and Williams, 1979)

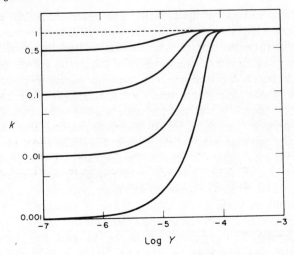

FIG. 8.14. As Fig. 8.13 but for k values $\leqslant 1$. (After Henderson and Williams, 1979)

Clearly, there are severe limitations to the applicability of a formulation such as this to igneous systems (see Carruthers (1975) for discussion). Strictly, the crystal growth rate should have been the same in each direction, and the values of all the parameters, δ, D and k_0, are unlikely to be available. However, as shown above, the use of the model by Burton *et al.* gives a semi-quantitative assessment of the effects of crystal growth rate on element distribution. Henderson and Williams (1979) have used the model in an attempt to estimate the growth rates of some igneous olivine crystals.

The evidence for the existence of diffusion-controlled growth processes in the petrogenesis of igneous rocks is not substantial. The presence of concentration profiles in the glassy groundmass adjacent to skeletal olivine crystals in some basaltic rocks has been cited as evidence that the growth was controlled by diffusion. Bottinga *et al.* (1966) suggested that the presence of oscillatory zoning in plagioclase is caused by diffusion-controlled growth coupled with a periodic change in the growth mechanism at the interface. There are other examples of this kind but much research into mechanisms of crystal growth in magmas is needed before an accurate assessment of the effect of this phenomenon can be given.

In the preceding discussions there has been an implicit assumption that the crystals have grown in a liquid of unchanging composition. However, if the fluid volume is such that the formation of crystals produces a compositional change then growth rates might alter as a result of the associated changes in degree of undercooling and diffusion rates. Furthermore, the possibility then arises of the solid phase becoming compositionally zoned if the diffusion rates in the solid are too slow to maintain an homogeneous crystal. An analogous situation occurs in a magma chamber where the early formed crystals sink (or perhaps float) and by their accumulation are effectively removed from the supernatant magma and prevented from re-equilibrating with it. The compositional changes in the successive solids and residual liquids that accompany this process may be estimated if the appropriate partition coefficients are known. Section 8.5 discusses the theory of this in the context of fractional crystallization of magmas, although it should be realized that much of the discussion is applicable to crystal zonation.

8.4 Transformations

8.4.1 Cation ordering

In Chapter 7 the temperature dependence of cation ordering was discussed but the kinetics of the process, which may be regarded as a transformation, were irrelevant to that treatment of equilibrium systems. However, the ordering process in some minerals takes place at measurable rates or is sufficiently slow to be of use in determining the cooling rates of some rocks. There is often a cut-off temperature below which the ordering rate is effectively zero; this serves to illustrate the temperature dependence of the ordering rate. In general the temperature dependence of a reaction rate (ξ) is given by an Arrhenius relationship

$$\xi = Ae^{-E/RT}, \tag{8.17}$$

where A is a rate constant and E is the activation energy of reaction.

Investigations into the kinetics of ordering behaviour in natural minerals have not been extensive, but attention is being given increasingly to the phenomenon. Mueller (1969) has developed a kinetic model which may be applied to the exchange of cations between the M1 and M2 sites in orthopyroxene. However, the ordering rate in orthopyroxene is probably too fast to be of much use in petrogenetic problems (at 500°C, 95% of the equilibrium ordering is attained in about 145 days). The kinetics of order–disorder reactions in anthophyllite show more promise as a petrogenetic indicator. The Fe^{2+}–Mg^{2+} ordering rates over the available sites (M1, M2, M3, M4) are relatively slow and provide for the possible determination of the cooling rate of a metamorphic rock (Seifert and Virgo, 1975).

8.4.2 Exsolution

The process of exsolving one mineral phase of different composition from its host occurs frequently in igneous and metamorphic minerals and is well exhibited by some pyroxenes, amphiboles and feldspars. The 'unmixing' or 'phase transformation' occurs because the originally homogeneous crystal becomes supersaturated with a dissolved phase, usually as a result of a reduction in temperature. The development of an exsolved mineral phase clearly causes a redistribution of elements and, as with ordering, may be used to give some indication of the rate of cooling.

The production of an exsolved phase may occur by one of two processes:

(a) Nucleation and growth of the phase, where the nucleation event may be described by classical nucleation theory. This is the only possible mechanism when the exsolved phase is of a fixed stoichiometric composition. The nucleation may be homogeneous or heterogeneous, but the latter is undoubtedly more common.

(b) Spinodal decomposition. This may occur when the exsolved phase is of a variable composition not very different from that of the host. The interface between the two phases is diffuse and the onset of crystallization of the exsolved phase is not abrupt. These two processes will be briefly discussed.

The expression for the nucleation of a spherical nucleus in a solid contains an additional energy term compared with equation (8.10) for nucleation in a liquid. This is a strain energy term so that the free energy change for nucleation is given by:

$$\Delta G = 4\pi r^2 \sigma + \frac{4}{3}\pi r^3 (\Delta G_v + \Delta G_\varepsilon), \qquad (8.18)$$

where ΔG_ε is the strain free energy. The ΔG_ε term is always positive, so that nucleation has a higher activation energy in the solid compared with the value for a related liquid. Equation (8.18) is of little practical use since the nucleus is rarely spherical, and also, the free energy terms are compositionally dependent. Furthermore, the nucleation of an exsolved phase is probably initiated at a crystal dislocation or grain boundary and is therefore heterogeneous but will often have a sufficiently high activation energy for there to be a frequent occurrence of metastable solid solution in igneous and metamorphic assemblages.

In spinodal decomposition there is no separate nucleation event. Consider the binary-phase diagram in Fig. 8.15, which contains a miscibility gap at low temperatures. At any temperature below T_s, the temperature at the peak of the solvus, the variation of free energy of the solid solution with change in composition will be of the form shown in Fig. 8.16. Within the area defined by

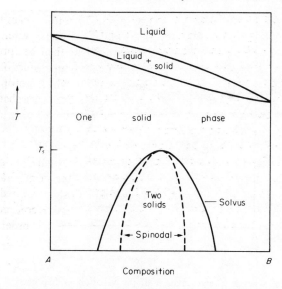

Fig. 8.15. Phase diagram for a binary solid-solution with a miscibility gap at low temperatures.

the solvus there are two inflexions, labelled S_1 and S_2, in the free energy curve. If a compound has an overall composition $A_x B_{1-x}$ lying between S_1 and S_2, and within it there is a small compositional fluctuation so that small local areas of the compounds has compositions $A_{x+dx} B_{1-(x+dx)}$ and $A_{x-dx} B_{1-(x-dx)}$ $(a-a'$ on Fig. 8.16), then the free energy of the compound is given by the weighted average of the free energies of the compositional fluctuations. Therefore, the free energy of the compound has been *decreased* by ΔG_s (in Fig. 8.16) compared with the compound without any compositional fluctuations. Thus a compound of any composition between S_1 and S_2 is unstable with respect to *small* fluctuations in composition and will undergo spontaneous decomposition at temperatures $< T_s$, without a nucleation event. The inflexion points S_1 and S_2 are called the *spinodes*, and their locus at all temperatures below T_s define the *spinodal* in a temperature–composition diagram, Fig. 8.15. Compositions which lie between the spinodal and the solvus (at temperatures below T_s) would suffer an increase in free energy through the occurrence of small compositional fluctuations and so the creation of two phases requires a nucleation event with large compositional fluctuation differences as shown by the line $b–b'$ in Fig. 8.16. A fuller discussion of spinodal decomposition may be found in Yund and McCallister (1970). Exsolution by spinodal decomposition has been observed in alkali feldspars, plagioclase feldspars, clinopyroxenes (augite-pigeonite) and some amphiboles.

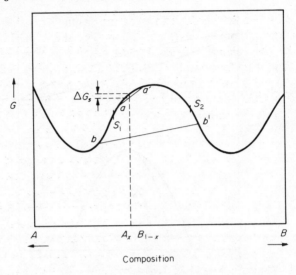

FIG. 8.16. Schematic representation of the variation in free energy, G, with composition, for the phase diagram depicted in Fig. 8.15, at low temperature.

The continued development of the exsolved phase requires the diffusion of atoms or ions through the solid phase, commonly by volume diffusion (see Section 8.2.4). The diffusivities of ions in most silicates are very low and determine the rate of growth of the exsolved phase.

The linear growth rate, Y, of a particle growing in a solid matrix is given approximately by equation (8.19) (Christian, 1975, p. 485):

$$Y \approx \frac{\Delta C}{C_a - C_b} \cdot \frac{D}{\delta},$$

(8.19)

where ΔC is the concentration change of an element in the matrix over the effective diffusion distance δ, and D is the volume diffusion coefficient in the matrix. C_a and C_b are the equilibrium concentrations of the element in the two phases. As the diffusion coefficient is temperature-dependent, so also is the growth rate Y. It is therefore convenient to describe exsolution behaviour with the aid of a time–temperature–transformation plot (sometimes known as a TTT plot). Usually this is presented with transformation boundaries plotted within temperature–log time space. The boundary curves are constructed from a number of experimental runs either under isothermal or continuous cooling conditions over a particular time interval.* Different runs are made at different temperatures or different cooling rates.

---

* The phase transformation boundaries on a TTT plot constructed from isothermal experiments may differ slightly in their position from those constructed from continuous cooling experiments.

Figure 8.17 is a schematic TTT plot for exsolution through spinodal decomposition or homogeneous nucleation. The shape of the plot is determined by the balance between the free-energy change for the phase separation and the diffusion rate of the elements through the solid host. The upper part of the curve (high temperatures) is dominated by the free-energy change, diffusion being rapid. At lower temperatures, diffusion is sluggish and becomes the rate-determining factor.

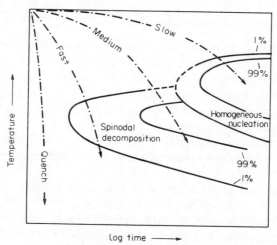

FIG. 8.17. TTT diagram showing the effective start (1 %) and finish (99 %) curves for spinodal decomposition and homogeneous nucleation for a hypothetical system. The four cooling curves show the effect that different cooling rates have on the operation of the different crystallizing processes.

The TTT plot (for continuous cooling) may be used to predict the behaviour of a mineral that is subjected to different rates of cooling. Four cooling curves are shown in the figure. At slow cooling rates, homogeneous nucleation of an exsolved phase will be favoured over spinodal decomposition and it can be seen that the exsolved phase reaches complete development over a relatively short time span. If cooling is fast, spinodal decomposition will occur, and it is possible for the exsolved phase to be incompletely formed if the cooling rate is sufficiently fast. The diagram helps to emphasize the point that spinodal decomposition is less dependent upon element diffusion and can occur, therefore, at lower temperatures than homogeneous, or heterogeneous nucleation. Heterogeneous nucleation is not shown on the TTT plot in Fig. 8.17 because it need not be a reproducible process in temperature-time space; its field will tend to lie at higher temperatures and earlier times than that of homogeneous nucleation.

Only a few time–temperature–transformation plots have been experimentally constructed for phase transformations in mineral phases. A good example can be found in Owen and McConnell's description (1974) of spinodal decomposition in alkali feldspar. (A related example is the exsolution behaviour of kamacite within taenite, which has been used to determine the cooling rate of some stony-iron and iron meteorites, see Chapter 1.)

8.4.3 Polymorphic transformations

Transformations that involve a change in crystal structure without, in general, the gain or loss of chemical components are of importance to the metamorphic petrologist as a guide to pressures and temperatures of rock formation. Polymorphic transformations which are geologically common, such as aragonite–calcite or kyanite–andalusite–sillimanite, rarely involve a significant change in trace or minor element distribution, although some polymorphs occasionally contain more of a trace element than others. For example, Cr is sometimes present in moderate amounts in kyanite but not in andalusite.

Trace constituents do not appear to have much influence on the rate of transformation. Aragonite may exist metastably at low pressures whether or not it is pure calcium carbonate, although the presence of an aqueous phase may facilitate the transformation. Earlier suggestions that trace Sr inhibits the transformation have not been substantiated.

8.4.4 Closed-system metamorphism

Metamorphism is 'the mineralogical and structural response of a rock to imposed conditions of temperature and pressure different from those of its origin' (Turner, 1968, page 2). The term 'closed-system metamorphism' is used here to describe rock transformation processes that take place without the introduction or removal of chemical components into or from the rock.

So far in Section 8.4 we have examined transformation behaviour of individual minerals. The metamorphism of pre-existing igneous or sedimentary rocks, with their aggregations of mineral grains and the possible presence of intergranular fluids, can entail additional transformation mechanisms. A study of these, as well as the rates of the transformations, could give us important clues to the metamorphic history of a terrain even in those systems where thermodynamic equilibrium either was not reached or has subsequently been disturbed.

Diffusion undoubtedly plays an important role in the formation of metamorphic rocks even in those cases where an apparently simple polymorphic transformation has occurred. Sometimes in rocks taken from a location near to an isograd, two polymorphs of the same composition may coexist, but not in contact with one another. It is possible that the production of one of the polymorphs at the isograd involved a number of reactants and reaction steps and was not simply a result of a structural inversion. For example, Carmichael (1969) has shown that in some pelitic rocks the generation of sillimanite at the expense of kyanite could have involved a two-step reaction with muscovite as an intermediate product:

$$3 \text{ kyanite} + 3 \text{ quartz} + 2K^+ + 3H_2O \rightleftharpoons 2 \text{ muscovite} + 2H^+$$

and

$$2 \text{ muscovite} + 2H^+ \rightleftharpoons 3 \text{ sillimanite} + 3 \text{ quartz} + 2K^+ + 3H_2O.$$

Frequently the kyanite is rimmed with muscovite in these rocks. Although other proportions of the reactants could be postulated, the essential point is that the sillimanite develops after a discrete nucleation event by the transference of chemical species (e.g. K^+) over short distances through the rock. The formation of sillimanite by such a reaction probably takes place with a lower activation energy than that for the inversion directly from kyanite. If the postulated reactions are correct then intergranular water is an essential ingredient, so that in totally dry systems such reactions would be prevented.

In a similar approach to metamorphic differentiation, Fisher (1970) showed that some andalusite–biotite–quartz segregations in a sillimanite–biotite–feldspar–quartz gneiss in Sweden developed (after the metamorphism) via reactions that were driven by the small free energy difference between andalusite and sillimanite. In these rocks the transformation of sillimanite to andalusite again involved diffusion of species through the rock fabric and not just a structural transformation of the polymorphs. (Water is a component of the postulated reactions.)

Compositional zoning in minerals can give a relative measure of the rate of development of the metamorphic fabric. Edmunds and Atherton (1971) have postulated that zoning in garnets within the Fanad aureole, Donegal, Ireland, is a result not only of depletion of components during crystal growth in the solid medium but also of diffusion-controlled growth. The compositional zoning may be removed or modified by subsequent volume diffusion.

A brief qualitative discussion of some processes causing element redistribution during metamorphism in open systems is given in Chapter 5 (Section 5.6) but there is, as yet, little information on their rates. Having reviewed the kinetics of crystal nucleation and growth within liquid and solid matrices, we can now turn our attention, in the following section, to the time-dependent processes of crystal fractionation and magma generation.

8.5 Crystal fractionation

8.5.1 Theory

The frequent existence of a density difference between crystals and their supernatant magma provides for the possible gravity accumulation of the crystals and thereby the effective separation of liquid and solid phases. The rate of cooling and the viscosity of the magma are two of the most important parameters which determine the rate and extent of the segregation process. Slowly cooled, basaltic magmas with their low viscosities give some of the best conditions for extensive fractional crystallization, for which ample evidence of its occurrence in nature exists. One of the best described examples is the Skaergaard layered intrusion, east Greenland (Wager and Brown, 1968). As seen in that intrusion, the accumulation of the early formed and subsequent crystals produced systematic changes in the successive liquid and solid phases to produce an extremely well-fractionated rock sequence from an initial magma of tholeiitic composition (see Chapter 5). There is evidence also for the occurrence of the process in the genesis of acid intrusive rocks, but examples are fewer (see Wager and Brown, 1968).

In this section a brief theoretical treatment is given of the compositional changes in solid and liquid phases during equilibrium crystallization and fractional crystallization.

Under conditions of complete equilibrium of solid with liquid phases during crystallization (equilibrium crystallization) the concentration of an element, M, in the two phases when proportion, F, of liquid remains is given by the mass balance equation:

$$FC_l + (1 - F)\overline{k}C_l = C_l^\circ, \tag{8.20}$$

where C_l° is the concentration of M in the initial liquid,

C_l is the concentration of M in the fractionated liquid,

$\overline{k}$ is the bulk solid/liquid partition coefficient C_s/C_l, and equals $\alpha^a k^a + \alpha^b k^b + \alpha^c k^c, \ldots, \alpha^n k^n$, where α^i is the mass proportion of a particular mineral with associated partition coefficient k^i.

Equation (8.20) on rearrangement gives:

$$\frac{C_l}{C_l^\circ} = [F(1 - \overline{k}) + \overline{k}]^{-1} \tag{8.21}$$

and for the solid:

$$\frac{C_s}{C_l^\circ} = \overline{k}/[F(1 - \overline{k}) + \overline{k}]. \tag{8.22}$$

Plots of compositional changes in solid and liquid for different $\bar{k}$-values are shown in Fig. 8.18.

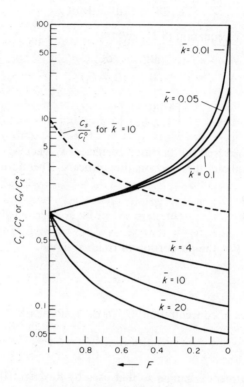

FIG. 8.18. Relative changes in the concentration of an element in a magma as a function of the proportion of remaining liquid (F) during conditions of equilibrium crystallization (as defined by equation 8.21) for various values of the bulk partition coefficient, $\bar{k}$, and for the coexisting solid when $\bar{k} = 10$ (equation (8.22)).

If the accumulated crystals do not remain in equilibrium with the residual liquid during subsequent cooling and further crystallization, then equations (8.20) to (8.22) are inapplicable. The following treatment deals with this situation, where crystals are effectively removed from their parent melt as soon as they have formed (fractional crystallization), it being further assumed that the separation of solid from liquid is complete.

A small amount of crystallization removes an amount dW from a melt of mass W. If the concentration of an element in the solid phase is C_s, then the total mass of this element in the solid phase is $C_s\,dW$ and this equals the amount removed from the liquid, i.e.:

$$C_s \, dW = d(C_l W) \qquad (8.23)$$

or $$C_s \, dW = C_l \, dW + W \, dC_l. \qquad (8.24)$$

Rearrangement of equation (8.24) gives

$$\frac{dW}{W} = \frac{dC_l}{(C_s - C_l)} \qquad (8.25)$$

and so

$$\int_{W^\circ}^{W} \frac{dW}{W} = \int_{C_i^\circ}^{C_l} \frac{dC_l}{(C_s - C_l)}. \qquad (8.26)$$

It will be assumed that the partition coefficient $\overline{k}$, where $\overline{k} = C_s/C_l$, remains constant throughout the fractionation process. Therefore equation (8.26) becomes:

$$\ln \frac{W}{W^\circ} = \frac{1}{(\overline{k} - 1)} \ln \frac{C_l}{C_i^\circ}. \qquad (8.27)$$

Since $W/W^\circ = F = $ the proportion of residual liquid,

$$F^{(\overline{k}-1)} = \frac{C_l}{C_i^\circ}. \qquad (8.28)$$

For the changes in concentration, C_s, in the solid phase:

$$kF^{(\overline{k}-1)} = \frac{C_s}{C_i^\circ}. \qquad (8.29)$$

The above treatment is similar to that used by Rayleigh (1896, 1902) in the analysis of the distillation of gases from liquids and so the fractionation process is sometimes referred to as the crystallization analogue of Rayleigh distillation.

Figure 8.19 shows the changes in C_l/C_i° (and indirectly in C_s/C_i°) as a function of the proportion of liquid remaining for different $\overline{k}$-values on the basis of the theoretical model expressed by equation (8.28). The relative concentration of an element in the liquid cannot rise above the line defined by $\overline{k} = 0$ (dashed line in Fig. 8.19).

In practice a number of effects can nullify the applicability of equation (8.28) and (8.29) in petrology. For a closed system these are:

(a) Variation in the partition coefficient, $\overline{k}$, as a result of changes in temperature, pressure, liquid composition, crystal growth rate or the proportions of the different crystallizing minerals.

(b) Variation in the amount of supernatant liquid trapped with the accumulated crystals.

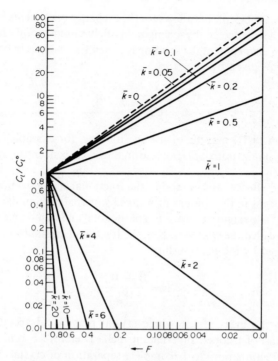

FIG. 8.19. Relative concentration changes in a magma as a function of proportion of remaining liquid (F) during fractional crystallization (equation 8.28), plotted on log–log scales for different values of the bulk partition coefficient ($\bar{k}$).

The first of these has been considered by Greenland (1970). He showed that equation (8.28) can be generalized to admit a varying partition coefficient, and as an example used a linear variation in $\bar{k}$ with variation in the proportion of residual liquid, such that $\bar{k} = a + bF$. Hence $\bar{k}$ varies from $a + b$ to a, and the integrated equation is:

$$\ln\left(\frac{C_l}{C_l^{\circ}}\right) = (a - 1)\ln F + b(F - 1). \tag{8.30}$$

It is improbable, however, that the partition coefficient will vary as a simple linear function of F. The frequent strong temperature dependence of the partition coefficient, usually of the form (see Chapter 5, Section 5.4.3):

$$\ln\bar{k} = A/T + B,$$

where A and B are constants for a particular composition will tend to dominate the partition coefficient variation during fractionation.

A partition coefficient that allows for the exchange of elements between melt and solid is less affected by changes in melt composition. If element a substitutes for element b in the solid phase, then the changes in concentration ratios with F is given by:

$$\frac{C_{l,a} \cdot C_{l,b}^{\circ}}{C_{l,b} \cdot C_{l,a}^{\circ}} = F^{k_a - k_b}. \tag{8.31}$$

This equation (8.31) may be of use in studying, for example, the effect of fractional crystallization of plagioclase on the Sr/Ca concentration ratio in the magma.

The second effect that can modify the fractional crystallization equations, that of variations in the amount of trapped supernatant liquid, needs further examination. The extent of separation of the solid and liquid phases from each other will be called the *efficiency*, E, of the fractional crystallization process. It may be defined by the relationship:

$$E = \frac{W_r - W_t}{W_r}, \tag{8.32}$$

where W_r is the mass of the whole crystal plus liquid accumulation and W_t is the mass of trapped supernatant melt. Therefore F ranges from zero (no crystal fractionation) to 1 (complete separation of crystals from liquid). The fractionation efficiency may change from one crystallization increment to the next; its effect on the change in concentration of an element in a melt is given by:

$$\frac{C_l}{C_l^{\circ}} = F^{E(k-1)} \tag{8.33}$$

and for the solid phase:

$$\frac{C_s}{C_l^{\circ}} = [E(k-1)+1]F^{E(k-1)} \tag{8.34}$$

for a constant partition coefficient. The effect on the concentrations of elements in the fractionated liquid and solid for different fractionation efficiencies of 0.6 and 1, with $k = 10$ and 0.01 is shown in Fig. 8.20.

It should be noted that if the logarithmic form of equation (8.33) for one element, a, is divided by that for another element, b, then E and $\ln F$ cancel to give:

$$\ln \frac{C_{l,a}}{C_{l,a}^{\circ}} / \ln \frac{C_{l,b}}{C_{l,b}^{\circ}} = \frac{(k_a - 1)}{(k_b - 1)} \tag{8.35}$$

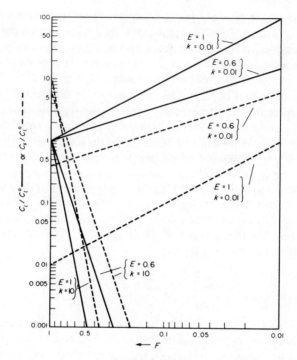

FIG. 8.20. Effect of fractionation efficiency (E) on the relative changes in melt and solid compositions during fractional crystallization for different values of the partition coefficient, k.

or

$$\frac{C_{l,a}}{C_{l,a}^{\circ}} = \left(\frac{C_{l,b}}{C_{l,b}^{\circ}}\right)^{(k_a-1)/(k_b-1)}. \qquad (8.36)$$

Equation (8.36) shows that the concentration ratio (C_l/C_l°) for one element may be determined from the ratio of another element irrespective of the value of fractionation efficiency and the proportion of residual liquid.

8.5.2 Application of theoretical models

There has been widespread use of the fractional crystallization equations in attempts to elucidate the petrogenesis of diversified igneous rocks. A good example of the use of equation (8.29) is to be found in the work of Irvine and Smith (1967) on the petrogenesis of the Muskox layered intrusion, Northwest Territories, Canada. In this intrusion, ultramafic rocks have formed by the fractional crystallization of basaltic magma, so that towards the base of the layered series a thick dunite sequence was developed. The concentrations of nickel determined in rocks taken from over a stratigraphic interval of 520 m define four cycles, each of which contains a steady decline in Ni concentration with stratigraphic height, followed by a relatively abrupt return to a higher concentration value (Fig. 8.21). The cyclic nature of the nickel concentration can be interpreted as resulting from four successive replenishments of magma during the accumulation of the dunite sequence, and the gradual decline in Ni abundance within each cyclic unit results from the fractional crystallization of olivine from a finite amount of magma. If the partition coefficient of nickel between the cumulus minerals and the magma is known, then it is possible to calculate by use of equation (8.29) the mass proportions of magma that gave

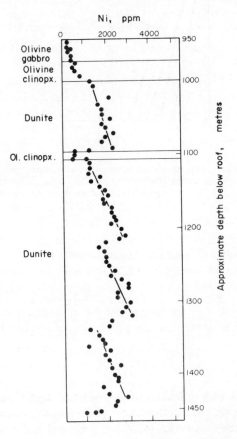

Fig. 8.21. Variation in the Ni content of rocks of part of the Muskox intrusion, Northwest Territories, Canada. (After Irvine and Smith, 1967)

each cyclic unit from the change in Ni concentration. For example, the unit at depths between 1220 m and 1320 m has nickel concentration changing from 3100 to 1700 $\mu g/g$ over a thickness of about 90 m. Irvine and Smith assumed nickel partition coefficients of 10 to 15, which are reasonable on the basis of published data for olivine/melt systems (see Chapter 5). Hence we may write, if $k = 15$, from equation (8.29):

$$\frac{C_s}{C_i^\circ/15} = 15 F^{14},$$

$$\frac{1700}{3100} = F^{14},$$

i.e. $\ln 0.548 = 14 \ln F$

and so $\qquad\qquad\qquad\qquad F = 0.960.$

Therefore this cyclic unit represents about 4 % by mass of the parent magma. The calculation has not allowed for a possible variation in k_{Ni} with change in temperature during fractional crystallization nor for a fractionation efficiency of less than 1, but the result is consistent with geological evidence (Irvine and Smith, 1967) and shows the applicability of the theoretical model to a relatively simple natural system.

Since the degree of fractional crystallization helps to determine the precise concentration trend of an element in both the solid and liquid phases during solidification, it would be advantageous to know the order of magnitude of E (equation (8.32)) during the formation of cumulates. Unfortunately, little quantitative work has been done on this aspect, although a discussion is given by Wager and Brown (1968, pages 64–68). In an attempt (Henderson, 1975) to establish the average fractionation efficiency during crystallization of part of the Skaergaard layered intrusion, use was made of the very low partition coefficient of uranium between cumulus minerals, such as olivine, pyroxene and plagioclase feldspar, and the parent magma. Variations in the uranium concentration in cumulates over small crystallization intervals reflect changes in the proportion of crystallized trapped liquid. An independent estimate of the uranium concentration in the parent magma at the appropriate horizon of the Skaergaard intrusion allowed the determination of the parameter E for a particular cumulate, simply from analysis for its uranium content. Results for rocks taken from a 350-m-long drill-core through part of the lower zone and hidden zone of the intrusion indicated significant variations in E from one cumulate to the next (Fig. 8.22) despite a general uniformity in the rock type. The average efficiency, E_{av}, was about 85 %. The result is interesting in that trapping of magma in a cumulate to the extent of about 15 % of the total cumulate assemblage can add considerably to the concentration of incom-

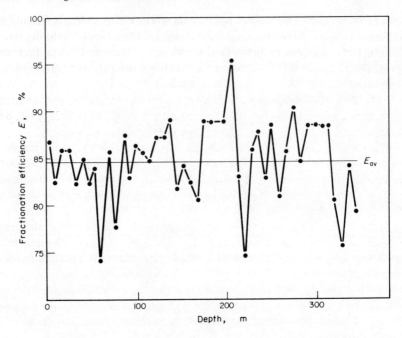

FIG. 8.22. Variation in fractionation efficiency, E, with depth as indicated by cumulates from a 350-m-long drill-core through part of the Skaergaard layered intrusion. The core was taken from the lowest exposed part, and down to about 200 m depth in the hidden part of the intrusion. (After Henderson, 1975)

patible elements in the resultant rock and also to the constituent minerals by virtue of crystallization of the trapped liquid during cooling, with the consequent addition of lower-temperature mineral zones around the initial cumulus phases.

At the present time, equations (8.28) to (8.34) may be used to give semi-quantitative estimates of the extent of possible fractionation mechanisms involved in the petrogenesis of a suite of rocks. Except in simple systems, they cannot be used to give us a quantitative determination of the parameters involved in a particular fractionation history because of a lack of data on the effects of some of the variables mentioned in the introduction to this chapter.

8.6 Equilibrium and fractional fusion

The generation of magmas involves the partial fusion of pre-existing solid mantle or crustal material or the tapping of intergranular fluid already within mantle layers. The compositional changes in solid and liquid that occur during different kinds of fusional process may be defined theoretically if the existence of chemical equilibrium between the phases, at least during part of the processes, is assumed, and furthermore that there is no contamination of the generated melt.

If during partial fusion the liquid that is produced continually reacts and re-equilibrates with the solid residue then equations (8.21) and (8.22) apply, but with C_l° replaced by C_s°. This process is called *equilibrium fusion*.

Equations that define the chemical changes involved in *fractional fusion*— that is, a fusion process in which the liquid, on forming, is immediately isolated—have been derived by Shaw (1970) in a modification of a treatment by Gast (1968). The derivations of the equations given below may be found in Shaw's (1970) paper.

In the fractional fusion process where the mineral phases melt in the same proportions as they occur in the rock, then C_l/C_s° for any element in a particular melting increment when F proportion of liquid has already been removed, is given by:

$$\frac{C_l}{C_s^\circ} = \frac{1}{k}(1 - F)^{(1/k)-1)}$$ (8.37)

and for the residual solid:

$$\frac{C_s}{C_s^\circ} = (1 - F)^{((1/k)-1)}.$$ (8.38)

If the liquids removed during fractional fusion accumulate in a well-mixed magma reservoir then the aggregate liquid composition $\overline{C}_l$ is given by:

$$\frac{\overline{C}_l}{C_s^\circ} = \frac{1}{F}[1 - (1 - F)^{1/k}].$$ (8.39)

Relative changes in concentration for different values of $\overline{k}$, according to equations (8.37), (8.38) and (8.39), are shown in Figs 8.23 and 8.24.

However, the melting of the mineral phases in the same proportion as they exist in the rock is a very improbable event for most polymineralic rocks. Equations (8.37) to (8.39) may be made more general (Shaw, 1970). If different minerals α, β, γ, . . . , etc., melt in proportions p^α, p^β, p^γ, . . . , etc., and if $P = p^\alpha k^\alpha + p^\beta k^\beta + p^\gamma k^\gamma + \ldots$ etc., then the expression analogous to equation (8.37) is:

$$\frac{C_l}{C_s^\circ} = \frac{1}{\overline{k}^\circ}(1 - PF/\overline{k}^\circ)^{(1/P-1)}$$ (8.40)

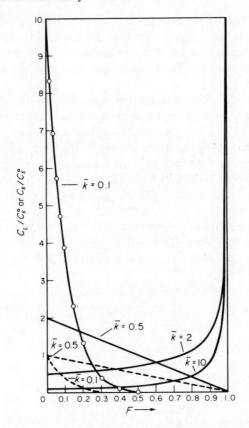

FIG. 8.23. Variation in relative concentration of an element in the liquid as a function of degree of fractional fusion (F), according to equation (8.37), for different bulk partition coefficients $\overline{k}$. The dashed lines are the variations shown by the coexisting solid when $\overline{k} = 0.5$ and 0.1 (equation (8.38)).

and to equation (8.38) is:

$$\frac{C_s}{C_s^\circ} = \frac{1}{(1-F)}(1 - PF/\overline{k}^\circ)^{1/P} \tag{8.41}$$

and to equation (8.39) is:

$$\frac{\overline{C}_l}{C_s^\circ} = \frac{1}{F}\left[1 - (1 - PF/\overline{k}^\circ)^{1/P}\right]. \tag{8.42}$$

In equations (8.40) to (8.42), $\overline{k}^\circ$ is the bulk-solid/liquid partition coefficient at the first moment of melting. These equations, however, are still limited in their

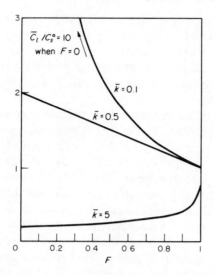

FIG. 8.24. Variation in relative concentration of an element in the aggregate liquid as a function of melt fraction (F) according to equation 8.39 for three values of the bulk partition coefficient, $\bar{k}$

use since they do not allow for the possibility of a mineral being used up, nor for variations in partition coefficients and melting proportions during the melting process. Formulations which encompass these possibilities have been derived by Hertogen and Gijbels (1976) and Apted and Roy (1981); the interested reader is referred to those papers for the details.

At the present time, uncertainties about many of the parameters involved in fractional fusion seriously restrict the reliable use of most of the theoretical models. We do not have an adequate knowledge of the variation of element partition coefficients with changes in temperature, pressure or composition, and usually we can only estimate the mineralogy of the source material. Furthermore, during melting, elements may be moved to or from the magma by fluids which could be hydrous and relatively rich in carbon dioxide and the halogens or by other contamination and mixing mechanisms. (Shaw (1978) has treated theoretically the behaviour of trace elements during anatexis in the presence of a fluid phase.) This is not to say that the models are useless, since in practice they have proved to be helpful in the testing of petrogenetic hypotheses for igneous rocks in many different tectonic settings. Where element concentrations in a melt are very sensitive to the extent of fusion of a wide range of likely source materials, such as shown by the incompatible elements (also see Figs 8.23 and 8.24), the theory can provide a good indication of the extent of fusion and whether the parent magmas of the members (e.g. lavas) of a rock suite are genetically related through different degrees of partial melting of the same source. One example, amongst many, of a discussion on the application of petrogenetic models can be found in a paper by Hanson (1978) on the use of trace elements in elucidating the petrogenesis of some granitic rocks.

Other mechanisms besides those described above could be important in magma genesis and in changing a melt's composition. A process equivalent to *zone refining* is one possibility. In this, a succession of 'zones' of magma advance through a solid by causing melting at each zone's front and with crystallization and deposition behind (Fig. 8.25). This process would strongly affect the distribution of incompatible elements since these will be preferentially partitioned into the melt and carried forwards by each advancing zone. Analytical expressions which describe changes in element concentrations have been derived for one zone pass and also for the ultimate distribution but not for an intermediate number of passes. Good evidence for the occurrence of zone refining in magma genesis is still wanting, but the mechanisms cannot yet be discounted. However, little attention has been given to it by petrologists.

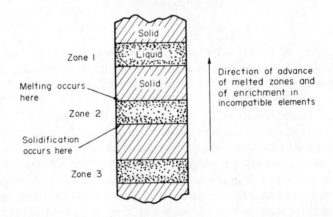

FIG. 8.25. Representation of zone refining showing three zones of liquid advancing through a solid.

Suggested further reading

ALLEGRE, C. J. and MINSTER, J. F. (1978) Quantitative models of trace element behavior in magmatic processes. *Earth Planet. Sci. Lett.* **38**, 1–25.
ATHERTON, M. P. (1976) Crystal growth models in metamorphic tectonites. *Phil. Trans Roy. Soc. London*, A, **283**, 255–270.

DOWTY, E. (1980) Crystal growth and nucleation theory and the numerical simulation of igneous crystallization. In *Physics of Magmatic Processes* (ed. R. B. HARGRAVES). pp. 419–485. Princeton Univ. Press.

FISHER, G. W. (1977) Nonequilibrium thermodynamics in metamorphism. In *Thermodynamics in Geology* (ed. D. G. FRASER), pp. 381–403. Reidal Publishing Co.

FYFE, W. S., PRICE, N. J. and THOMPSON, A. B. (1978) *Fluids in the Earth's Crust*. Elsevier Scientific Publishing Co. Chapter 5. 383 pp.

HOFMANN, A. W. (1980) Diffusion in natural silicate melts: a critical review. In *Physics of Magmatic Processes* (ed. R. B. HARGRAVES), pp. 385–417. Princeton Univ. Press.

KIRKPATRICK, R. J. (1975) Crystal growth from the melt: a review. *Am. Mineral.* **60**, 798–814.

PUTNIS, A. and MCCONNELL, J. D. C. (1980) *Principles of Mineral Behaviour*. Blackwell Scientific Publishing. Chapters 5 to 8. 157 pp.

9

Geochemical Applications of Isotope Distribution

9.1 Introduction

The advent of the mass spectrometer was the enabling factor that allowed geologists to determine the ratios of isotope abundances in rocks and minerals. Initially the aims were to determine the age of the Earth and the age relationships of rocks. For both purposes radioactive nuclides with a sufficiently wide range of half-lives occur naturally, and include ^{238}U with a half-life of 4.47 Ga, which decays through a chain of daughter products ending in the stable product ^{206}Pb, and ^{14}C with a half-life of 5730 a, which may be used to date certain, very recent deposits. However, more than the ages of rocks and minerals may be established from the use of isotopes. The abundance ratios of both stable and radioactive nuclides are of particular value in helping to solve a wide range of petrogenic problems and in giving insights into some geochemical processes. These include identification of magma sources, the extent to which rocks may have interacted with natural waters, geothermometry and as indicators of the provenance of ore-forming fluids. As the sensitivity and precision of the analytical methods has improved, so has the field of application of isotope studies expanded to become one of the more exciting and fruitful areas of research.

The purpose of this chapter is to review briefly some of the applications of isotope studies to the general problems of rock composition and element distribution. The topic of geochronology will receive only passing attention; there are a number of excellent texts on this subject, some of which are cited at the end of this chapter.

The field of study of isotope geochemistry falls naturally into two parts, one dealing with radioactive nuclides and their daughter products, the other with stable isotopes. The main principles behind each are clearly different. The first is essentially concerned with determining and interpreting the abundances of the parent and daughter products and requires a knowledge of radioactive decay rates. The second, through determination of the abun-

228

dance ratios of stable isotopes of one and the same element, sets out to explain how natural processes cause isotopic fractionation and the uses to which this may be put. There are, of course, areas where the two studies overlap, especially in topics on the mixing of naturally occurring materials. Here the essential requirement is that each contributor to the final product has sufficiently distinctive characteristics for use as an indicator in some way.

9.2 Radioactive nuclides: some principles and occurrences

The decay of a radioactive substance follows the exponential relationship (9.1)

$$N = N_0 e^{-\lambda t}, \tag{9.1}$$

where N is the number of unchanged atoms at time t and N_0 is the number present when $t = 0$ and λ is a constant, which is characteristic of the particular radioactive species. λ is called the decay constant and has the dimensions of reciprocal time.

The rate of a radioactive decay is conveniently expressed in terms of the half-life $t_{\frac{1}{2}}$, which is the time required for an initial number of atoms to be reduced to half that number by the decay process, i.e. when $t = t_{\frac{1}{2}}$, $N = N_0/2$. Therefore,

$$t_{\frac{1}{2}} = \ln 2/\lambda. \tag{9.2}$$

In many cases a radioactive species decays directly to a stable 'daughter' product, but in others a chain of radioactive daughter products may be involved before a final stable one is reached. Such is the case with the important naturally occurring nuclides of uranium and thorium: ^{238}U, ^{235}U and ^{232}Th, Tables 9.1 to 9.3.

Some other naturally occurring radioactive nuclides are listed in Table 9.4. The Table includes ^{14}C, which is produced in the upper atmosphere as a result of cosmic ray bombardment. The others in Table 9.4 were generated during nucleosynthetic events prior to the formation of the Earth (see Chapter 2).

The short half-life of ^{14}C makes it a useful nuclide in the dating of archaeological deposits; Faure (1977) gives a brief review of ^{14}C dating. The principal nuclides which are used in petrology and geochemistry are ^{40}K, ^{87}Rb, ^{147}Sm, ^{232}Th, ^{235}U and ^{238}U. Rhenium-187 has been used with limited success in age studies of iron meteorites. There has been a problem with the use of ^{87}Rb in geochronology and related studies because of difficulties in the accurate measurement of its decay constant, resulting from the large number of very low-energy beta particles emitted during decay. This problem now appears to have been overcome; the value given in Table 9.4 is that determined by Davis *et al.* (1977).

TABLE 9.1. *Decay chain of* ^{238}U

Element	Uranium – 238 series						
Uranium 92	^{238}U 4.47 x 10⁹ a	(99.85%)	^{234}U 2.44 x 10⁵ a				
Protactinium 91	α β	^{234}Pa 1.18 m / $\frac{235$Pa}{$ }$ 6.75 h β β	α				
Thorium 90	^{234}Th 24.10 d		^{230}Th 7.7 x 10⁴ a				
Actinium 89			α				
Radium 88			^{226}Ra 1600 a				
Francium 87			α				
Radon 86			^{222}Rn 3.824 d				
Astatine 85			α	^{218}At ~ 2 s			
Polonium 84			^{218}Po 3.05 m β	α	^{214}Po 164 μs β		^{210}Po 138.4 d β
Bismuth 83			(99.98 %) α	^{214}Bi 19.8 m	(99.96%) α	^{210}Bi 5.01 d β	α
Lead 82			^{214}Pb 26.8 m	α β	^{210}Pb 22.3 a	(5x10⁻⁵ %) α β	^{206}Pb Stable
Thallium 81				^{210}Tl 1.3 m	β	^{206}Tl 4.20 m	β

The use of radioactive nuclides in geological studies usually entails a modified form of equation (9.1):

$$N_d = N(e^{\lambda t} - 1), \qquad (9.3)$$

where N_d, the number of daughter atoms, is equal to $N_0 - N$. Hence the age (t)* of a mineral can, in principle, be determined from (9.3) if the numbers of daughter, or *radiogenic*, atoms and remaining radioactive parent atoms are known, together with the appropriate decay constant. In practice the mineral

* t is the period of time during which radioactive decay has given N_d atoms; it can represent the time that a rock has been in existence. Hence, t is a positive number.

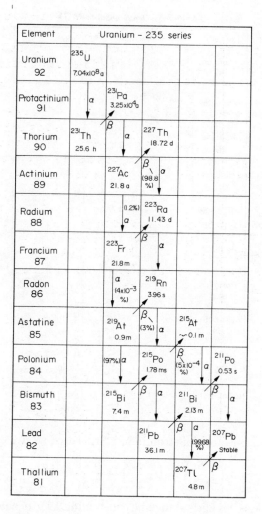

Element	Uranium – 235 series			
Uranium 92	^{235}U 7.04x10^8 a			
Protactinium 91	α	^{231}Pa 3.25x10^4 a		
Thorium 90	^{231}Th 25.6 h	β α	^{227}Th 18.72 d	
Actinium 89		^{227}Ac 21.8 a	β (98.8 %) α	
Radium 88		(1.2%) α	^{223}Ra 11.43 d	
Francium 87		^{223}Fr 21.8 m	β α	
Radon 86		α (4x10^{-3} %)	^{219}Rn 3.96 s	
Astatine 85		^{219}At 0.9 m	β (3%) α	^{215}At ~0.1 m
Polonium 84		(97%) α	^{215}Po 1.78 ms	β (5x10^{-4} %) α	^{211}Po 0.53 s
Bismuth 83			^{215}Bi 7.4 m	β α	^{211}Bi 2.13 m	β
Lead 82			^{211}Pb 36.1 m	β α (99.68 %)	^{207}Pb Stable
Thallium 81				^{207}Tl 4.8 m	β

may have incorporated some previously formed daughter atoms in its structure at the time of its formation. Thus, using ^{87}Rb as an example, we have (9.4)

$$^{87}Sr_p = {}^{87}Sr_0 + {}^{87}Rb(e^{\lambda t} - 1) \qquad (9.4)$$

where $^{87}Sr_p$ and $^{87}Sr_0$ are the numbers of ^{87}Sr atoms present now and initially (i.e. t years ago) respectively.

Also, it is often more convenient in mass spectrometric analysis to determine the ratios of isotopes rather than their absolute abundances. In the example

TABLE 9.3. *Decay chain of* ^{232}Th

Element	Thorium – 232 series				
Thorium 90	^{232}Th 1.39 × 10^{10} a		^{228}Th 1.913 a		
Actinium 89	α	^{228}Ac 6.13 h	β ↗ α		
Radium 88	^{228}Ra 5.78 a	β	^{224}Ra 3.64 d		
Francium 87			α		
Radon 86			^{220}Rn 55.6 s		
Astatine 85			α		
Polonium 84			^{216}Po 0.15 s (66.3%)		^{212}Po 0.3 μs
Bismuth 83			α ^{212}Bi 60.55 m	β ↗ α	
Lead 82			^{212}Pb 10.64 h	β (33.7%) α	^{208}Pb Stable
Thallium 81				^{208}Tl 3.054 m	β

above, the stable isotope of strontium, ^{86}Sr, is used to give equation (9.5)

$$\left(\frac{^{87}\text{Sr}}{^{86}\text{Sr}}\right)_p = \left(\frac{^{87}\text{Sr}}{^{86}\text{Sr}}\right)_0 + \frac{^{87}\text{Rb}}{^{86}\text{Sr}}(e^{\lambda t} - 1). \qquad (9.5)$$

^{86}Sr is suitable as it is fairly abundant, constituting 9.9 % of natural strontium, and is not the product of any known radioactive decay series. Equation (9.5) may be solved simultaneously for the age, t, of the rock and the initial ratio of strontium $(^{87}\text{Sr}/^{86}\text{Sr})_0$ at t years ago, by the analysis for $(^{87}\text{Sr}/^{86}\text{Sr})_p$ and $^{87}\text{Rb}/^{86}\text{Sr}$ of two or more minerals with differing rubidium contents, from the same rock system, Fig. 9.1.

9.3 Radioactive nuclides: petrogenesis

The chemical fractionation processes operating in the Earth lead to wide variations in the concentrations of the radioactive elements. For example,

TABLE 9.4. *Naturally occurring radioactive nuclides*

Active nuclide	Disintegration mode	Decay const. $\lambda(a^{-1})$	Half-life $t_{\frac{1}{2}}$ (a)	Isotopic abund., %	Stable products
^{14}C	β	1.21×10^{-4}	5730	—	^{14}N
^{40}K	β,EC	5.480×10^{-10}	1.27×10^9	0.018	^{40}Ca, ^{40}Ar
^{50}V	β,EC	1.16×10^{-16}	6×10^{15}	0.24	^{50}Cr, ^{50}Ti
^{87}Rb	β	1.419×10^{-11}	4.89×10^{10}	27.85	^{87}Sr
^{115}In	β	1.39×10^{-15}	5×10^{14}	95.72	^{115}Sn
^{123}Te	EC	5.78×10^{-14}	1.2×10^{13}	0.87	^{123}Sb
^{138}La	EC,β	6.30×10^{-12}	1.1×10^{11}	0.089	^{138}Ba, ^{138}Ce
^{142}Ce	α	$\sim 1.4 \times 10^{-17}$	$\sim 5 \times 10^{16}$	11.07	^{138}Ba
^{144}Nd	α	2.89×10^{-16}	2.4×10^{15}	23.85	^{140}Ce
^{147}Sm	α	6.54×10^{-12}	1.06×10^{11}	14.97	^{143}Nd
^{152}Gd	α	6.30×10^{-15}	1.1×10^{14}	0.20	^{148}Sm
^{176}Lu	β	1.98×10^{-11}	3.5×10^{10}	2.59	^{176}Hf
^{174}Hf	α	3.5×10^{-16}	2×10^{15}	0.18	^{170}Yb
^{187}Re	β	1.61×10^{-11}	4.3×10^{10}	62.5	^{187}Os
^{190}Pt	α	1×10^{-12}	7×10^{11}	0.013	^{186}Os
^{232}Th	chain	4.990×10^{-11}	1.39×10^{10}	100	^{208}Pb
^{234}U	chain	2.794×10^{-6}	2.44×10^5	0.0055	^{206}Pb
^{235}U	chain	9.8485×10^{-10}	7.04×10^8	0.72	^{207}Pb
^{238}U	chain	1.5512×10^{-10}	4.47×10^9	99.28	^{206}Pb

EC = electron capture.

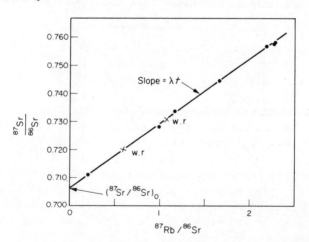

FIG. 9.1. Rb–Sr isochron plot. The age of the rock system is obtained from the slope of the line (isochron).

uraniferous ore deposits may contain several weight per cent of uranium in contrast to (say) dunites which contain only about $10 \, ng \, g^{-1}$. Hence, the production of radiogenic lead will also show considerable variation. If the rocks or ore deposits are subsequently reworked or remobilized and contribute to another, new, deposit, the derived isotopic ratios may be useful in identifying the source or sources of the younger material, provided that the genetic process has been simple. The rubidium–strontium method also has been used in this way.

Rubidium has an ionic charge and size that make it an incompatible element during fractionation of a basalt magma (see Chapter 6). It tends to reside in the melt phase until it is concentrated in the final, more siliceous products. A related distribution pattern for the element is found in the Earth as the crust contains significantly more Rb than the upper mantle (see Chapter 4). Hence the rate of generation of radiogenic ^{87}Sr will be greater in the crust than in the upper mantle. The rate of increase of the $^{87}Sr/^{86}Sr$ ratio of any system which is closed to Rb and Sr is clearly proportional to its Rb/Sr ratio. The value of $^{87}Sr/^{86}Sr$ of the upper mantle at the time of formation of the Earth was probably close to that recorded from basaltic achondrites, i.e. 0.698 98 $\pm 0.000\,03$ (Papanastassiou and Wasserburg, 1969) and is considered to have evolved to a present-day value of 0.7037 ± 0.0002 as indicated by the Sr-isotopic composition of oceanic tholeiites (Cox *et al.*, 1979). These two values define the Rb/Sr ratio of the upper mantle as 0.024 ± 0.001, Fig. 9.2.

A magma extracted from the mantle at any time will have a $(^{87}Sr/^{86}Sr)_0$ ratio that lies along the growth curve OP of Fig. 9.2 if the mantle is iso-

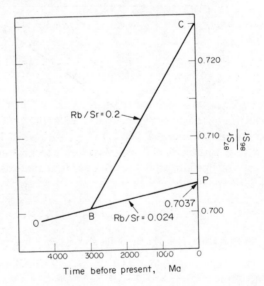

FIG. 9.2. Diagrammatic representation of the variation in $^{87}Sr/^{86}Sr$ isotope ratio with time. The line OP shows the ratio change for an isotopically homogeneous mantle. Line BC shows the change for a rock derived from the mantle 3000 years ago and with a Rb/Sr concentration ratio of 0.2.

topically homogeneous, but the Rb/Sr of this liquid is likely to be different from the mantle's value. A high Rb/Sr value will lead to the relatively rapid growth of $^{87}Sr/^{86}Sr$ as depicted by line B–C on Fig. 9.2. Similarly, the sialic crust which differentiated from the mantle at some early time in the Earth's history will now have a relatively high $^{87}Sr/^{86}Sr$ ratio.

These simple considerations concerning the evolution of isotopic ratios have been of use in helping to establish the provenance of granitic magmas. A granite could be derived from the upper mantle either by partial melting of mantle material or from fractional crystallization of a mantle-derived basic magma, or it could come from the fusion of sialic crust. Granites derived by crustal anatexis could have a relatively high initial $^{87}Sr/^{86}Sr$ ratio. A low initial ratio (0.704 or less) indicates that the granite was derived from the mantle or from a recent mantle-derived precursor.

The application of the above reasoning contains a number of assumptions. Three important ones are: the mantle is isotopically homogeneous; granites are derived from the mantle or crust through simple, single-stage processes; and isotopic ratios in rocks are not affected by post-crystallization processes. Evidence suggests that at least the first two assumptions may be invalid, but it seems as though the broad differences in isotopic ratios of possible source materials are sufficient to allow strontium isotopes to be used in many situations as a petrogenetic indicator (Cox *et al.*, 1979). Hence, Blaxland *et al.* (1978) have suggested a mantle derivation of the Proterozoic granites and syenites of the Gardar Province, southwest Greenland, on the basis of their low initial ratios $[(^{87}Sr/^{86}Sr)_0 = 0.702$ to $0.704]$ and Gunner (1974) claims that some quartz monzonites with high initial $^{87}Sr/^{86}Sr$ ratio of 0.734 from the Martin Dome of the Miller Range in the Transantarctic Mountains were produced from a magma which either was derived by partial melting of the granitic basement of the area or assimilated large amounts of old sialic rocks. On the other hand, Faure and Powell (1972) have shown that a few large Californian granite batholiths of North America, which from other evidence were almost certainly produced as a result of crustal anatexis, have relatively low initial $^{87}Sr/^{86}Sr$ ratios of about 0.707.

Granites produced through the fusion of sediments and other crustal materials may contain very different trace-element concentrations from those derived from a mantle source. Some of the physico-chemical conditions of melt generation and crystallization (e.g. P_{H_2O}) are also likely to have been different and may be reflected in the nature of element partition between mineral phases. Isotope studies in combination with other geochemical methods should help our understanding of the processes affecting element distribution.

The possibility of mantle heterogeneity alluded to above has been the subject of a number of studies including, recently, the use of Nd isotopes. Samarium-147 decays by alpha emission to ^{143}Nd, and it is possible to apply the same principles used in the Rb–Sr method to these isotopes. Relationship (9.6)

$$\left(\frac{^{143}Nd}{^{146}Nd}\right)_p = \left(\frac{^{143}Nd}{^{146}Nd}\right)_0 + \frac{^{147}Sm}{^{146}Nd}(e^{\lambda t} - 1) \tag{9.6}$$

may be used to determine the age of a rock, but the long half-life of ^{147}Sm (Table 9.4) precludes the use of this method except for Archaean ($> 2.5\,Ga$) rocks. However, the isotope ratios can give some information on the sources of magmas of more recent rocks. A source region with a chondrite-normalized rare-earth concentration pattern that shows light REE depletion will undergo a more rapid growth in its $^{143}Nd/^{146}Nd$ ratio because of its higher Sm/Nd ratio than one with light REE enrichment and a similar Nd content. For example, small but none the less significant differences in the $^{143}Nd/^{146}Nd$ ratios of some oceanic alkali basalts (values $= 0.7080$) and tholeiite basalts (values $= 0.7083$) studied by Richard *et al.* (1976) indicate that the two basalt types were derived from mantle regions with different Nd/Sm ratios. These regions must have been chemically distinct for a time long enough to create the observed isotopic difference.

9.4 Stable isotopes: some principles

Most of the naturally occurring elements consist of more than one stable isotope. The isotopes of any one element show the same general chemical behaviour but there are subtle and important differences in isotope partition as a result of differences in the isotopic masses. This effect is more pronounced when the mass differences are of a significant magnitude in relation to the mass of the isotope (usually 5% or greater). Fractionation of isotopes through natural processes has been detected in elements up to a mass of about 40 (i.e. calcium) and possibly in the case of selenium (where fractionations of ^{76}Se from ^{82}Se should be similar to those occurring in sulphur). Table 9.5 lists the abundances of the naturally occurring isotopes of the light elements which have been used to varying extents in stable isotope geochemistry.

The fractionation of isotopes arises in the first place from the fact that the modes of molecular motion and hence the energy of a compound depend upon the masses of the constituent atoms. The vibrational, translational and rotational components of molecular motion are all affected by differences in the masses of the atoms and therefore affect isotope fractionation. Only the vibrational component of the fractionation is temperature-dependent and this leads to the use of stable isotopes in geothermometry. As translational and rotational motions do not usually occur in solids, it is the vibrational component that controls isotope distribution in minerals.

TABLE 9.5. *Abundances (%) of selected stable isotopes*

isotope	Abundance, %	Isotope	Abundance, %
^{1}H	99.9844–99.987	^{28}Si	92.27
^{2}D	0.013–0.0156	^{29}Si	4.68
		^{30}Si	3.05
^{6}Li	7.52		
^{7}Li	92.48	^{32}S	95.02
		^{33}S	0.75
^{10}B	18.98	^{34}S	4.21
^{11}B	81.02	^{36}S	0.02
^{12}C	98.893	^{39}K	93.10
^{13}C	1.107	^{41}K	6.88
^{14}N	99.64	^{40}Ca	96.97
^{15}N	0.36	^{42}Ca	0.64
		^{43}Ca	0.145
^{16}O	99.763	^{44}Ca	2.06
^{17}O	0.0375	^{46}Ca	0.0033
^{18}O	0.1995	^{48}Ca	0.185
^{24}Mg	78.8		
^{25}Mg	10.15		
^{26}Mg	11.06		

The main processes causing isotopic fractionation in nature are:

(a) isotopic exchange reactions;

(b) kinetic processes in which the reaction rates are a function of the isotopic compositions of the reactants and products;

(c) physico-chemical processes including evaporation/condensation, melting/crystallization and diffusion.

This chapter is concerned principally with processes of type (a).

In geochemical studies the fractionation of isotopes is measured as the relative difference in the ratio of the isotopes in the sample under study to that in a standard. This relative difference is given the symbol δ, and for any sample, A, is defined as:

$$\delta_A = \left(\frac{R_A - R_{std}}{R_{std}} \right) 10^3, \tag{9.7}$$

where R_A = isotopic ratio in A, the heavier isotope as the numerator (e.g. ^{18}O/^{16}O; ^{34}S/^{32}S). Therefore, the delta-value, δ (or 'del value' as it is sometimes called) is in parts per thousand or per mill (‰). It is now possible to measure δ-values with precisions of better than ± 0.5‰ for hydrogen and often ± 0.05‰ for most other light elements.

Two relationships which are of use are the difference between the δ-values of two phases A and B:

$$\Delta_{A-B} = \delta_A - \delta_B, \tag{9.8}$$

and the isotopic fractionation factor, α, between A and B, defined as:

$$\alpha_{A-B} = R_A/R_B \tag{9.9}$$

or

$$\alpha_{A-B} = (1000 + \delta_A)/(1000 + \delta_B). \tag{9.10}$$

The values of fractionation factors are usually close to unity and show variation in the third decimal place. Therefore, it is possible to use the approximation $10^3 \ln 1.00X \simeq X$, which is convenient in isotope geothermometry (see below).

The reference standards are agreed internationally and are given in Table 9.6. They are either natural or hypothetical samples, with their isotopic composition being taken by convention or definition. Experimentally used standards are defined in relation to an internationally accepted one. For example, SMOW (see Table 9.6) was originally defined in terms of a National Bureau of Standards reference water NBS-1 as follows:

$$({}^{18}O/{}^{16}O)_{SMOW} = 1.008\,({}^{18}O/{}^{16}O)_{NBS-1}.$$

TABLE 9.6. *International Reference Standards for Reporting Stable Isotope Fractionations*

Isotopic Ratio	Standard	Abbreviation
${}^{18}O/{}^{16}O$	Standard Mean Ocean Water Hypothetical	SMOW
${}^{2}D/{}^{1}H$	Standard Mean Ocean Water	SMOW
${}^{13}C/{}^{12}C$	Carbonate of belemnite from Peedee formation (Cretaceous) of Carolina. Now exhausted	PDB
${}^{34}S/{}^{32}S$	Troilite from Cañon Diablo meteorite (by convention ${}^{34}S/{}^{32}S = 0.045\,004\,5$)	CD

9.5 Isotope geothermometry

The equilibrium value of the fractionation factor, α (equation (9.9)), is a function of temperature but is unaffected by pressure variations except in systems involving a gas phase. However, the fractionation of hydrogen isotopes is dominated by variations in the proportions of common elements, especially Mg, Al and Fe, in the mineral phases and it is not possible to use these isotopes as a geothermometer. Most isotope geothermometric studies have used oxygen or sulphur isotope fractionations and the following account concentrates on these.

The measurement by Urey *et al.* (1951) of temperatures of some marine waters of the Mesozoic from data on oxygen isotopes in fossil belemnites was the first important work in the use of isotopic fractionation in geo-thermometry. Urey and his co-workers were able to show, with the use of an empirical temperature calibration and isotope data from the individual growth rings of a Mesozoic belemnite guard from the Isle of Skye, Scotland, that there were cyclic variations between 25° and 20° C in the temperature of the water in which the belemnite lived. They showed that the oxygen isotope composition of the shell carbonate was established under equilibrium conditions and that subsequent alteration was absent. These workers were concerned with temperatures at which isotope fractionation is relatively marked, but even so their work was a considerable achievement. Later technological developments in isotope measurement, together with experimental work on temperature calibrations, has meant that under favourable conditions oxygen isotope geothermometry now can be used up to 1200°C.

Isotope geothermometry is based mainly on empiricially established temperature scales, despite the problem of sluggish reaction rates, as theoretical calibrations are often insufficiently precise. The temperature relationship is of a simple form (9.11), at the high temperatures of igneous and metamorphic systems:

$$\ln \alpha \propto T^{-2}, \tag{9.11}$$

where T is in kelvins. Because of the useful mathematical fact that $10^3 \ln (1.00 X) \simeq X$, and also because the plotting of experimentally determined fractionation factors as $10^3 \ln \alpha$ against T^{-2} often yields straight lines, expression (9.12) is used:

$$10^3 \ln \alpha = A(10^6 T^{-2}) + B, \tag{9.12}$$

where A and B are experimentally determined constants for each isotope exchange system. Therefore, $10^3 \ln \alpha$ is the per mil fractionation, and as temperature increases so its value tends to zero (i.e. α tends to 1). Minerals with OH groups often show a more complicated temperature dependence than that given by (9.12).

Figure 9.3 shows mineral–water fractionations for oxygen isotopes as a function of temperature as determined from hydrothermal experiments and, in some cases, from theoretical considerations. These data are useful not only in their own right but because mineral–mineral fractionations may be readily computed from them, as the per mil fractionation, $10^3 \ln \alpha$, is very well approximated by the Δ value:

$$\Delta_{A-B} = \delta_A - \delta_B \simeq 10^3 \ln \alpha_{A-B}. \tag{9.13}$$

For example, subtraction of the fractionation expression for magnetite–water from that for quartz–water gives that for quartz–magnetite. The resulting calibration may then be applied to the co-existing mineral pair in a rock to determine the temperature of formation of the minerals or, more correctly, the temperature at which isotopic equilibrium was reached and below which no further isotopic exchange between the minerals took place. The interpretation of the determined temperatures is likely to require care and a good understanding of the nature of formation of the mineral assemblage.

Mineral pairs which show strongly diverging fractionation lines in relation to water (Fig. 9.3) will make good geothermometers. Hence, Fig. 9.3 shows that the quartz–magnetite mineral pair is one of the most sensitive. As early as 1962, Taylor and Epstein showed that, of a number of minerals, quartz has the greatest tendency to concentrate ^{18}O, while magnetite has the lowest. Later work has shown the order to be: rutile, quartz, dolomite, alkali-feldspar,

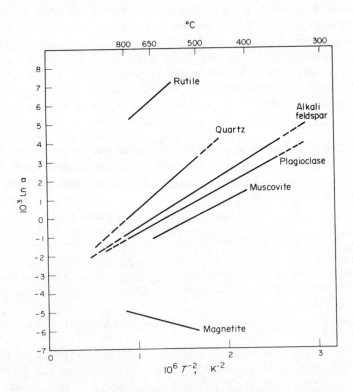

FIG. 9.3. Mineral-water fractionations for oxygen isotopes as a function of temperature. (The fractionation factor, α, is defined by equation (9.9).) (Data mainly from Friedman and O'Neil, 1977)

calcite, intermediate plagioclase, muscovite, anorthite, pyroxene, hornblende, olivine, garnet, biotite, chlorite, ilmenite, magnetite. This sequence is reflected in the fractionation of oxygen isotopes as a function of temperature for selected mineral pairs in Fig. 9.4. These fractionations were established either from fractionation data on mineral–water isotope exchange or from direct experiment. Anderson *et al.* (1971) used the plagioclase–magnetite pair to establish temperature relationships of some mafic igneous rocks and they estimated that the uncertainty in the temperature determination is $\pm 25°C$ at $800°C$ and $\pm 50°C$ at $1150°C$ if the analytical error is $\pm 0.1\%$.

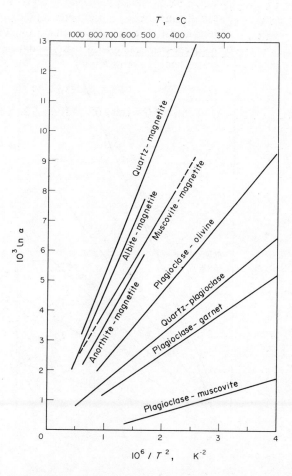

FIG. 9.4. Oxygen isotope fractionation for mineral pairs as a function of temperature. (Data from Friedman and O'Neil, 1977)

Figure 9.5 shows experimentally determined sulphur isotope fractionations between some mineral sulphides. It can be seen that the order of sensitivity is pyrite–galena > sphalerite–galena > pyrite–chalcopyrite > pyrite–sphalerite. A sulphate–sulphide mineral pair is more sensitive than any sulphide–sulphide pair. Temperatures may be determined to within ± 40°C (Rye and Ohmoto, 1974), but the sequence of crystallization has to be suitable and well understood. For example, the mineral pair pyrite–galena rarely yields a reliable temperature determination as pyrite can be precipitated over a much larger temperature range than the galena.

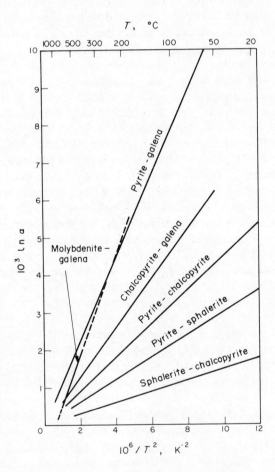

FIG. 9.5. Sulphur isotope fractionation for selected mineral pairs as a function of temperature. (Data from Friedman and O'Neil, 1977)

If three or more minerals coexist then it is possible to check for concordancy in the temperatures determined from each mineral pair. Concordancy is a necessary condition of equilibrium but its attainment does not appear to be common in many igneous and metamorphic rocks (Deines, 1977). Isotope exchange during rock cooling is an important process and probably takes place more readily between some minerals than others. Before isotope geothermometry can be applied with confidence, it is necessary to confirm that there is concordancy within, at least, a mineral triplet.

The application of isotope geothermometry has already been quite extensive. As some examples, the following are given:

(a) A study (Shieh and Taylor, 1969) of temperature variation across a contact metamorphic aureole in carbonate rocks at Birch Creek, California, USA. The temperatures determined from the quartz–biotite mineral pair show a systematic decrease from about 540°C with increasing distance from the contact.

(b) Oxygen isotope fractionations in quartz–magnetite and plagioclase–magnetite pairs from granulites of a metamorphic aureole around an intrusive charnockitic adamellite in the Musgrave Ranges, Central Australia yield concordant temperatures of about 550°C (Wilson *et al.*, 1970) but the temperature from the pyroxene–magnetite pair is discordant, presumably as a result of retrogression.

(c) The $^{18}O/^{16}O$ isotope ratios in ordinary chondrite meteorites increases systematically from H through L to LL types (see Chapter 1 for meteorite classification). The isotope fractionations between plagioclase–pyroxene and between pyroxene–olivine give a concordant temperature estimate of about 950°C $\pm$ 100°C (Onuma *et al.*, 1972a). This is a reasonable temperature of metamorphism if the system contained no water. However, studies by Onuma *et al.* (1972b) on carbonaceous chondrites have shown that the oxygen isotope distribution was not at equilibrium. The authors suggest that the isotope ratios in the various phases of these meteorites were established by equilibration at different temperatures for each phase with the cooling gas of the solar nebula. From an estimated $\delta^{18}O$ value of $-1 \pm 2\%$ (SMOW) for the solar nebular gas, estimates were made of the temperatures of formation of four groups representing different stages in the removal of solids from the nebular gas:

(i) $\delta^{18}O$: -6 to -12%. Iron-poor olivine and pyroxene in C2 meteorites and Ca, Al-silicate aggregates in C3 meteorites, condensed at temperatures > 1000 K.

(ii) $\delta^{18}O$: -2 to $+2\%$. Iron-rich olivine and pyroxene in the chondrules of C2 and C3 meteorites formed at temperatures of 530–620 K.

(iii) $\delta^{18}O$: $+4$ to $+6°/_{oo}$. A group comprising the Earth, Moon, achondrites and the chondrules of ordinary chondrites, which separated from the nebular gas at 450–470 K.

(iv) $\delta^{18}O$: $+11$ to $+27°/_{oo}$. Hydrous silicates in C2 matrix and bulk of Cl meteorites, formed in temperature range 350–400 K.

(d) The zinc–lead deposits of the Upper Mississippi Valley, USA, provide excellent material for the application of sulphur isotope geothermometry. There, banded ores of a low temperature of deposition are in a readily determinable paragenetic sequence and have not suffered any metamorphism. The co-precipitated minerals—sphalerite, galena and marcasite—yield temperatures of 150° to 260°C for the early deposits and 100°C or less for the later ones (Pinckney and Rafter, 1972). These results are in agreement with temperature determinations from fluid inclusions within the ore minerals.

Sulphur isotopes also can provide important information on some other chemical conditions of ore deposition. This aspect is discussed below.

Isotope geothermometry, in combination with the thermodynamic principles of Chapter 7, provides a powerful tool for helping to establish the nature of element distribution in rocks and ore-deposits. The ability to check for isotopic equilibrium through temperature concordancy offers a useful guide to the relevance of applying other chemical equilibria principles to the system in question. Furthermore, as isotopic fractionation in solids and liquids is insensitive to pressure, isotope geothermometry can provide an independent temperature estimate, so allowing the application of other thermodynamic geobarometers that are not totally unaffected by temperature (see Chapter 7).

9.6 Further applications of stable isotopes

9.6.1 *Magma genesis*

The distribution of stable isotopes during magmatic differentiation is subject to the same controlling factors, described in Chapter 8, that affect element distribution. These factors include the type and sequence of crystallization and the efficiency of fractionation. However, during differentiation of a basic magma the oxygen isotope ratio undergoes little change except during the late stages. The lack of change results mainly from the high temperatures of crystallization. There is a tendency for the $\delta^{18}O$ values to increase with increase in silica content of an igneous rock. Thus, unaltered ultramafic rocks have $\delta^{18}O$ values of $+5$ to $+7°/_{oo}$; gabbros, basalts, andesites and syenites, $+5.5$ to $+7.5°/_{oo}$, and granites $+7$ to $+13°/_{oo}$. The sequence of mineral crystallization can affect isotope ratios. For example, the early crystallization of magnetite will tend to enrich the magma in ^{18}O compared with a magma undergoing crystal fractionation under such a low oxygen fugacity that magnetite formation is prevented.

Taylor (1968, 1974) has reported on oxygen and hydrogen isotope analyses of a substantial number of igneous rock and mineral samples from various localities throughout the world. Most volcanic and plutonic rocks are very uniform in $^{18}O/^{16}O$ (commonly with $\delta^{18}O$ values of 5.5 to $10.0^o/_{oo}$) and D/H ratios (commonly δD values of -50 to $-80^o/_{oo}$), as seen in Figs 9.6 and 9.7. The uniformity stems from the similarity in the majority of magma sources and the frequently close relationship of one igneous rock type to another, occasioned by such processes as fractional crystallization. Stable isotopes can tell us little about the petrogenetic details of a rock that has the common isotopic ratios; their use tends to be restricted to situations where the ratios deviate from the isotopic norm defined by the majority of igneous rocks of a similar kind.

Igneous rocks with isotope ratios that lie outside the common ranges are likely to have been derived from relatively unusual sources, or were contaminated by material of dissimilar isotopic composition, or have undergone hydrothermal alteration. This last possibility is discussed separately below (Section 9.6.2).

The sialic crust tends to be richer in $\delta^{18}O$ than are igneous rocks derived from the mantle. Therefore, the genesis of the silica-rich igneous rocks with abnormally high ^{18}O concentrations is likely to have involved assimilation or melting of especially ^{18}O-rich sialic material. One particularly interesting study is that by Taylor and Turi (1976) on the Tuscan magmatic province, Italy, where the calc–alkaline rocks of Plio–Pleistocene age have unusually high δ^{18} values (11.2 to $16.4^o/_{oo}$). The rhyolites and quartz latites have some of the highest values (15.3 to 16.4) for primary igneous rocks and this feature can be explained only by an origin involving the melting or large-scale assimilation of ^{18}O-rich argillaceous rocks. This interpretation is supported by the strontium isotope data ($^{87}Sr/^{86}Sr = 0.713$ to 0.720) and the presence of cordierite in the rocks. It is to be expected that the trace element contents of these rocks could be very different from those produced by fractional crystallization of a basic magma.

9.6.2 Rock-water interaction

The interaction of water with igneous rocks has been identified, through stable isotope studies, as a very important petrogenetic process. Interaction at almost any of the stages of rock genesis appears to be possible and a range of

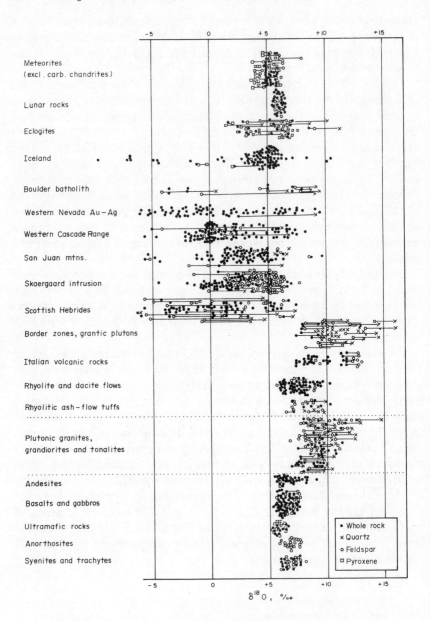

FIG. 9.6. Composition of $\delta^{18}O$ analyses of minerals and igneous rocks from a variety of localities. (After Taylor, 1974)

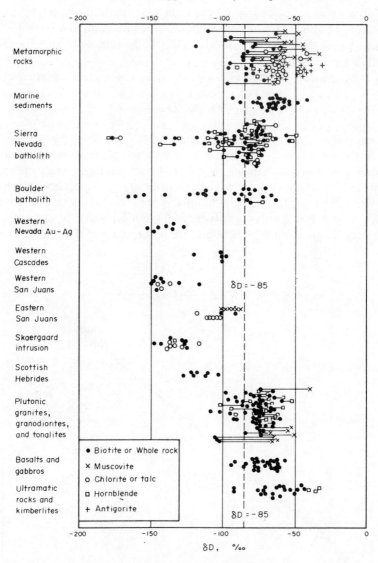

FIG. 9.7. Compilation of δD analyses of minerals from igneous, metamorphic and hydrothermally altered rocks from a variety of localities. (After Taylor, 1974)

natural waters may be involved—meteoric, magmatic or sea water. The isotopic ratios of natural waters are often sufficiently distinctive to leave their mark on an igneous rock with which they have interacted.

The isotopic ratios in rain or snow vary systematically with latitude or elevation. The higher the latitude or elevation the lower are the δD and $\delta^{18}O$ values. Values of $\delta^{18}O$ vary from about $0\%_{oo}$ at the equator to about $-55\%_{oo}$ (SMOW) at the poles.

A linear relation between δD to $\delta^{18}O$ values exists and is approximately

$$\delta D = 8\delta^{18}O + 5 \qquad (9.14)$$

as indicated by the data in Fig. 9.8 (Taylor, 1974). The relationship arises from the fact that condensation of water from the atmosphere is an equilibration process, is temperature-dependent, and D/H fractionation is directly proportional to $^{18}O/^{16}O$ fractionation. Thus, meteoric ground waters normally have a lower $\delta^{18}O$ value than those of normal igneous rocks. Interaction of such water with a hot igneous intrusion could bring about, through oxygen exchange, a considerable reduction in the $\delta^{18}O$ values of the rocks.

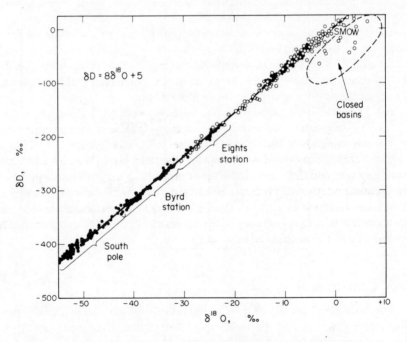

FIG. 9.8. Plot of δD against $\delta^{18}O$ for various meteoric surface waters. The data define the line shown, with $\delta D = 8\,\delta^{18}O + 5$. Eights Station, Byrd Station and South Pole are localities in Antarctica. The closed basins represent lakes where there has been strong evaporation. (After Taylor, 1974)

There is now a sizeable body of evidence to show that such an effect does occur. It is most evident where an intrusion is emplaced in permeable country rocks so that convective circulation of the ground water is readily set up. Heated water penetrates into and exchanges oxygen with the intrusion as well as with the surrounding rocks, the Tertiary igneous complexes of western Scotland being exemplars of this (Taylor, 1968, 1974, 1978; Taylor and Forester, 1971).

In the Isle of Mull the basalts show a systematic decrease in $\delta^{18}O$ values from $+9.1$ to $-6.5\%_{oo}$ towards the central plutonic complex. Depletion in $\delta^{18}O$ is observed over an area of $500\,km^2$ and is interpreted (Forester and Taylor, 1976) as arising from interaction of the rocks with heated ground waters, which are estimated to have had initial $\delta^{18}O$ values of -11 to $-12\%_{oo}$.

In Mull, as in the other centres, the oxygen exchange appears to have taken place predominantly after solidification of the intrusive magma. This is shown by the differential $\delta^{18}O$ depletion of minerals; the feldspar, being susceptible to alteration, shows the greatest change, while quartz often retains a typical igneous $\delta^{18}O$ value. However, the temperature of interaction is estimated to have been relatively high (of the order of 300°C and greater) to produce the observed ^{18}O depletion in rocks and minerals where oxygen exchange was complete. Figure 9.3 shows that if a feldspar is to have a $\delta^{18}O$-value $\simeq -5\%_{oo}$ (see points for Scottish Hebrides on Fig. 9.6) when in isotopic equilibrium with water of $\delta^{18}O \simeq -11\%_{oo}$ (i.e. $10^3 \ln \alpha \simeq 6$), a relatively high temperature ($\sim 200-300°C$) is required. Low temperatures, such as in weathering, would give rise to a much higher equilibrium value.

The extent of $\delta^{18}O$ depletion in these intrusive complexes is sufficiently large to require a considerable volume of ground water, probably at least as great as the volume of affected rock (Taylor, 1974). There is the possibility that the interaction of such a large volume of water would leach some of the trace and minor elements from the rocks. There is some evidence of such an effect (see Section 9.6.3) as some ore-deposits associated with igneous activity do have low $\delta^{18}O$ values, indicating the involvement of meteoric water in their genesis. These waters may have been the carriers of some of the ore material.

Not all natural waters available for interaction with an igneous intrusion are necessarily low in ^{18}O. Primary magmatic waters have $\delta^{18}O$ values in the range $+5$ to $+9\%_{oo}$, i.e. similar to that of many unaltered igneous rocks. Some metamorphic waters may be high in ^{18}O. For example, one of the grano-diorites that forms part of the Southern California batholith shows evidence of some exchange between the margins of the pluton and high ^{18}O metamorphic water derived from the surrounding pelitic schists (Taylor, 1978). Quartz from near the margin of the pluton has $\delta^{18}O$ values of about $10\%_{oo}$ and occasionally up to $\sim 19\%_{oo}$, in contrast to values of about $8.7\%_{oo}$ for quartz from the central zone.

Furthermore, as mentioned above and as indicated by equation (9.12) and the data in Fig. 9.3, low-temperature interactions could lead to ^{18}O enrichment even though the water itself is not ^{18}O-enriched. The low temperature ($\sim 4°C$) alteration of ocean-floor basalts by seawater ($\delta^{18}O \simeq 0$) leads to considerable enrichment in ^{18}O, i.e. from the $\delta^{18}O$ value $\sim +6°/_{oo}$ of fresh basalt to $+17°/_{oo}$ or higher (Muehlenbachs and Clayton, 1972). Evidence from the deep drilling of the ocean floor indicates that cold seawater can penetrate, and interact with, the oceanic crust to a depth of at least 600 m (Muehlenbachs and Clayton, 1976). Similarly, at least some ophiolitic rocks which are considered to be fragments of the oceanic crust and upper mantle show ^{18}O-enrichment through the low-temperature interaction of the initially basaltic material with seawater (Spooner *et al.*, 1974). The implications of these findings to chemical oceanography are discussed in Chapter 11.

9.6.3 Genesis of ore-deposits

In addition to their use as a geothermometer, stable isotopes can be indicators of other conditions of ore formation, such as the pH and oxygen fugacity, as well as the provenance, of the mineralizing fluid. Both sulphur and oxygen isotopes have been successfully applied in this way, particularly to hydrothermal sulphide deposits. Carbon isotopes are also useful genetic indicators where the ore-deposit contains carbonates and/or graphite (Ohmoto, 1972).

The range of $\delta^{34}S\,(CD)$ values of sulphide minerals in hydrothermal ore deposits is commonly from -7 to $+15°/_{oo}$. In sulphates the range is about $+8$ to $+25°/_{oo}$. However, $\delta^{34}S$ variations in a mineral are a function of the composition of the hydrothermal fluid from which it was derived, especially with respect to the H_2S/SO_4^{2-} ratio. For example, for a system in which the fluid has a $\delta^{34}S$ value $= 0°/_{oo}$ and contains H_2S and SO_4^{2-} in the molar proportions 0.1:0.9 respectively, the $\delta^{34}S$ in galena would be $-33.3°/_{oo}$ at 200°C. If the fluid composition is changed to equal proportions of H_2S and SO_4^{2-}, $\delta^{34}S$ of the galena changes to $-20.5°/_{oo}$, at the same temperature (Rye and Ohmoto, 1974).

The likely important sulphur species in hydrothermal fluids at less than 500°C are: H_2S, HS^-, S^{2-}, SO_4^{2-}, HSO_4^-, KSO_4^- and $NaSO_4^-$, and the most important equilibria area:

$$H_2S_{aq} = H^+ + HS^-,$$
$$HS^- = H^+ + S^{2-},$$
$$2H^+ + SO_4^{2-} = H_2S_{aq} + 2O_2,$$
$$HSO_4^- = H^+ + SO_4^{2-},$$
$$KSO_4^- = K^+ + SO_4^{2-},$$
$$NaSO_4^- = Na^+ + SO_4^{2-}.$$

These show the importance of pH and oxygen fugacity on the H_2S/SO_4^{2-} ratio in the fluid and hence on the sulphur isotope ratio of the minerals forming from it. Ohmoto (1972) has used these equilibria and constraints of mass balance to deduce the systematic relationships between isotope ratios and the chemistry of the hydrothermal fluids. The value of this approach lies in the fact that from the S isotope ratios of the ore minerals it should be possible to establish the T, pH, f_{O_2}, as well as the $\delta^{34}S$ value of the ore forming fluid. With these data we will be nearer to a proper understanding of the conditions of ore deposition. The $\delta^{34}S$ value of the ore-bearing fluid has significance and use, as the origin, in broad terms, of the sulphur in the fluid may be established from it. Three main sulphur sources have been recognized on this basis (Rye and Ohmoto, 1974):

(a) A $\delta^{34}S$ value in the fluid of near zero indicates that the sulphur is probably of magmatic origin.

(b) A $\delta^{34}S$ value of about $+20^o/_{oo}$ (CD), which is that of seawater, shows that the sulphur has come from seawater or marine evaporites.

(c) Intermediate values of $\delta^{34}S$ are likely when the sulphur is from the surrounding rocks.

It is also possible to identify the source of the water of the hydrothermal fluid from the oxygen isotope ratios of the fluid inclusions in the ore minerals. The famous stratabound Fe–Cu–Pb–Zn sulphide deposits in Kuroko, Japan have been subjected to studies of this kind. There the sulphur isotope values of the minerals indicate that the $\delta^{34}S$ of the ore-bearing fluid was close to $+20^o/_{oo}$ (CD) and the $\delta^{18}O$ values of water in fluid inclusions in the pyrite and chalcopyrite fall within a narrow range of -1.6 to $+0.3^o/_{oo}$ (Ohmoto and Rye, 1974). These data suggest that a significant proportion of the ore-forming fluids was predominantly of seawater origin. It has been suggested (Ohmoto and Rye, 1974) that the genesis of the ore involved the high-temperature convective circulation of seawater through hot volcanic rocks of the area. Here is further evidence of the importance of rock–water interaction and of the leaching of elements from igneous rocks.

Suggested further reading

Economic Geology (1974) Special Issue on *Stable Isotopes as Applied to Problems of Ore Deposits.* Vol. 69, no. 6.
FAURE, G. (1977) *Principles of Isotope Geology.* J. Willey & Sons. 464 pp.
HOEFS, J. (1973) *Stable Isotope Geochemistry.* Springer-Verlag, New York. 140 pp.
JAGER, E. and HUNZIKER, J. C. (editors) (1979) *Lectures in Isotope Geology.* Springer-Verlag, New York. 329 pp.

PART THREE

AQUEOUS SYSTEMS

10

Continental Waters

10.1 Chemical weathering

The weathering of rocks is both a mechanical and a chemical process. The thermodynamic instability of igneous and metamorphic rocks brought into the conditions prevailing at the Earth's surface provides the driving force for mineral dissolution and breakdown reactions. Rarely does the weathering process reach equilibrium as the systems are nearly always open, mixing is incomplete and changes in the variables are usually rapid compared with the weathering rates.

The weathering process entails the interaction of an aqueous solution (and/or gas) with rock material to produce a solution of different composition from the reactant one, a residue of insoluble solids of the initial rock, and other solids that are secondary mineral phases. The weathering medium transports the products and in so doing fractionates the material into accumulations of differing sediments and natural waters.

In this section we examine chemical weathering and start by considering the solubilities and stabilities of some of the principal rock-forming minerals. This is followed by an account of the main types of weathering reactions and the effects of pH, redox potential and rock composition.

Minerals may dissolve in the active aqueous reactant phase either by simple congruent dissolution reactions such as:

$$SiO_2 + 2H_2O \rightleftharpoons H_4SiO_4$$
$$\text{quartz} \qquad\qquad \text{silicic acid}$$

$$(10.1)$$

or by an incongruent one such as:

$$MgCO_3 + 2H_2O \rightleftharpoons Mg(OH)_2 + HCO_3^- + H^+.$$
$$\text{magnesite} \qquad\qquad \text{brucite}$$

$$(10.2)$$

In these two examples the aqueous reactant phase is shown as pure water. Often it is the soil waters or ground waters that are the active weathering agents and frequently these are mildly acid from uptake of CO_2 produced via the respiration of organisms. Rain water is also mildly acid (pH ≈ 5–6) as it dissolves some atmospheric CO_2 to give a weak carbonic acid solution. Stream and river waters vary slightly in pH but often are close to neutral. (The interaction of seawater with rock material is discussed in the next chapter.)

10.1.1 Congruent dissolution

The congruent dissolution of many minerals is strongly dependent upon pH. The solubility of SiO_2 will be taken as an example. The relevant equilibrium constants (Stumm and Morgan, 1970) are as follows:

$$SiO_2 + 2H_2O \; \rightleftharpoons \; H_4SiO_4 \qquad \log K = -3.7 \, (25°C), \qquad (10.3)$$
$$\text{quartz} \qquad\qquad\qquad \text{silicic}$$
$$\text{acid}$$

$$SiO_2 + 2H_2O \; \rightleftharpoons \; H_4SiO_4 \qquad \log K = -2.7, \qquad (10.4)$$
$$\text{amorphous}$$
$$\text{silica}$$

$$H_4SiO_4 \rightleftharpoons H_3SiO_4^- + H^+ \qquad \log K = -9.46. \qquad (10.5)$$

These equilibria allow the calculation of the solubility either of quartz or amorphous silica as a function of pH. The calculated curve (Fig. 10.1) for amorphous silica shows that at pH less than 8 the solubility is constant (at

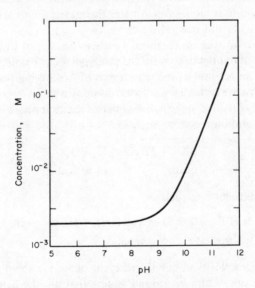

FIG. 10.1. Calculated solubility of amorphous silica, at 25° C, as a function of pH.

$10^{-2.7}$ M) with decreasing pH, while it rises rapidly at higher pH values. The equilibrium constant for reaction (10.3) indicates that the solubility of quartz is an order of magnitude lower than that of amorphous silica. At pH values of greater than 11 the curve shown in Fig. 10.1 will be modified slightly as the following equilibria would begin to play a role:

$$H_3SiO_4^- \rightleftharpoons H_2SiO_4^{2-} + H^+ \qquad \log K = -12.56, \quad (10.6)$$

$$4H_4SiO_4 \rightleftharpoons H_6Si_4O_{12}^{2-} + 2H^+ + 4H_2O \ \log K = -12.57. \quad (10.7)$$

These increase the calculated solubility, but such pH values are outside those found in most natural waters (at 25°C).

Many carbonates undergo chemical weathering by means of a congruent chemical reaction. Here acid attack is the principal weathering mechanism, with the acid being formed by the reaction of carbon dioxide, from the atmosphere or from the respiratory cycle of organisms in the soil, with water:

$$CO_2 + H_2O \rightleftharpoons H^+ + HCO_3^-. \qquad (10.8)$$

The congruent dissolution reaction may then be written (using calcite as an example):

$$CaCO_3 + H^+ \rightleftharpoons Ca^{2+} + HCO_3^-. \qquad (10.9)$$

The sum of the two reactions means that every doubly charged calcium is balanced by two singly charged bicarbonate anions.

10.1.2 Carbonate equilibria

The above reactions are important because they show that there is a relationship between the partial pressure of carbon dioxide in the atmosphere and the maximum solubility of calcite in water at equilibrium with the atmosphere. A short digression will be made here to show how the concentration of certain ionic species may be calculated from a knowledge of the equilibrium constant (at 25°C) for the appropriate reactions as follows:

$$\begin{array}{llll} CaCO_3 & \rightleftharpoons Ca^{2+} + CO_3^{2-} & \log K = -8.3, & (10.10) \\ H_2CO_3 & \rightleftharpoons H^+ + HCO_3^- & \log K = -6.4, & (10.11) \\ HCO_3^- & \rightleftharpoons H^+ + CO_3^{2-} & \log K = -10.3, & (10.12) \\ H_2O & \rightleftharpoons H^+ + OH^- & \log K = -14, & (10.13) \\ pCO_2 + H_2O & \rightleftharpoons H_2CO_3 & \log K = -6.48 & (10.14) \\ \text{atmosphere} & \text{solution} & & \end{array}$$

The following treatment closely follows that by Garrels and Christ (1965) in which equations (10.10) to (10.14) are used, together with a charge balance equation (10.15):

$$2m_{Ca^{2+}} + 2m_{H^+} = 2m_{CO_3^{--}} + m_{HCO_3^-} + m_{OH^-}, \qquad (10.15)$$

to calculate the concentrations of the various ions in solutions at equilibrium with an atmosphere of constant CO_2 partial pressure of about 32 Pa ($\sim 10^{-3.5}$ atm). The assumption is made that the activity coefficients of all ionic species are equal to one. Since p_{CO_2} for the atmosphere is given then, from equation (10.14)

$$[H_2CO_3] = 10^{-6.48} \times 32 \approx 10^{-5.0}, \qquad (10.16)$$

and substituting this result into the equilibrium relationship for equation (10.11):

$$[H^+][HCO_3^-] = 10^{-6.4} \times 10^{-5.0}. \qquad (10.17)$$

Therefore,
$$[HCO_3^-] = 10^{-11.4}/[H^+]. \qquad (10.18)$$

From (10.12) and (10.18):

$$[CO_3^{2-}] = 10^{-21.7}/[H^+]^2. \qquad (10.19)$$

From (10.13):

$$[OH^-] = 10^{-14.0}/[H^+]. \qquad (10.20)$$

From (10.10) and (10.19):

$$[Ca^{2+}] = 10^{13.4}[H^+]^2. \qquad (10.21)$$

Substitution of equations (10.18) to (10.21) into equation (10.15), together with rearrangement, yields:

$$10^{13.7}[H^+]^4 + [H^+]^3 - 10^{-11.4}[H^+] = 10^{-21.4}. \qquad (10.22).$$

The solving (by an iterative method) of this last equation, and substitution into equations (10.18) to (10.21) yields:

$$[H^+] = 10^{-8.4} \qquad [CO_3^{2-}] = 10^{-4.9},$$
$$[Ca^{2+}] = 10^{-3.4} \qquad [HCO_3^-] = 10^{-3.0},$$
$$[H_2CO_3] = 10^{-5.0} \qquad [OH^-] = 10^{-5.6}.$$

The calculated pH is lower (a value of 8.4 compared with about 10) and the Ca^{2+} concentration is about three times greater in the solution in contact with the atmosphere than in a solution with no external CO_2 pressure. Garrels and Christ (1965) also discuss carbonate equilibria under differing conditions, including the case where the pH of a solution in contact with the atmosphere is determined by factors other than that of the carbonate equilibrium alone.

The lowering of the pH of natural water through the interaction with the atmosphere will also affect the nature of the weathering reactions of silicates. This may be illustrated by the hydrolysis of forsterite olivine in the presence of carbonic acid:

$$MgSiO_4 + 4H_2CO_3 \rightarrow 2Mg^{2+} + 4HCO_3^- + H_4SiO_4. \tag{10.23}$$

10.1.3 Incongruent dissolution

A number of silicates, including the feldspars and micas, undergo incongruent dissolution reactions during weathering. This type of reaction may be illustrated by the case of potassium feldspar which, on dissolution, gives kaolinite as the secondary mineral and a basic solution. The overall reaction, which undoubtedly proceeds by a number of steps with intermediary products, is:

$$4KAlSi_3O_8 + 11H_2O \rightarrow 4K^+ + 4OH^- + Al_4Si_4O_{10}(OH)_8 + 8H_4SiO_4. \tag{10.24}$$

In the presence of a CO_2-rich reactant solution the overall reaction is:

$$4KAlSi_3O_8 + 4H_2CO_3 + 18H_2O \rightarrow 4K^+ + 4HCO_3^- + Al_4Si_4O_{10}(OH)_8 \\ + 8H_4SiO_4 \tag{10.25}$$

and this is likely to be the more common of the two.

The production of OH^- ions in reaction (10.24) shows that the pH of solutions involved in the weathering of silicates can be increased by the process. Solutions containing dissolved CO_2 become less acid during weathering. In the absence of CO_2 the pH can rise to a value of about 9 where it is then buffered by the dissociation of silicic acid.

From experimental and theoretical considerations, Helgeson (1968, 1971 and 1972) and Helgeson *et al.* (1969) have been able to determine, with a reasonable degree of certainty, the reaction steps in the dissolution of potassium feldspar at 25°C:

(a) Feldspar hydrolysis by exchange at the mineral surface, of H^+ for K^+. This is followed by
(b) the development of gibbsite as an intermediate product, which in turn is followed by
(c) development of kaolinite.

The rate of dissolution is rapid at first and then decreases. There has been an interesting series of papers discussing the possible cause of this decrease. Early experimental results by Lagache (1965) and her co-workers on the kinetics of the dissolution were interpreted as showing that the rate is controlled by the concentration of components in the reacting solution. Other workers (namely, Helgeson, 1971; Pačes, 1973; and Wollast, 1967) have interpreted experimental data as indicating the presence of a secondary mineral or amorphous layer built up on the weathering feldspar, and with the dissolution rate controlled by the diffusion of components through the layer. Recent experiments by Lagache (1976) and X-ray photoelectron spectroscopy by Petrović *et al.* (1976) suggest that such a layer does not exist and so the rate must be controlled by the interface reaction, as initially thought. Confirmatory evidence for this has come from further experimental work (Holdren and Berner, 1979) and a study of feldspars from some soils (Berner and Holdren, 1979). This particular problem serves to illustrate some of the difficulties encountered in elucidating the reaction mechanisms of dissolution.

Although a reaction-product layer around weathering feldspars does not occur, it is clear that with some other minerals, coatings of insoluble oxides or hydroxides (especially those of iron) produced during the initial stages of weathering can act as protective skins against the alteration process.

10.1.4 Redox reactions

These are the third type of weathering reaction and they can have a marked influence on the pH of the product solutions. Many of these reactions are complex and probably proceed via a number of steps. For example, the redox weathering of pyrite may be expressed by reaction (10.26):

$$4FeS_2 + 15O_2 + 14H_2O \rightarrow 4Fe(OH)_3 + 16H^+ + 8SO_4^{2-} \qquad (10.26)$$

but probably proceeds by the steps (10.27) to (10.29):

$$2FeS_2 + 7O_2 + 2H_2O \rightarrow 2Fe^{2+} + 4SO_4^{2-} + 4H^+, \qquad (10.27)$$

$$4FeS_2 + O_2 + 4H^+ \rightarrow 4Fe^{3+} + 2H_2O, \qquad (10.28)$$

$$Fe^{3+} + 3H_2O \rightarrow Fe(OH)_3 + 3H^+. \qquad (10.29)$$

The reactions release a large number of hydrogen ions; for every mole of pyrite 4 moles of H^+ ions are produced, so the weathering solution becomes very acid: pH values of 2 or lower may be reached. Hence, most mine waters are distinctly acid and the associated weathering is often deep.

Other examples of redox reactions are given in the following equations:

$$PbS + 4Mn_3O_4 + 12H_2O \rightarrow Pb^{2+} + SO_4^{2-} + 12Mn^{2+} + 24OH^-, \qquad (10.30)$$

$$2MnCO_3 + O_2 + 2H_2O \rightarrow 2MnO_2 + 2H_2CO_3, \qquad (10.31)$$

$$2Fe_2SiO_4 + O_2 + 4H_2O \rightarrow Fe_2O_3 + H_4SiO_4. \qquad (10.32)$$

10.1.5 Mineral stability

The relative resistance of minerals to weathering at the surface of the Earth was shown early on by Goldich (1938) to be:

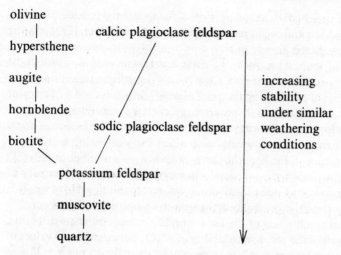

A more complete list has been drawn up by Ollier (1975), and is given in Table 10.1 but it is important to stress that the sequence is only a guide as local conditions could change the precise order. The list is based upon the observed occurrence or absence of minerals in different sediments, so the order shows the relative resistance of minerals to the integrated effects of mechanical as well as chemical weathering.

TABLE 10.1. *Relative resistance of minerals to weathering*

Most resistant	8 Magnetite	16 Zoisite
1 Zircon	9 Staurolite	17 Augite
2 Tourmaline	10 Kyanite	18 Sillimanite
3 Monazite	11 Epidote	19 Hypersthene
4 Garnet	12 Hornblende	20 Diopside
5 Biotite	13 Andalusite	21 Actinolite
6 Apatite	14 Topaz	22 Olivine
7 Ilmenite	15 Sphene	Least resistant

10.1.6 Rock weathering

As might be expected, the mineralogy of a rock mass is the dominant factor controlling the release of many elements from the rock during chemical weathering. In a study of the weathering profiles of five granitic rocks, Harriss and Adams (1966) ascertained that calcium and sodium were some of the first elements to be lost during weathering. Potassium, rubidium and thorium are examples of elements mobilized during the intermediate and final stages of the weathering process, while lithium, copper, manganese and zinc are some that are found to be enriched in the soil portion of the weathered residue. The early loss of much sodium and calcium was also observed by Goldich (1938) during his study of weathered granite gneiss.

In the case of more basic rocks, calcium and magnesium are readily lost during the early weathering stages. Clearly this is a simple reflection of the relative stabilities of minerals as discussed above. The precise order of element loss, therefore, will depend on the mineralogy, rock texture and also upon the climatic conditions as well as any biological control. The presence of some organic complexing compounds may also affect the weathering sequence.

It is impracticable to list here the changes shown by a number of different rock types during weathering, especially as many of the observations may be predicted from a knowledge of individual mineral stabilities. However, it is appropriate to state the generalized changes in composition that occur.

Aluminium is usually one of the least mobile elements during weathering, mainly as a result of the very low solubility of Al_2O_3 between the pH values of 5 and 8. It has become customary, therefore, to treat the aluminium loss as negligible and to use the alumina concentrations in the various *in-situ* weathered products as a reference against which the loss or gain of other elements may be determined. This assumption of the total inertness of alumina is not strictly valid, for some is lost into solution, but it is a useful and reasonable approximation provided that the weathering solutions did not attain pH values of greater than 8 or less than 5 at any stage during the weathering process.

In relation to constant alumina, the composition, expressed as the element oxides, of a silicate rock changes on weathering as follows:

SiO_2 some decrease
Fe_2O_3 significant increase;
FeO   almost total removal by oxidation to Fe_2O_3;
MgO   significant decrease;

$\left.\begin{array}{l} CaO \\ Na_2O \end{array}\right\}$ almost total removal;

K_2O decrease;
H_2O great increase.

Some researchers have suggested that the constancy of the alumina content during weathering may be used to help define a scale of rock alteration (e.g. Ruxton (1968) proposed the use of a silica to alumina mole ratio to quantify the degree of weathering in humid regions). The relative immobility of alumina and of iron is well demonstrated by the existence of bauxites and laterites—those final, insoluble residues of an extended weathering process. Chesworth (1973) has proposed that the components SiO_2, Al_2O_3, Fe_2O_3 and H_2O can be used to define a 'residua system' of weathering. In the form of a

SiO_2–Al_2O_3–Fe_2O_3 projection, it can be used to express the chemistry and mineralogy of most common residual products from the weathering of silicate rocks.

10.2 Spring, river and lake waters

The nature of aqueous solutions that are produced or modified by the processes of weathering is determined by several factors, including chemical controls such as reaction rate, solubility and interface reactions, as well as environmental controls such as climate, geology and the hydrologic cycle.

The solutions from weathering may mix with other waters that effectively have not been involved in a weathering process. In turn, the mixed waters may be modified by further reactions such as by some cation exchange with clay or other mineral phases, or by the activities of man. It is, therefore, a difficult matter to predict the composition of a given water at a particular stage of its hydrologic evolution, or to establish the precise influence of each of the factors which determine its composition. However, it is possible to make some comments about the general geochemistry of fresh and saline waters of two types:

(a) surface waters: spring, river and lake waters;
(b) ground waters (i.e. subsurface waters that are in the zone of water saturation).

The second of these is covered in the next section.

Although there is a very great variation in the concentrations of dissolved materials in stream and river water, nonetheless an extensive amount of available data allowed Livingstone (1963) to estimate the mean composition of world river water (Table 10.2).

TABLE 10.2. *Mean chemical contents ($\mu g/g$) of world river water*

HCO_3^-	58.4	Ca	15	Fe	0.67
SO_4^{2-}	11.2	Mg	4.1	SiO_2	13.1
Cl^-	7.8	Na	6.3	Sum	120
NO_3^-	1	K	2.3		

Livingstone also reviewed the trace-element contents of river waters and was able to give, for some elements, an approximate value for the world mean river water. These are summarized in Table 10.3. Data on the compositions of major rivers throughout the world may be found in Livingstone (1963). The total amount of dissolved solids carried annually to the oceans by rivers is estimated to be about 3.905×10^{12} kg.

Lake waters also vary greatly in composition, not only from lake to lake but

TABLE 10.3. *Mean trace element contents in world river water*

Halogens

F	less than 1 $\mu g/g$
Br	about 0.02 $\mu g/g$. World average for fresh water
I	about 2 ng/g. Mean for lakes and rivers

Transition elements

V	significantly less than 1 ng/g
Ni	about 10 ng/g. Ordinary fresh water
Cu	about 10 ng/g. Ordinary fresh water

Others

B	about 13 ng/g
Rb	about 1 ng/g
Ba	about 50 ng/g
Zn	about 10 ng/g
Pb	between 1 and 10 ng/g
U	about 1 ng/g

often within a lake where marked temperature and compositional stratifications can occur. Reducing conditions often exist in the lower, more saline level of stratified lakes and these give rise to relatively high concentrations of nitrite, ammonia and Fe^{2+} in the water. The reducing conditions may also lead to the production of hydrogen sulphide gas and the precipitation of some metal sulphides (including iron sulphides). Silica and phosphorus may be released from the sediments.

The existence of a zone (the thermocline) characterized by a large change in temperature through its depth and which separates the upper and lower levels of a thermally stratified lake, helps to maintain the reducing conditions in the lower level. This zone prevents the rapid diffusion of atmospheric oxygen through to the anaerobic (i.e. reducing) water. Atmospheric gases determine the pH and redox conditions of the upper layer and surface waters. Pure water which is brought to equilibrium with the atmosphere will be slightly acid through the dissolving of CO_2 to give carbonic acid. However, most lake waters contain more dissolved CO_2 than that corresponding to the equilibrium condition. This results from the respiratory action of organisms and the presence of dissolved carbonates. The importance of carbonate equilibria in controlling the pH and composition of natural waters has been discussed in Section 10.1.

In a perceptive study of the origin of some spring and lake waters, Garrels and Mackenzie (1967, 1972) used the idea that the reaction process could be treated as operating in a closed system. The spring and lake waters are those of the Sierra Nevada, USA where the rocks consist mainly of a range of gneisses from quartz diorite to quartz microcline types. The major minerals are plagioclase feldspar, alkali feldspar, and quartz, while biotite and hornblende are minor. The weathered residue consists of aluminosilicates—mainly kaolinite—together with small amounts of authigenic gibbsite, mica and montmorillonite.

The method that Garrels and Mackenzie adopted to see if the system could be treated as a closed one was to calculate whether a mass balance would exist when the spring water and solid reaction phases were back-reacted to produce the original minerals of the gneisses. The basis of the calculation is:

(a) The aluminosilicate residue, plus all the Na^+ and Ca^{2+} ions in the water, are used to give the initial plagioclase ($\sim An_{35}$).

(b) All the Mg^{2+} ions, some K^+, some HCO_3^-, some of the dissolved SiO_2, and authigenic kaolinite are used to produce the biotite.

(c) The remaining small amount of K^+, enough HCO_3^-, and dissolved SiO_2 are used to give the initial potassium feldspar.

The simple calculations show a surprisingly good balance and indicate the effectively closed nature of the weathering system. Garrels and Mackenzie were able to draw a number of conclusions:

(a) Silica in the water came from the breakdown of the silicates and an insignificant amount from the direct solution of quartz, i.e. dissolved silica represents the dissolved CO_2.

(b) The CO_2-rich water is the active component that produces the kaolinite from the original minerals.

(c) About 80% of the dissolved constituents of the spring water can be accounted for by the breakdown of plagioclase feldspar alone. Potassium feldspar shows little breakdown.

(d) The high Na/K concentration ratio in the water is, therefore, related to this differential weathering rate rather than to the adsorption upon the clay minerals of the K^+ after release from the feldspar.

These conclusions are of wider applicability than just to the Sierra Nevada spring system, and they show that in appropriate cases a quantitative approach to chemical weathering processes is justified and can produce a new insight. Furthermore, it was possible, for the Sierra Nevada area, to make quantitative estimates of the reacting capacity of the waters and of the weathering rate from knowledge of the annual rainfall.

10.3 Ground waters

The hydrogeochemistry of ground waters reflects the source of the water, the lithology of the aquifer, and the local chemical conditions such as temperature, pressure and redox potential. The source* may be:

(a) magmatic,

(b) meteoric (e.g. precipitated and surface water),

* A genetic classification of ground waters is given in White *et al.* (1963).

(c) connate (i.e. water trapped in the pore spaces of a sediment at the time of deposition),

(d) oceanic.

The importance of the lithology in determining the composition of ground waters can be gauged from analyses of ground water from a granite and from a serpentinite (Table 10.4). However, it must be emphasized that these are examples and not necessarily 'typical' or average analyses. The reader is referred to White *et al.* (1963) for extensive compilations of ground-water compositions.

TABLE 10.4. *Constituents of ground waters from different rock types*

	Granite	Serpentinite	Shale
Cations or oxide			
SiO_2	39	31	5.5
Al	0.9	0.2	0
Fe	1.6	0.06	3.5
Ca	27	9.5	227
Mg	6.2	51	29
Na	9.5	4.0	12
K	1.4	2.2	2.7
Anions			
HCO_3	93	276	288
CO_3	0	0	0
SO_4	32	2.6	439
Cl	5.2	12	24
F	0	0	0
NO_3	7.5	6.8	0.9
PO_4	0	0	0

Granite: drilled well, Howard County, Maryland, USA.
Serpentinite: well, Baltimore County, Maryland.
Shale: drilled well, Onondaga County, NY, USA.

Data source: White *et al.* (1963).

The influence of local chemical conditions can be illustrated by the example of pressure solution. The amount of dissolved SiO_2 (and some other components) can be increased by this effect. Where the grains of a sediment are in contact and under pressure, the mineral at the points of stress will more readily dissolve in any interstitial water than where the stress is absent. This can, for example, cause the migration in solution of some SiO_2 from chert-rich sandstones to quartz-rich ones (Lerbekmo and Platt, 1962). The dissolution will be promoted further by higher pH values as indicated by Fig. 10.1.

Ground waters play an important role in the generation of ores, for they act as a medium for the dissolution and transportation of heavy metals to the place of selective precipitation. While some ore bodies within a sedimentary sequence were probably generated from ore-bearing fluids of an easily determinable origin, such as from an adjacent igneous intrusion, it is clear that the genesis of others was complex and involved considerable migration, mixing and reaction of the parent fluids. From a detailed study of the Mississippi Pb–Zn–baryte–fluorite deposits, (White, 1968) showed that brines (of 4 to 10 times greater salinity than seawater) had been involved in the genesis of the ores. The fluid inclusions have consistently high concentrations of Na^+, Ca^{2+} and Cl^- ions but are low in K^+, Mg^{2+} and SO_4^{2-} ions. For example, a fluid inclusion in galena had the composition ($\mu g/g$): Ca 20,000; Mg 4000; Na 55,000; K 3100; Cl 115,000. White suggested that the heavy metals were transported in sulphide-deficient brines of connate origin from deep sedimentary basins and then later mixed with sulphide-rich brines from shallow depths, so causing the precipitation of the sulphide ores.

Another interesting example is to be found in the brine waters of the Salton Sea Geothermal System (White, 1968). This occurs at the south-east edge of the Salton Sea on the northern slope of Colorado River delta. A very saline, slightly acid (pH is about 5.5 at 25°C) brine permeates the deltaic sedimentary rocks of the system at depths below about 1000 m. The brine is rich in heavy metals (Table 10.5) that occur dominantly as chloride complexes. Clearly, this brine is capable of producing a significant ore deposit. White (1968) considers that the brine formed largely from local meteoric water which had gained its high salinity from the dissolution of local evaporites and had subsequently reacted with host sediments.

TABLE 10.5 *Analysis of brine from Salton Sea Geothermal System (units $\mu g/g$)*

SiO_2	400	Ba	235	NH_4	409
Al	4.2	Cu	8	HCO_3	150
Fe	2290	Pb	102	SO_4	5.4
Mn	1400	Zn	540		
As	12	Na	50,400	Cl	155,000
Sb	0.4	K	17,500	F	15
Ca	28,000	Li	215	Br	120
Mg	54	Rb	135	I	18
Sr	400	Cs	14	B	390

In some cases it seems clear that the ground waters are modified seawater. Brines in the Coal Measures of north-east England have heavy metal contents that are 1 to 7 times richer than present-day seawater (Edmunds, 1975). The source of the brine was, at least partly, from flushing of the sediments by seawater during the marine transgressions of Coal Measures times. Although the brines are depleted in magnesium and enriched in calcium relative to seawater, this can be explained by ground-water–sediment interactions such as dolomitization during early diagenesis. However, in other cases the Ca/Mg ratio of a brine may be increased by the precipitation of sulphate minerals (Lerman, 1970).

10.4 Oxidation-reduction

The composition of a natural aqueous solution, the oxidation state of the ions in solution, and the nature of the solid phases in equilibrium with the solution, all depend on the reducing or oxidizing (redox) condition of the system. Under reducing conditions certain dissolved ions will exist in lower oxidation states (e.g. Fe^{2+}, Cu^+, Mn^{2+}) than under oxidizing conditions. Conversion to higher oxidation states may involve a marked change in solubility (e.g. Fe III and Mn IV compounds are usually much less soluble than those of the $2+$ oxidation state).

In order to put the stability relations of the different oxidation states onto a quantitative basis we use the concept of the *oxidation potential* or *electrode potential* (E). Consider the reaction (10.33):

$$\underset{\text{metal}}{Fe} + \underset{\text{solution}}{Cu^{2+}} = \underset{\text{solution}}{Fe^{2+}} + \underset{\text{metal}}{Cu} . \tag{10.33}$$

The reaction involves the flow of electrons, so it could be used as the basis of a galvanic cell which would have an electromotive force or electrode potential (E) characteristic of the reaction. This potential is a measure of the driving force of the redox reaction, and is the sum of the individual potentials of the half-reactions (10.34) and (10.35):

$$Fe^{2+} + 2e^- = Fe, \tag{10.34}$$
$$Cu^{2+} + 2e^- = Cu. \tag{10.35}$$

The individual potentials for (10.34) and (10.35) clearly cannot be determined from the electrode potential for reaction (10.33). For the potentials to be usable as a quantitative indicator of the ease of oxidation or reduction it is necessary to have a reference half-reaction. This, by convention, is the reduction of hydrogen ion to hydrogen gas (10.36),

$$H_{aq}^+ + e^- = \tfrac{1}{2}H_{2\,gas}, \tag{10.36}$$

which is assigned an electrode potential of zero at standard temperature and pressure. This follows from the convention that the standard Gibbs free energy ($G°$) of the hydrogen half-cell is zero. Thus, the half-reactions (10.35) and (10.36) may be combined to give reaction (10.37) and the resultant potential is that for the copper reduction half-reaction:

$$H_{2\,gas} + Cu_{aq}^{2+} = 2H_{aq}^+ + Cu_{metal}. \tag{10.37}$$

This potential is now referred to as the relative electrode potential*; or electrode potential; or redox potential, and is given the symbol E_H—the suffix, H, showing that the hydrogen cell is the reference zero.

The redox potential may also be used to determine the free-energy charge of the half-cell from relationship (10.38)

$$\Delta G^\circ = -n_e E_H^\circ F, \tag{10.38}$$

where E_H° is the standard redox potential of the reaction when all components are at unit activity, n_e is the number of electrons involved (e.g. two in reaction (10.37)), and F is the Faraday constant, equal to $9.648\,456 \times 10^4$ coulombs per mole ($C\,mol^{-1}$). (This value is suitable for use in equations where the free energy, ΔG, values are given in joules as 1 joule = 1 volt-coulomb. Where G values are expressed in units of kilocalories per mole, F is $2.306\,036 \times 10^2\,kcal\,mol^{-1}$.)

Relationship (10.38) shows that a positive E_H° value for a reaction indicates that it will tend to proceed from left to right (negative ΔG°).

Lists of redox potentials are given in the conventional form with the oxidized species on the left-hand side, e.g.

$$
\begin{array}{ll}
 & E_H^\circ,\ \text{volts} \\
Cu^{2+} + 2e^- = Cu & +0.34 \\
Fe^{2+} + 2e^- = Cu & -0.44.
\end{array}
$$

These values may be used to calculate the standard redox potential, E_H°, for reaction (10.33), as $+0.78$ volts at $T = 298$ K.

For the calculation of redox potentials, E_H, at some non-standard state, the Nernst equation (10.39) is used:

$$E_H = E_H^\circ + \frac{RT}{n_e F} \ln \frac{a_D^d \cdot a_E^e}{a_B^b \cdot a_C^c} \tag{10.39}$$

---

* The relative electrode potential of an electrode (or half-cell) is defined as the electromotive force of a cell where the electrode on the *left* is a *standard hydrogen electrode* and that on the right is the electrode in question. For example, for the zinc electrode (written as $Zn^{2+}|Zn$) the cell is:

$$Pt, H_2|H^+|Zn^{2+}|Zn$$

which implies the reaction:

$$H_2 + Zn^{2+} \rightarrow 2H^+ + Zn.$$

The reaction taking place at the electrode is $Zn^{2+} + 2e^- \rightarrow Zn$. In the standard state the electromotive force of this cell has the value -0.763 V. Therefore, the standard electrode potential is -0.763 V. The above is the convention agreed by the International Union of Pure & Applied Chemistry (IUPAC). Readers should note, however, that not all authors have adopted it. Electromotive forces of cells where the standard hydrogen electrode is on the *right* should not be called electrode potentials.

for the conventionally written half-reaction $dD + eE + n_e e^- = bB + cC$.

Reaction (10.40) will be used to illustrate the use of the Nernst relationship

$$MnO_2 + 4H^2 + 2Fe^{2+} = Mn^{2+} + 2H_2O + 2Fe^{3+}.$$

$$\text{solid} \quad \text{aq.} \quad \text{aq.} \qquad \text{aq.} \quad \text{liq.} \quad \text{aq.}$$

(10.40)

E_H° values for the half-reactions are:

$$Fe^{3+} + e^- = Fe^{2+} \qquad\qquad\qquad +0.77\text{volt}, \qquad (10.41)$$

$$MnO_2 + 4H^+ + 2e^- = Mn^{2+} + 2H_2O \quad +1.23 \text{ volts.} \qquad (10.42)$$

Therefore E_H° for reaction (10.40) is $+0.46$ volt. A non-standard state will be taken in which reaction (10.40) occurs in an aqueous medium with pH $= 3.5$, with the other ions at unit activity and at $T = 298$ K. Hence, application of the Nernst equation to the half-reaction (10.42) gives:

$$E_H = E_H^\circ + \frac{RT}{n_e F} \ln \frac{a_{H^+}^4}{a_{Mn^{2+}}}, \qquad (10.43)$$

$$E_H = 1.23 + \frac{8.314 \times 298}{2 \times 9.65 \times 10^4} \ln (10^{-3.5})^4$$

$$= 1.23 + 0.012\,84 \times 2.303 \log 10^{-14}$$

$$= 0.82 \text{ volt.}$$

Therefore, the value of E_H for reaction (10.40) when the hydrogen ion activity is $10^{-3.5}$ is $0.82 - 0.77 = 0.14$ volt. The Nernst equation could be applied directly to the full reaction (10.40), but care must be used over the sign conventions. In the application to half-reactions, the activities of the oxidized species must always be placed as the numerator, i.e.

$$E_H = E_H^\circ + (RT/n_e F) \ln [\text{ox.}]/[\text{red}]. \qquad (10.44)$$

In the same way that it is possible to define a relative hydrogen ion activity of a solution in the form of the pH scale, so may a relative electron activity be defined:

$$p_\varepsilon = -\log a_e, \qquad (10.45)$$

where a_e is the relative activity of the electron. By this definition p_ε is then related to E_H by

$$p_\varepsilon = \frac{F}{2.303\,RT} E_H, \qquad (10.46)$$

p_ε is dimensionless*.

---

* This is because the Faraday constant has units of C mol^{-1}; E_H is in volts, or the equivalent J C^{-1}; R is in J K^{-1} mol^{-1}; and T is in K. The units cancel in relationship (10.46).

Both p_ε and E_H are in current usage by authors of research papers and books dealing with redox reactions. The book by Stumm and Morgan (1970) develops the use of p_ε, while that by Garrels and Christ (1965) does the same with E_H. In the succeeding parts of this chapter E_H is used. It has the advantage of being the value of the electromotive force that is actually measured. p_ε has the advantage of making some calculations simpler.

Discussions on the measurement of electromotive forces and of the problems encountered may be found in Garrels and Christ (1965) and Stumm and Morgan (1970). Tables of redox potentials are in the reference works by Latimer (1952), Charlot (1958), Sillen and Martell (1964, 1971), and in the texts of Krauskopf (1967) and Stumm and Morgan (1970). (Readers should remember that sign conventions for half-reactions may differ in some reference books and papers from that of the IUPAC convention.)

10.5 $E_H - $ pH diagrams

The effect of changes in pH on the redox potential of reaction (10.40) illustrates the important interconnecting roles that E_H and pH have on mineral and ionic equilibria. The stability limits of minerals and ionic species may be conveniently displayed on diagrams relating these two master variables in aqueous systems.

In natural systems the ranges of pH and E_H are restricted. Only very rarely are pH values of less than 2 or greater than 12 encountered. In the case of E_H, upper and lower limits, as a function of pH, can be defined. The upper limit is taken to be that of water stability at 1 atmosphere (101,325 Pa) and $T = 298$ K, i.e.

$$O_{2_g} + 4H_{ap}^+ + 4e^- = 2H_2O_{liq}, \quad E_H^\circ = 1.23 \text{ V}. \tag{10.47}$$

Use of the Nernst equation (10.39) gives

$$E_H = 1.23 - 0.059 \text{ pH} \tag{10.48}$$

for a partial pressure of oxygen of 1 atmosphere, and

$$E_H = 1.22 - 0.059 \text{ pH} \tag{10.49}$$

for the observed oxygen partial pressure of about 0.2 atmosphere (see Table 4.9). In defining the lower limit the hydrogen electrode reaction is used, since there is no known naturally occurring reaction causing the decomposition of water, to give hydrogen gas.

$$2H^+ + 2e^- = H_{2\,gas} \quad E_H^\circ = 0.00 \text{ V}, \tag{10.50}$$

E_H for this reaction is given by

$$E_H = E_H^\circ - \frac{0.059}{2} \log p_{H_2} - 0.059 \, \text{pH}. \tag{10.51}$$

The partial pressure of hydrogen gas (p_{H_2}) can be put at a maximum (i.e. most reducing) condition of 1 atmosphere. E_H°, by convention, is zero so we have:

$$E_H = -0.059 \, \text{pH}. \tag{10.52}$$

Both (10.49) and (10.52) define the dashed lines plotted in Fig. 10.2.

The construction of the upper and lower stability limits of water shows the method of establishing stability fields on E_H–pH diagrams. As a further example the stability relationships of the iron oxides Fe_3O_4 and Fe_2O_3 will be determined (based on the treatment by Garrels and Christ, 1965).

The half-cell reaction involving oxidation of iron to magnetite written in the conventional form is:

$$Fe_3O_{4_s} + 8H_{aq}^+ + 8e^- = 3Fe_s + 4H_2O_{liq}. \tag{10.53}$$

for which ΔG° is $+65.3$ kJ (15.6 kcal) and so $E_H^\circ = -0.084$ V (using

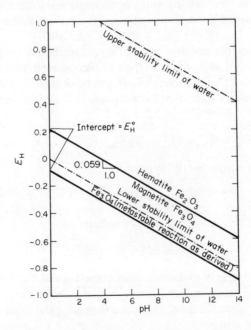

Fig. 10.2. Boundary for iron-magnetite and magnetite-hematite, as a function of E_H and pH, at 25° C and 1 atm total pressure. (After Garrels and Christ, 1965)

relationship (10.38)). Use of the Nernst equation for the half-reaction (10.53) gives:

$$E_H = -0.084 + \frac{0.059}{8} \log \frac{a_{Fe_3O_4} \cdot a_{H^+}^8}{a_{Fe}^3 \cdot a_{H_2O}^4}. \tag{10.54}$$

Substituting the activities of $Fe_3O_{4_s}$, Fe_s and H_2O_{liq} as being each unity, and $-pH$ for $\log a_{H^+}$, gives:

$$E_H = -0.084 - 0.059 \; pH. \tag{10.55}$$

Thus, the iron-magnetite reaction defines a straight line on an E_H–pH plot. The half-reaction for oxidation of magnetite to hematite is

$$3Fe_2O_{3_s} + 2H^+ + 2e^- = 2Fe_3O_{4_s} + H_2O_{liq}, \tag{10.56}$$

for which $\Delta G°$ is -42.7 kJ $(-10.2$ kcal) and so $E_H° = 0.221$ V. Following the same treatment as before:

$$E_H = 0.221 - 0.059 \; pH. \tag{10.57}$$

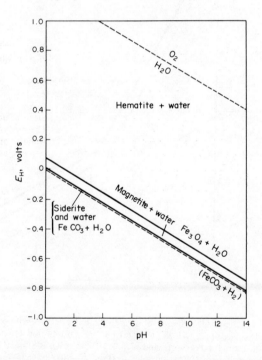

FIG. 10.3. Stability fields of iron oxides and siderite as functions of E_H and pH, at 25° C and 1 atm total pressure. Compare with Fig. 10.2.

This relationship defines a line parallel to that for (10.55); both are given on Fig. 10.2. The diagram shows that metallic iron is unstable in the presence of water at 25°C as the Fe/Fe_3O_4 reaction boundary lies below the lower stability limit of water.

The determination of the stability fields of the iron oxides in Fig. 10.2 is for a pure water–iron system, in the absence of any other component, such as CO_2, that is likely to exist in natural systems. These may clearly affect the stability fields of oxidation states of dissolved species and of mineral phases. The effect of atmospheric carbon dioxide, $p_{CO_2} = 0.000\,31\ (= 10^{-3.5})$ atm (or 31.4 Pa), on the stability of some iron compounds will be considered. The relevant reactions and E_H relationships are:

$$2FeCO_{3_s} + H_2O = Fe_2O_{3_s} + 2CO_{2_g} + 2J^+_{aq} + 2e^-, \tag{10.58}$$

siderite   hematite

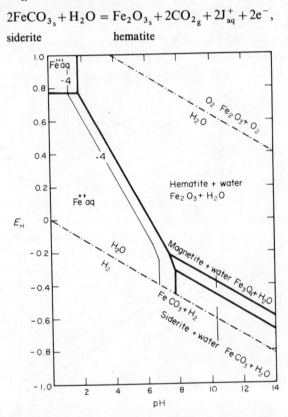

FIG. 10.4. Composite diagram showing the stability fields of iron oxides, siderite, and of dissolved ionic species Fe^{2+} and Fe^{3+}, for a total ion activity of 10^{-6}, at 25° C and 1 atm total pressure, and $p_{CO_2} = 10^{-2}$ atm. An additional boundary line for an ion activity of 10^{-4} is included to show the effect of changing activity. The stability fields are shown only within the boundaries of water stability. (After Garrels and Christ, 1965)

for which

$$E_H = 0.286 + 0.059 \log p_{CO_2} - 0.059 \text{ pH}, \qquad (10.59)$$

$$3FeCO_{3_s} + H_2O = Fe_3O_{4_s} + 3CO_{2_g} + 2H^+_{aq} + 2e^- \qquad (10.60)$$

magnetite

for which

$$E_H = 0.319 + 0.0885 \log p_{CO_2} - 0.059 \text{ pH}. \qquad (10.61)$$

The substitution of $p_{CO_2} = 10^{-3.5}$ gives the equations for the two reaction lines shown in Fig. 10.3, at 25°C and a total pressure of 1 atm.

Composite diagrams may also be constructed. These can show the stability fields of ionic species in equilibrium with the solids depicted on the diagram for a chosen value of the sum of the ion activities: an example is shown in Fig. 10.4. Note that the p_{CO_2} for this diagram is $10^{-2.0}$ atm, so giving a larger stability

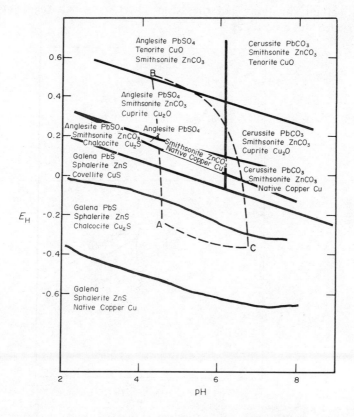

FIG. 10.5. Composite diagram of stability of some metal sulphides and their oxidation products, at 25° C and 1 atm total pressure. (After Garrels, 1954)

field to siderite compared with that in Fig. 10.3. Garrels and Christ (1965) give a detailed account of the construction and uses of composite diagrams.

Figure 10.5 shows the stability fields of the various oxidation products of common metal sulphides (Garrels, 1954). It acts as a good example of the use of such diagrams, for its shows, quantitatively, the roles of E_H and pH in the oxidation of ores undergoing a change in natural environment. An ore initially in environment C on Fig. 10.5 (i.e. before exposure of the primary sulphides at the surface of the Earth) is brought to environment B–exposed at the surface. Oxidation leads to production of acid ground waters which on progressive neutralization follow the path B–C in a carbonate rock system, or B–A–C in an igneous rock system, and precipitate the depicted phases at successive redox and pH conditions. This diagram (Fig. 10.5) was the result of one of the earlier attempts at applying E_H–pH relationships to weathering processes. Since then E_H–pH diagrams have become more refined by being better representations of likely natural situations, but the basic tenets on which the figure was constructed still hold today.

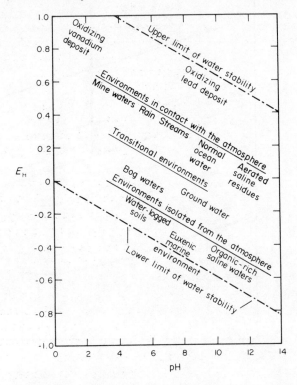

FIG. 10.6. Approximate position of some natural environments on an E_H–pH plot.
(After Garrels and Christ, 1965)

A diagram of the approximate pH–E_H positions of some natural environments was constructed by Garrels and Christ (1965) and is reproduced as Fig. 10.6. It may be used in conjunction with composite E_H–pH diagrams, such as Fig. 10.4, to help predict likely mineral occurrences. However, success in this has been somewhat limited. Composite E_H–pH diagrams can only represent the stable phases for the given chemical composition of a system at equilibrium. A natural situation, especially one involving weathering, may be some way from an equilibrium condition and is likely to have a chemical make-up which is more complex than one depicted on a composite diagram.

Suggested further reading

BERNER, R. A. (1971) *Principles of Chemical Sedimentology*. McGraw-Hill. 240pp.
GARRELS, R. M. and CHRIST, C. L. (1965) *Solutions, Minerals and Equilibria*. Harper. 450 pp.
GARRELS, R. M. and MACKENZIE, F. T. (1971) *Evolution of the Sedimentary Rocks*. Norton Chapters 10 and 11. 397 pp.
LERMAN, A. (1979) *Geochemical Processes, Water and Sediment Environments*. Wiley. 481 pp.
LOUGHNAN, F. C. (1969) *Chemical Weathering of the Silicate Minerals*. American Elsevier. 154 pp.
STUMM, W. and MORGAN, J. J. (1970) *Aquatic Chemistry. An Introduction Emphasizing Chemical Equilibria in Natural Waters*. Wiley. Chapters 7 and 8. 583 pp.

11

Chemical Oceanography

11.1 General data and composition

The oceans account for a little over 70% of the Earth's surface and comprise 98% of the hydrosphere. They represent a relatively well-mixed system of considerable mass and potential economic use. Despite their overall size, the oceans are sufficiently uniform to make description of their chemical nature relatively straightforward. However, organic activity plays a significant role in controlling the distribution of some elements, and this inevitably leads to greater complexity within the system. This chapter is concerned with those inorganic aspects of chemical oceanography that are helpful in giving an understanding of the broad chemical nature of the oceans and the effect that seawater has on the chemistry of parts of the lithosphere. Hence, this chapter discusses the chemical composition of seawater; the speciation and reactivity of the components; the factors affecting element flux into and from the oceans; possible changes in ocean composition with time; and the effect of seawater interaction with rocks of the ocean crust. We start by giving some general data on the ocean system and its chemical composition.

The general, physical data given in Table 11.1 do not call for much comment except that the pH value of 8.1 clearly needs some explanation.

TABLE 11.1. *The Oceans: General physical data*

Temperature,	average	5°C
	range	1°C (deep water) to $\sim$ 30°C (tropical surface water)
Pressure,	average	200 bars (20 MPa)
	common range	$\sim$ 1 bar (surface) to 540 bars (at 5500 m depth)
Density,	average	1.024 g ml^{-1}
Depth,	average	3730 m
	greatest	10850 m? (Mariana trench)
Total volume		1.36×10^9 km^3
Total mass		1.40×10^{24} g
Total area		3.62×10^8 km^2
pH		8.1 ± 0.2

At equilibrium with the atmospheric partial pressure of CO_2, the CO_3^{2-}–HCO_3^-–CO_2 system produces a pH of about 8. This system is undoubtedly the major determining factor of the pH value in ocean waters, and it is controlled, in part, by biological activity. It has been argued (e.g. Sillen, 1961) that certain mineral reactions also help to buffer the pH at the observed value. These may be of the following type (11.1):

$$3Al_2Si_2O_5(OH)_4 + 4SiO_2 + 2K^+ + 2Ca^{2+} + 9H_2O$$
$$\text{solid} \qquad\qquad \text{solid} \quad = 2KCaAl_3Si_5O_{16}(H_2O)_6 + 6H^+. \qquad (11.1)$$

Other types of reactions, possibly involving borate species, also may provide a small control on the pH.

The concentrations of the major constituents are given in Table 11.2. Eleven elements make up 99.9 % of the dissolved salts and the majority of these show little variation in concentration throughout the ocean: nitrogen is considered as a minor element since most of it occurs as the dissolved gas rather than salts. Silicon shows wide variation in its concentration (see below) and is also classified as a minor constituent.

As early as 1884 Dittmar noticed that the *ratios* of the concentrations of the majority of the major constituents that he determined did not change from place to place within the open ocean. He did note that calcium showed some variation with depth. We now know that the concentration ratios of seven major constituents are constant: Cl, Na, S, K, Br, Sr and B. Calcium is enriched in middle and deeper waters through biological activity, which also affects C concentrations, and Mg and F show small departures from constancy. However, the general constancy of the ratios of most of the major constituents has led to the use of *chlorinity* alone as a measure of the composition of seawater samples, as the concentration of the halogens are readily determinable. The chlorinity ($Cl°/_{oo}$) may be defined as the concentration (per ml) of chloride in seawater on the basis that all bromide and iodide have been replaced by chloride.* The chlorinity is related to the total salt content or *salinity* ($S°/_{oo}$) by the empirical relationship:†

$$S°/_{oo} = 1.80655 \ Cl°/_{oo}. \qquad (11.2)$$

Seawater normally has a salinity of about 35 °/$_{oo}$ with the range commonly from 34 to 36 °/$_{oo}$. The concentrations of elements in Tables 11.2 and 11.3 are

---

* Chlorinity is strictly defined as the mass of silver necessary to precipitate the halogens in 328.5233 g of seawater and is 1.000 43 times the mass in grams of chlorine equivalent to the mass of halogens contained in 1 kg of seawater (Pytkowicz, 1975).

† The electrical conductivity of seawater is almost directly proportional to the salinity. Hence, conductivity measurements can be used as a basis for the determination of salinity and chlorinity. A discussion on the concept and definition of salinity may be found in Wilson (1975).

TABLE 11.2. *Concentrations of the major, dissolved constituents of sea-water*

Element	Concentration	
	Molar	$mg\,l^{-1}$
Chlorine	5.46×10^{-1}	1.94×10^4
Sodium	4.68×10^{-1}	1.077×10^4
Magnesium	5.32×10^{-2}	1.29×10^3
Sulphur	2.82×10^{-2}	9.05×10^2
Calcium	1.03×10^{-2}	4.12×10^2
Potassium	1.02×10^{-2}	3.8×10^2
Bromine	8.4×10^{-4}	67
Carbon	2.3×10^{-3}	28
Strontium	9.1×10^{-5}	8
Boron	4.1×10^{-4}	4.44
Fluorine	6.8×10^{-5}	1.3

Source: Brewer (1975) with minor modifications.

TABLE 11.3. *Geochemical parameters of seawater* $(S = 35\,{}^{\circ}\!/_{\circ\circ})$

Element	Concentration		Principal species	Residence time (years)
	molar	$mg\,l^{-1}$		
He	1.7×10^{-9}	6.8×10^{-6}	He gas	
Li	2.6×10^{-5}	0.18	Li^+	2.3×10^6
Be	6.3×10^{-10}	5.6×10^{-6}	$BeOH^+$	
B	4.1×10^{-4}	4.44	H_3BO_3, $B(OH)_4^-$	1.3×10^7
C	2.3×10^{-3}	28	HCO_3^-, CO_3^{2-}	
N			N_2, NO_3^-, NO_2^-	
O			O_2 gas	
F	6.8×10^{-5}	1.3	F^-, MgF^+	5.2×10^5
Ne	7×10^{-9}	1.2×10^{-4}	Ne gas	
Na	0.468	1.077×10^4	Na^+	6.8×10^7
Mg	5.32×10^{-2}	1.29×10^3	Mg^{2+}	1.2×10^7
Al	7.4×10^{-8}	2×10^{-3}	$Al(OH)_4^-$	1.0×10^2
Si	7.1×10^{-5}	2	$Si(OH)_4$	1.8×10^4
P	2×10^{-6}	6×10^{-2}	HPO_4^{2-}, $MgPO_4^-$	1.8×10^5
S	2.82×10^{-2}	9.05×10^2	SO_4^{2-}, $NaSO_4^-$	
Cl	0.546	1.94×10^4	Cl^-	1×10^8
Ar	1.1×10^{-5}	0.43	Ar gas	
K	1.02×10^{-2}	3.8×10^2	K^+	7×10^6
Ca	1.03×10^{-2}	4.12×10^2	Ca^{2+}	1×10^6
Sc	1.3×10^{-11}	6×10^{-7}	$Sc(OH)_3$	4×10^4
Ti	2×10^{-8}	1×10^{-3}	$Ti(OH)_4$	1.3×10^4
V	5×10^{-8}	2.5×10^{-3}	$H_2VO_4^-$	8×10^4
Cr	5.7×10^{-9}	3×10^{-4}	$Cr(OH)_3$, CrO_4^{2-}	6×10^3
Mn	3.6×10^{-9}	2×10^{-4}	Mn^{2+}, $MnCl^+$	1×10^4
Fe	3.5×10^{-8}	2×10^{-3}	$Fe(OH)_2^+$	2×10^2
Co	8×10^{-10}	5×10^{-5}	Co^{2+}, $CoCO_3$	3×10^4
Ni	2.8×10^{-8}	1.7×10^{-3}	Ni^{2+}	9×10^4
Cu	8×10^{-9}	5×10^{-4}	$CuCO_3$, $CuOH^+$	2×10^4

TABLE 11.3. *(cont.)*

Element	Concentration		Principal species	Residence time (years)
	molar	mg l^{-1}		
Zn	7.6×10^{-8}	4.9×10^{-3}	$ZnOH^+$, Zn^{2+}, $ZnCO_3$	2×10^4
Ga	4.3×10^{-10}	3×10^{-5}	$Ga(OH)_4^-$	1×10^4
Ge	6.9×10^{-10}	5×10^{-5}	$GeO(OH)_3^-$	—
As	5×10^{-8}	3.7×10^{-3}	$HAsO_4^{2-}$	5×10^4
Se	2.5×10^{-9}	2×10^{-4}	SeO_3^{2-}	2×10^4
Br	8.4×10^{-4}	67	Br^-	1×10^8
Kr	2.4×10^{-9}	2×10^{-4}	Kr gas	—
Rb	1.4×10^{-6}	0.12	Rb^+	4×10^6
Sr	9.1×10^{-5}	8	Sr^{2+}	4×10^6
Y	1.5×10^{-11}	1.3×10^{-6}	$Y(OH)_3$	—
Zr	3.3×10^{-10}	3×10^{-5}	$Zr(OH)_4$	—
Nb	1×10^{-10}	1×10^{-5}	—	—
Mo	1×10^{-7}	1×10^{-2}	MoO_4^{2-}	2×10^5
Tc	—	—	—	—
Ru	—	—	—	—
Rh	—	—	—	—
Pd	—	—	—	—
Ag	4×10^{-10}	4×10^{-5}	$AgCl_2^-$	4×10^4
Cd	1×10^{-9}	1×10^{-4}	$CdCl_2$	—
In	8×10^{-13}	1×10^{-7}	$In(OH)_2^+$	—
Sn	8.4×10^{-11}	1×10^{-5}	$SnO(OH)_3^-$	—
Sb	2×10^{-9}	2.4×10^{-4}	$Sb(OH)_6^-$	7×10^3
Te	—	—	$HTeO_3$	—
I	5×10^{-7}	6×10^{-2}	I^-, IO_3^-	4×10^5
Xe	3.8×10^{-10}	5×10^{-5}	Xe gas	—
Cs	3×10^{-9}	4×10^{-4}	Cs^+	6×10^5
Ba	1.5×10^{-7}	2×10^{-2}	Ba^{2+}	4×10^4
La	2×10^{-11}	3×10^{-6}	$La(OH)_3$	6×10^2
Ce	1×10^{-11}	1×10^{-6}	$Ce(OH)_3$	—
Pr	4×10^{-12}	6×10^{-7}	$Pr(OH)_3$	—
Nd	1.9×10^{-11}	3×10^{-6}	$Nd(OH)_3$	—
Sm	3×10^{-13}	5×10^{-8}	$Sm(OH)_3$	—
Eu	7×10^{-14}	1×10^{-8}	$Eu(OH)_3$	—
Gd	4×10^{-12}	7×10^{-7}	$Gd(OH)_3$	—
Tb	9×10^{-13}	1×10^{-7}	$Tb(OH)_3$	—
Dy	6×10^{-12}	9×10^{-7}	$Dy(OH)_3$	—
Ho	1×10^{-12}	2×10^{-7}	$Ho(OH)_3$	—
Er	4×10^{-12}	8×10^{-7}	$Er(OH)_3$	—
Tm	1×10^{-12}	2×10^{-7}	$Tm(OH)_3$	—
Yb	5×10^{-12}	8×10^{-7}	$Yb(OH)_3$	—
Lu	1×10^{-12}	2×10^{-7}	$Lu(OH)_3$	—
Hf	4×10^{-11}	7×10^{-6}		—
Ta	1×10^{-11}	2×10^{-6}	—	—
W	5×10^{-10}	1×10^{-4}	WO_4^{2-}	1.2×10^5
Re	2×10^{-11}	4×10^{-6}	ReO_4^-	—
Os	—	—	—	—
Ir	—	—	—	—
Pt	—	—	—	—
Au	2×10^{-11}	4×10^{-6}	$AuCl_2^-$	2×10^5
Hg	1.5×10^{-10}	3×10^{-5}	$HgCl_4^{2-}$, $HgCl_2$	8×10^4

IG - J*

TABLE 11.3. (*cont.*)

| Element | Concentration | | Principal species | Residence time (years) |
	molar	$mg\,l^{-1}$		
Tl	5×10^{-11}	1×10^{-5}	—	—
Pb	2×10^{-10}	3×10^{-5}	$PbCO_3$, $Pb(CO_3)_2^{2-}$	4×10^2
Bi	1×10^{-10}	2×10^{-5}	BiO^+, $Bi(OH)_2^+$	—
Po	—	—	—	—
At	—	—	—	—
Rn	2.7×10^{-21}	6×10^{-16}	Rn gas	—
Fr	—	—	—	—
Ra	3×10^{-16}	7×10^{-11}	Ra^{2+}	—
Ac	—	—	—	—
Th	4×10^{-11}	1×10^{-5}	$Th(OH)_4$	60
Pa	2×10^{-16}	5×10^{-11}	—	—
U	1.4×10^{-8}	3.2×10^{-3}	$UO_2(CO_3)_3^{4-}$	3×10^6

Based on Brewer (1975) and Stumm and Brauner (1975) and other sources.

for the $35\,\%_{oo}$ value. It should be noted, however, that in some work a standard chlorinity of $19\,\%_{oo}$ is adopted, which gives a standard salinity of $34.32\,\%_{oo}$ on the basis of relationship (11.2) above.

As a result of biological activity many of the minor elements do not exhibit constant ratios in their concentrations and so are not necessarily related in abundance to salinity. The average concentrations of the minor elements, together with the major constituents, are given in atomic number sequence in Table 11.3. The biological activity is greatest in the surface layer of ocean water and so those elements acting as nutrients will tend to be depleted at the surface as they are removed by dead, sinking organisms. Hence, the biological activity in surface waters is controlled by the availability of nutrient elements.

Elements may be classified into three categories on the basis of their involvement in biological activity:

Biolimiting elements: those which are almost totally depleted in surface water relative to deep water. There are only three known: N, P and Si.

Biointermediate elements: those which are partially depleted. Four elements are known in this category but others may later be found to be biointermediate. The four are Ba, Ca, C, Ra.

Biounlimited elements: no depletion in surface waters so that their concentration ratios are constant in all samples, whether from surface or deep waters. Examples are B, Mg, Sr, S, the alkali metals, the halogens, and the inert gases.

Upwelling, the flux of which is estimated to be about twenty times that of river water, brings back biolimiting elements to the surface. The concentrations of some of the dissolved gases are affected by biological activity. Oxygen concentration is determined by both physical and biological processes but the concentration of many gases is principally a function of temperature, pressure and salinity.

The ocean is an open system in which a number of processes contribute to the incoming and outgoing fluxes of material (Fig. 11.1). However, there is evidence, as discussed below, to suggest that the system is at a steady state and has been so for a significant part of geological time. Therefore, we must view the fluxes of material in this light.

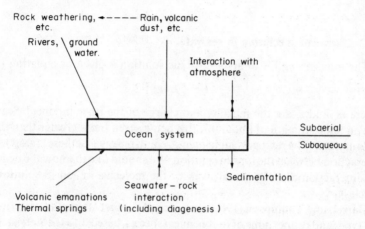

Fig. 11.1. Diagrammatic representation of influx and efflux to and from ocean water.

The influx of dissolved material in river waters is estimated (Livingstone, 1963) to be about 3.9×10^{12} kg a^{-1}, and the amount of suspended material is about 10^{13} kg a^{-1} (Garrels and Mackenzie, 1971). Most of the solid material is deposited (along with precipitation and adsorption of a little of the dissolved material) as sediments on the continental shelves and in sedimentary basins. Tectonic processes provide a mechanism for the efflux of these deposits from the ocean system.

Subaerial and subaqueous volcanic activity is an important contributor to the fluxes of both particulate and gaseous material but the extent of the process is not known. Interaction of newly created sea-floor rocks at ocean ridges with seawater causes exchange reactions that are significant in the mass balance of elements. Diagenesis of sediments is also a contributor in this way. Effluxes include the loss of material as aerosols created at the atmosphere–ocean interface, and by evaporation including some elements or compounds besides water. For example, iodide ions are oxidized at the surface and may be removed as elemental iodine.

As has been indicated, the biomass plays a big role in the geochemical behaviour of some elements. Discussion of this role is beyond the scope of this book, and readers interested in the topic are referred to the volumes edited by Riley and Skirrow (1975–1978) on Chemical Oceanography. This chapter emphasizes the inorganic chemical controls on the seawater chemistry and starts by considering the possible inorganic species that are likely to exist, of the dissolved constituents.

11.2 Chemical speciation in seawater

The ionic strength I of an electrolytic solution is given by equation (11.3):

$$I = \tfrac{1}{2} \Sigma m_i z_i^2 \tag{11.3}$$

where m_i and z_i are the molality and charge of the ith component. Seawater has an ionic strength of about 0.7, this being seven times greater than that of stream water. The relatively high ionic strength means that there is a significant interaction between the ions in solution. We should like to know the degree of the interaction and the actual form of the molecule or ion in solution (i.e. species).

Garrels and Thompson (1962) made one of the first attempts at determining the types and proportions of the chemical species, based on a theoretical model with the use of dissociation constants and individual ion activity coefficients of the major components. At that time Garrels and Thompson were forced to make assumptions about the values of some activity coefficients because the required data were not available. Although there may be some inaccuracies in the calculated proportions of the species, which can be corrected when there is a better understanding of activity–composition relationships in seawater, none the less many of the conclusions made by Garrels and Thompson have been confirmed by later experimental work. Their pioneering work, using an approach which has been adopted by many other workers, has greatly aided our understanding of the chemistry of ocean waters.

Dissociation constants for the appropriate components are given in Table 11.4. Study of this table reveals that there can be only eighteen important species present in seawater, as given in Table 11.5.

In order to be able to determine the proportions of these species it is necessary to use mass balance equations of the kind (11.4) (e.g. for K)

$$m_{\mathrm{K}^+} + m_{\mathrm{KSO_4^-}} = m_{\mathrm{K}} \text{ total} \tag{11.4}$$

for all species. Furthermore, to be able to use quantitatively the dissociation

TABLE 11.4. *Dissociation constants* $(-\log K)$ *for major species* $(25°C)$

	OH^-	HCO_3^-	CO_3^{2-}	SO_4^{2-}	Cl^-
H^+	14.0	6.4	10.33	2	—[†]
K^+	—	—	—	0.96	—
Na^+	−0.7	−0.25	1.27	0.72	—
Ca^{2+}	1.30	1.26	3.2	2.31	—
Mg^{2+}	2.58	1.16	3.4	2.36	—

[†] A dash denotes no measurable association.

TABLE 11.5. *Likely major species in seawater*

K^+	Ca^{2+}	$MgCO_3$
KSO_4^-	$CaHCO_3^+$	$MgSO_4$
Na^+	$CaCO_3$	HCO_3^-
$NaCO_3^-$	$CaSO_4$	CO_3^{2-}
$NaHCO_3^\circ$	Mg^{2+}	SO_4^{2-}
$NaSO_4^-$	$MgHCO_3^+$	Cl^-

constants in Table 11.4, values of activity coefficients (γ) are required as shown by relationship (11.5) using potassium as an example.

$$K_{KSO_4^-} = \frac{a_{K^+} a_{SO_4^{2-}}}{a_{KSO_4^-}} = \frac{m_{K^+} \gamma_{K^+} m_{SO_4^{2-}} \gamma_{SO_4^{2-}}}{m_{KSO_4^-} \gamma_{KSO_4^-}}. \tag{11.5}$$

It is at this point that assumptions about the γ-values had to be made because of lack of data for some species at an ionic strength of 0.7. Uncharged species were assumed to have a γ-value of 1.13 (that of H_2CO_3). The activity coefficients of $NaCO_3^-$, $MgHCO_3^+$, $CaHCO_3^+$, KSO_4^- were assumed to be the same as that of HCO_3^-, i.e. 0.68. Coefficients for the remaining species were calculated from available data or from experiment.

The activity coefficients and measured molalities (m_i total) are substituted into equations of the type (11.4) and (11.5), but their solution is not possible by standard methods because there are too many unknowns. As a first step in solving the equations, it can be assumed that the *cations* exist predominantly as dissociated species. This approximation is reasonable in the light of the dissociation constants (Table 11.4) and reduces the unknowns to a number that allows solution of the equations. The initial, approximate results so obtained for the associated *anions* are then used in a repeat of the calculations but without any assumptions regarding the nature of the cation species. Further iterations lead to the final calculated proportions in Table 11.6. It can be seen that only a small proportion of the cations exist as ion-pairs. Furthermore, calcium and magnesium exist more as ion pairs, especially with SO_4^{2-}, than do Na^+ and K^+.

TABLE 11.6. *Major dissolved species in seawater*
(*Garrels and Thompson*, 1962)

Metal ion Me	Conc. total, molal	Free ion %	$MeSO_4$ %	$MeHCO_3$ %	$MeCO_3$ %
Ca^{2+}	0.010	91	8	1	0.2
Mg^{2+}	0.054	87	11	1	0.3
Na^+	0.48	99	1.2	0.01	—
K^+	0.010	99	1	—	—

Ligand L	Conc. total, molal	Free ion %	CaL %	MgL %	NaL %	KL %
SO_4^{2-}	0.028	54	3	21.5	21	0.5
HCO_3^-	0.0024	69	4	19	8	—
CO_3^{2-}	0.00027	9	7	67	17	—
Cl^-	0.56	100	—	—	—	—

A similar approach to that adopted by Garrels and Thompson has been used to estimate the proportions of different species of the minor constituents (e.g. for Ba, Hanov, 1969). An interesting study of the speciation of Cd, Cu, Pb and Zn by Long and Angino (1977) showed the effect on speciation of mixing fresh-water with seawater. For zinc, chloride becomes an important ligand after the addition of only small amounts of seawater to fresh-water. As the amount of chloride in solution increases, so higher-order complexes (e.g. $ZnCl_2$, $ZnCl_3^-$ and $ZnCl_4^{2-}$) occur, but $ZnCl^+$ is always the commonest species in the zinc associations with chlorine (Fig. 11.2). The importance of pH is well shown in Fig. 11.2. At the pH of seawater (i.e. 8.1) the $Zn(OH)_2$ dissolved species is the commonest. A similar situation occurs in the case of copper, but with lead and cadmium the commonest species at pH 8.1 are $PbCl^-$ and $CdCl^-$ respectively. At this pH, MCO_3 associations are also common for Zn, Pb and Cu.

In a review of speciation in seawater, Stumm and Brauner (1975) point out that the high electronegativity of fluorine indicates some association of F^- with calcium and magnesium; the estimated proportions being F^- (51%), MgF^+ (47%) and CaF^+ (2%). Consideration of the speciation of elements such as F, Si, and B is important because of their relatively high concentrations and their possible association with other major constituents. Table 11.3 lists the probable main species of many of the constituents.

Improvements are continually occurring to our knowledge of chemical speciation in seawater. New methods of approach are being tested but it is interesting to note that in a consistency test for single ion activity coefficients in seawater, the model by Garrels and Thompson (1962) yields activity ratios within 10% of the thermodynamic consistent values (Nesbitt, 1980).

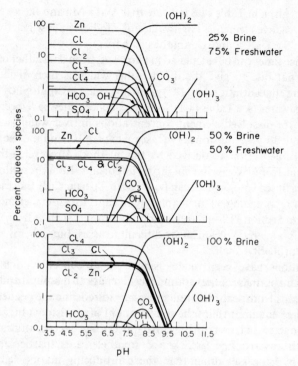

FIG. 11.2 Zinc chemical speciation in brine–freshwater mixtures. (After Long and Angino, 1977)

11.3 Residence times

There is evidence (see Section 11.7) that the composition of seawater has undergone no systematic change with time during, at least, much of the Phanerozoic and, therefore, that the ocean may be treated as a steady-state system in which the influx of elements is balanced by their efflux. This probable balance led Barth (1952) to use the concept of residence time, which he defined by equation (11.6):

$$\tau = A/(\mathrm{d}A/\mathrm{d}t) \tag{11.6}$$

where A is the total amount of an element dissolved in the ocean; and $\mathrm{d}A/\mathrm{d}t$ is the rate of influx or efflux of that element. The residence time can be seen as a measure of the reactivity of an element in the ocean system. If τ has a low value then the element on its entry must react relatively quickly and be removed by precipitation, exchange, uptake or adsorption processes involving other dissolved elements, the biomass or rocks and particulate matter. Residence time values given in Table 11.3 indicate that Al, Ti, Mn and Ba are examples of relatively reactive elements. High values of τ indicate a low reactivity such as shown by Na, Cl, Ca and U, amongst others.

Residence times can be calculated in two ways, from data either on input rate or on output rate to give the dA/dt value which is then divided into the appropriate concentration (mg l^{-1}) given in Table 11.3, multiplied by the total mass of seawater (see Table 11.1). Many residence times have been calculated from data for dissolved element concentrations in river water (such as those data complied by Livingstone, 1963; see Chapter 10) and most of the values given in Table 11.3 were derived by this method. However, other influxes besides that from river water might be important. Influx resulting from the interaction of seawater with ocean floor rocks—particularly those at centres of ocean floor spreading is probably significant for some elements and is discussed in Section 11.5. The effects of other inputs such as from dust storms or from sub-aqueous volcanism are difficult to assess but, other than on a local scale, are probably small.

Calculations have been made using element concentrations in massive sediments but as these concentrations represent both dissolved and particulate contributions from seawater, it is necessary to redefine the residence time by making A of equation (11.6) the total amount of an element in dissolved *and* particulate states in the ocean. Because appropriate data for values of A of this new definition are often lacking and some elements (particularly Mg) are removed by processes other than those producing marine sediments (see Section 11.5), the calculation of residence times by this method cannot be recommended.

There is a pattern to the residence time values listed in Table 11.3. The alkali and alkaline earth elements, together with magnesium, have long residence times ($\geqslant 10^6$ years). The halogens tend to have high τ values, while Si, P, As, Se and most of the transition elements have intermediate values. Al, Cr and Fe have relatively short residence times, probably because of their propensity to form insoluble hydroxides. Thorium also readily forms an insoluble hydroxide and this element has a particularly low τ value. However, not a lot of significance can be attached to the differences in the stated values of τ when these are small, since with these cases there are special difficulties involved in their measurement and interpretation.

It is implicit within the concept of residence times, as defined by relationship (11.6), that the ocean system is compositionally homogeneous, i.e. that seawater is well mixed. However, the 'mixing time' of the upper 100 m of ocean water is about 10a and that of deep ocean water is in the order of 1000a. Only elements with residence times greater than 1000a will be distributed homogeneously, provided that they are not nutrients. Of the elements for which data are available, Al, Fe, La, Pb and Th all have residence times of less than 1000a and will not, therefore, be homogeneously distributed. For such elements it becomes a very difficult, if not impossible, task to obtain a realistic assessment of the value of A in equation (11.6), and so alternative approaches in determining residence times have to be adopted. In the case of Th it is possible to recognize the short residence time from consideration of the rate of decay of isotopes in the ^{238}U series (see Table 9.1). In this the first steps are:

$$^{238}U \xrightarrow[t_{1/2} = 4.47 \times 10^9 \, a]{\alpha} {}^{234}Th \xrightarrow[24.1 \text{ days}]{\beta} {}^{234}Pa \longrightarrow {}^{234}U.$$

The daughter product, ^{234}Th, has a considerably shorter half-life than the parent, ^{234}U, so that a steady-state condition can be achieved between the rate of production of ^{234}Th and its decay provided the system is undisturbed. This condition is called secular equilibrium and is defined by equation (11.7):

$$\lambda_{^{238}U} \cdot N_{^{238}U} = \lambda_{^{234}Th} \cdot N_{^{234}Th}, \tag{11.7}$$

where N_i is the number of atoms of isotope i. Hence, if the decay constants are known and the concentration of ^{238}U is determined, then the secular equilibrium concentration of ^{234}Th can be calculated. Uranium-238 has a nearly uniform concentration in open ocean waters but the ^{234}Th concentration is found to be below its secular equilibrium value, and so thorium must be being removed quite rapidly after its formation. Estimates based on evidence of this kind give a residence time to thorium in deep ocean water of about 60 years.

Where the residence time exceeds 1000a, the element concentration will tend to be uniform throughout the bulk of ocean water, but highly variable concentrations have been recorded for some elements (e.g. Co, Robertson, 1970) in regional surface waters. Some elements show a systematic variation in concentration with depth as a result of biological controls such as the downward flux of calcareous tests and other organic detritus. Hence calcium shows enrichment with depth and so do the nutrient elements phosphorous and nitrogen. Variations in some trace-element concentrations are controlled biologically, including those shown by a number of the transition elements (e.g. Ni) but we still lack an adequate knowledge of the controls.

As residence time is a measure of element reactivity it should be possible to correlate a residence-time value with the process (or processes) by which an element is removed from seawater. A first step in this direction is to investigate the balance between the measured input and measured removal of an element. Such a task has proved not to be simple for any element yet studied but needs to be pursued if a fuller understanding of element reactivity is to be obtained.

11.4 Mass balances

If the ocean is in a steady state with respect to composition then the influx of any element should be balanced exactly by the removal of that element, and it is clearly desirable to check if this is true. No convincing demonstration of an exact mass balance for any element has yet been made, but this failure probably stems more from our ignorance about influx and removal processes than from the possibility of the ocean system not being at a steady state. Two attempts (Burton and Liss, 1973; Wollast, 1974) to establish details of the balance of silica in seawater serve to illustrate the problems that are encountered.

The major process for the removal of silica is considered by Burton and Liss to be the generation of marine biogeneous oozes. On the basis of previously published estimates of the areal extent and rates of deposition of the oozes, they calculated that this process removed a maximum of 1.9×10^{14} g $SiO_2 a^{-1}$. Three influx processes are significant: inflow of river water; release of silica from sediment pore waters; and the weathering of Antarctica. Exchange reactions between seawater and rocks were not considered to be significant but the loss of some silica through reaction of river water with amorphous material during passage through estuaries has an effect on the silica balance. The evidence for the contribution of silica from weathering of the Antarctic land mass came from the fact that the deep water supply to the oceans from the region is higher in dissolved silica than the inflowing waters. Furthermore, earlier estimates suggest that almost 14×10^{14} g SiO_2 a^{-1} are eroded from the Antarctic land mass, of which only about 5×10^{14} g a^{-1} are deposited as glacial marine sediments.

The tentative 'budget' for silica as calculated by Burton and Liss is given in Table 11.7, and it can be seen that there is a large excess in influx over the estimated removal. Other removal processes must operate. Sillen (1961) had postulated that silica is involved to a significant extent in the production of marine authigenic minerals—a process excluded from the silica budget of Burton and Liss. Wollast (1974) attempted to estimate the extent of this effect

TABLE 11.7. *Estimated 'budget' for silica in the ocean*

	Influx (g a^{-1})			Removal (g a^{-1})	
	Burton and Liss (1973)	Wollast (1974)		Burton and Liss	Wollast
Rivers	4.3×10^{14}	4.3×10^{14}	Sediments	$< 1.9 \times 10^{14}$	7.5×10^{14}
Pore water	1×10^{14}	3.2×10^{14}	Estuarine	0.9×10^{14}	
Antarctica	$\sim 7 \times 10^{14}$				
	$\sim 12.3 \times 10^{14}$	7.5×10^{14}		$< 2.8 \times 10^{14}$	7.5×10^{14}

which he tentatively put at about 3.5×10^{14} g a^{-1}. His estimate of the balance is given in Table 11.7, which shows equal totals for influx and efflux. Of the 7.5 $\times 10^{14}$ g a^{-1} silica initially deposited, 6.7×10^{14} g a^{-1} are dissolved, of which 3.5×10^{14} g a^{-1} react with clay minerals and thereby are retained by the sediment. The remainder $(3.2 \times 10^{14}$ g a$^{-1})$ in the interstitial water of the sediments is returned to the oceans.

The marked discrepancy between the two estimates shows that much more research is needed into the production of marine authigenic minerals and of the return of silica to the ocean from interstitial water. Our knowledge of mass balances is still very sketchy, but in the case of some elements—notably, magnesium—some significant advances in our understanding are being made.

An early assessment (Drever, 1974) of the modes of removal of magnesium from seawater indicated that less than 50% of the estimated influx by river waters could be balanced by the removal of the element in carbonate and glauconite formation, ion-exchange and burial of pore-fluids. Either the amount of magnesium in seawater is increasing or the estimates of influx and/or removal are seriously in error.

Experimental investigations of the interaction of seawater with rock has shown that one of the products of hydrothermal alteration of basalt is a smectite (generalized formula: $(\frac{1}{2}Ca, Na)_{0.7}(Al, Mg, Fe)_4[(Si, Al)_8 O_{20}](OH)_4 \cdot nH_2O)$. Elderfield (1977) suggests that at high temperatures (of the order of 300° C) the smectite will be Mg-rich, but at low temperatures ($< 100°C$) it is likely to be Fe-rich, and that the basalt–seawater system provides a major sink for magnesium. The following section discusses rock–seawater interaction and although quantitative evidence is still wanting, it seems possible that this process could account for the apparent disparity between magnesium inflow and removal rates.

For many elements the major influx appears to be from the weathering of continental rocks and the transport of the dissolved constituents by river waters. The total amount of dissolved sodium in seawater can be accounted for by this process. Other elements are too abundant to have been derived solely from weathering. This is particularly so in the cases of Cl, Br, and S, all of which are likely to have been contributed in large measure from volcanic emanations. For this reason they have been referrred to as the 'excess volatiles'. Besides Mg, it also seems probable that interaction of seawater with ocean floor rocks could be a significant contributor to the mass balances of calcium, silicon and a number of minor constituents.

11.5 Interaction of seawater with rocks

The change in chemical conditions accompanying the movement of particulate matter from river water to seawater and then to the location of sedimentation provides a driving force for solid–seawater interactions and re-equilibration. Similarly, the igneous rocks of the ocean floor, when fresh, are not in equilibrium with seawater and tend to undergo alteration either at the relatively low temperatures existing near the surface of much of the oceanic crust or at the higher temperatures encountered at places where new sea-floor is being generated. Recent research has led to a rapid development in our knowledge of the effects of these interactions especially those involving seawater and basalt. This section emphasizes some of the recent work.

Reaction of seawater with solid material whether it be as a particulate suspension, as bottom sediments or as igneous rocks of the ocean crust, is termed *halmyrolysis* (but some authors restrict use of the term to sediments alone). In sediments, the process is considered to take place before the action of diagenesis; i.e. it occurs when the particulate matter is in direct contact with the seawater. Clay minerals with marked cation exchange properties, such as the smectites (montmorillonite group) and especially montmorillonite itself $[(Na)_{0.7} (Al_{3.3} Mg_{0.7}) Si_8 O_{20}(OH)_4 \cdot nH_2O]$, are important in halmyrolysis as they take up Mg^{2+} and Na^+ from solution in exchange for Ca^{2+} and K^+. Furthermore, formation of montmorillonite and the clay mineral, illite, is favoured under alkaline conditions and so may be produced by the conversion of a 'cation-free' clay such as kaolinite, (11.8):

$$3Al_2Si_2O_5(OH)_4 + 2K^+ + 2HCO_3^- = 2KAl_3Si_3O_{10}(OH)_2 + 2CO_2$$

kaolinite   illite $+ 5H_2O$ (11.8)

Reactions of this kind, which are a form of 'reverse weathering' as new minerals are formed and CO_2 is returned to the atmosphere, were predicted by Mackenzie and Garrels (1966) to occur, for they help to remove the excess HCO_3^- ions brought in by rivers. Rivers provide more than two bicarbonate ions for every calcium ion so that not all the HCO_3^- can be removed by carbonate precipitation (11.9):

$$2HCO_3^- + Ca^{2+} = CaCO_3 + CO_2 + H_2O. \qquad (11.9)$$

However, there does not seem to be any evidence for the existence of reaction (11.8) in the ocean system, and it is probably necessary to consider the role of HCO_3^- in the formation of montmorillonite during the alteration of ocean floor basalt, as well as in the formation of authigenic alkali feldspars (Kastner, 1974) by reactions of the type (11.10):

$$Al_2Si_2O_5(OH)_4 + 2Na^+ + 2HCO_3^- + 4H_4SiO_4 = 2NaAlSi_3O_8 + 2CO_2$$

kaolinite   feldspar
                                                    $+ 11H_2O$ (11.10)

The nature of the interaction of seawater with oceanic igneous rocks is sensitive to temperature. At low temperatures ($\leqslant 25°C$) the alteration of crystalline or glassy basalt by seawater produces the changes given qualitatively in Table 11.8. The rates of change are probably slow, and the overall changes depend to some extent upon which secondary minerals develop. One point of particular interest is that low-temperature alteration causes addition of magnesium to seawater.

TABLE 11.8. *Qualitative changes in rock composition during interaction with seawater*

	Low temperature ($\leqslant 25°C$)	High temperature
SiO_2	Decrease (small)	Decrease
TiO_2	Inconsistent	Increase
Al_2O_3	Decrease (small)	Inconsistent
ΣFe	Decrease	Decrease
Fe_2O_3/FeO	Increase (large)	Increase (large)
CaO	Decrease (large)	Decrease (large)
MgO	Decrease	Increase (large)
K_2O	Increase (large)	Inconsistent
Na_2O	Inconsistent	Inconsistent
P_2O_5	Increase	Inconsistent
H_2O	Increase (large)	Increase (large)
Trace elements		
Li	Increase (large)	
B	Increase	Redistributed
Cl	Increase	
Cr	No change	No change
Mn	Inconsistent	
Co	No change	Redistributed
Cu	Inconsistent	Decrease
Rb	Increase	
Sr	Inconsistent	Decrease
Y	No change	No change
Cs	Increase	
Ba	Inconsistent	
U	Increase	

Table based on data for variably altered basalt specimens collected from the sea-floor, as reported by R. Hart (1970); Thompson (1973); S. R. Hart *et al.* (1974); Shido *et al.* (1974); Storzer and Selo (1976); Humphris and Thompson (1988 a, b).

Quantitative chemical and detailed mineralogical studies of low-temperature rock alteration are needed before much can be said of the relative significance of the phenomenon to mass balances of the ocean system. Such studies are made difficult by the slow reaction rates—a difficulty which does not arise to the same extent in high-temperature alteration. Numerous experimental studies of the alteration of basalt by seawater at elevated temperature have been made in recent years. These, for the most part, have been within the temperature range 70° to 350°C and at pressures of 500 to 1000 bars (5.10^7 to 10^8 Pa), i.e. at pressures commensurate with those occurring at the sea-floor. The general chemical changes experienced by this saline solution during these experiments are given in Table 11.9, and examples of the rates of change in Fig. 11.3 for a basalt and Fig. 11.4 for a diabase. The relevance of

TABLE 11.9. *Qualitative changes in the solution during experimental interaction of seawater with basalt*
(T: 70° to 350°C; P: 500 to 1000 bars; water/rock mass ratio $\leqslant$ 10)

pH	slightly alkaline to slightly acid		
Eh	oxidizing to reducing		
Composition:			
Major elements		Minor elements	
Na	Decrease	Fe	Increase
Mg	Almost total removal	Mn	Increase
$SO_4^=$	Almost total removal	Cu	Increase
Ca	Increase (large)	Ba	Increase
K	Increase at $T \geqslant 150°C$		
	Decrease at $T = 70°C$		
Sr	Inconsistent		
B	Increase		

Based on data in: Bischoff and Dickson (1975); Hajash (1975); Bischoff and Seyfried (1978); Mottl and Holland (1978); Seyfried and Bischoff (1979); Mottl *et al.* (1979).

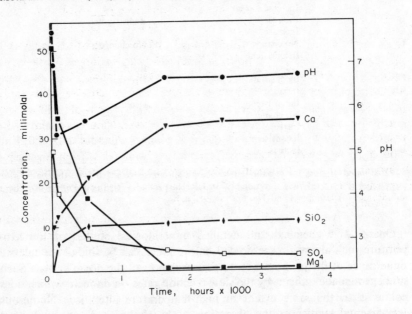

FIG. 11.3. Concentrations of Ca, Mg, SiO₂, SO_4^{2-} and pH in seawater during reaction with basalt glass at 150°C, 500 bars and water/rock mass ratio of 10. (After Seyfried and Bischoff, 1979)

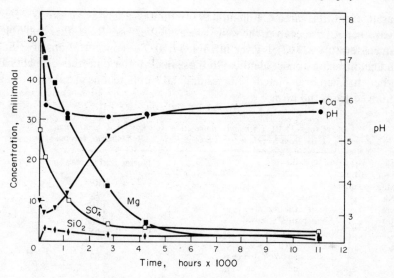

FIG. 11.4. Concentrations of Ca, Mg, SiO$_2$, SO$_4^{2-}$ and pH in seawater during reaction with diabase at 150° C, 500 bars and water/rock mass ratio of 10. (After Seyfried and Bischoff, 1979)

these experiments to quantitative estimation of the effects of seawater–rock interaction on seawater composition and mass balances hinges on the mass ratio of the interacting water and rock. The experimental work has often used water/rock ratios of $\leqslant 10$ and these are probably much lower than those involved in alteration processes at the sea-floor. Change in pH will be very small when a large volume of seawater reacts with a sea-floor basalt and so the resultant mineral assemblage and chemical changes will tend to be different. This could be one of the causes of magnesium being taken from basalt into seawater during low-temperature alteration but being almost quantitatively removed from solution during the experiments at high temperature (Tables 11.8 and 11.9).

However, the experimental work does reproduce the chemical changes seen in natural systems. Compositional differences between the fresh and altered portions of pillow basalts dredged from the Mid-Atlantic Ridge, which have a mineral assemblage of the greenschist facies of metamorphism, are as predicted from the experiments (Tables 11.8 and 11.9). Seawater is the source of the solutions in the wells of the geothermal area of Reykjanes Peninsula, Iceland, but its composition has been changed dramatically through inter-action with basalt at elevated temperatures (200–300° C). Tómasson and Kristmannsdóttir (1972) have described the mineral assemblages in the altered basalt from this area. Holland (1978) quotes analyses of some of the hydrothermal brines from the wells; one example, from Reykjanes drill hole 2 (at temperature 221°C), is given in Table 11.10. The concentrations of sodium and chlorine are almost identical to those in seawater but other elements are either enriched or depleted. Magnesium and sulphate ions are almost totally removed from the brine.

TABLE 11.10. *Chemical composition* $(mg\,kg^{-1})$ *of geothermal brine from Reykjanes drill hole 2, Reykjanes Peninsula, Iceland*

		Normalized to seawater
Na	10,440	0.97
K	1382	3.64
Ca	1812	4.40
Mg	8	0.006
SO_4^{2-}	72	0.03
S^{2-}	51	—
Cl	20,745	1.07
SiO_2	374	93.5
B	11.6	2.61
Fe	0.48	240

Source: quoted in Holland (1978).

Many of the secondary minerals found in the altered basalts of Reykjanes are produced in the experimental mineral assemblages. These include anhydrite, montmorillonite, albite, pyrite, chalcopyrite, pyrrhotine and hematite. However, chlorite and epidote, both recorded at Reykjanes and in greenschist-facies metabasalts of the ocean floor, have not been identified in the experimental products, probably as a result of nucleation and kinetic factors (Mottl and Holland, 1978).

The implications of all these findings for studies on seawater chemistry, ore-forming fluids and oceanic crust composition are immense. As mentioned in Section 9.6.2, the depth to which cold seawater can penetrate the sea-floor is at least $\frac{1}{2}$km, and there may be more than 1 km of oceanic crust that is hydrothermally altered. Holland (1978) has calculated on the basis of a feasible value for the flux of seawater through oceanic crust of about $10^{17}\,g\,a^{-1}$, the contribution that seawater–rock interaction makes on the mass balances of some elements (Table 11.11). It is clear that the process is a very significant one for many elements and could be the dominant one for a few. It is reasonable to conclude that this hydrothermal alteration of new oceanic crust can accommodate the apparent deficit for magnesium in the mass balance calculation made by Drever (1974) (see Section 11.4).

The importance of rock–water interactions to ore formation has been discussed already in earlier sections of this book. In addition, it is noteworthy

TABLE 11.11. *Flux of dissolved substances between the oceans and mid-ocean ridges as a result of seawater cycling*

	Percentage of annual river flux	
	Max. $T = 50°C$	Max. $T = 250°C$
Mg^{2+}	-63	-68
Ca^{2+}	$+48$	$+17$
Na^+	-23	-16
K^+	~ 0	$+100$
SO_4^{2-}	-24	-50
SiO_2	$+26$	$+15$

Source: Holland (1978).

that in the median valley of the Red Sea are hot brines, rich in dissolved metals, and which are contributing to the formation of metal-rich sediments containing iron and manganese oxides, sulphides and sulphates. The metal contents of the brines are derived partly by leaching of elements from sediments and volcanic rocks (Degens and Ross, 1969; Bignell *et al.*, 1976).

11.6 Kinetic aspects of some mineral deposition processes

As in any steady-state system, kinetics plays a very important role in the chemistries of the oceans and ocean-floor deposits. Two examples—calcium carbonate precipitation and dissolution, and the formation of ferro-manganese oxide deposits—illustrate some of the kinetic controls and complexities involved in mineral formation and deposition.

Much of the surface water of the oceans is supersaturated with calcium carbonate. The nucleation of calcite is inhibited by the presence of magnesium, probably because of its association with carbonate ions (see Section 11.2) and thereby reducing the carbonate activity below that required for calcite nucleation. When crystallization does occur it is aragonite, which is metastable under all oceanic conditions, that is precipitated first. The removal of calcium carbonate from solution is largely by the construction of calcareous skeletons (of calcite or aragonite) by organisms, and probably most of the calcium carbonate in calcareous sediments is biogenic.

The solubility of $CaCO_3$ increases with decrease in temperature and increase in pressure. Therefore, calcite is more soluble at greater depths in the ocean, the solubility being about twice as great at 5000 m depth (at $2°C$) as it is at the surface (at $2°C$). At a certain depth, seawater is no longer saturated with respect to $CaCO_3$. This depth, called the saturation level, is variable but for calcite occurs at about 4200 m depth in the Atlantic Ocean and down to about

3000 m in the Pacific Ocean. These variations result, amongst other factors, from differences in total dissolved carbon within and between the two oceans (see Fig. 11.5) and the downward flux of calcareous skeletal tests. The rate of dissolution of calcite below the saturation level is slow and is probably affected by protective coatings on the particles and by the inhibiting effect of specifically adsorbed species (e.g. PO_4^{3-}). The slow rate means that calcite can exist at some considerable depth below the saturation level. The depth at which the rate of carbonate sedimentation is balanced by the slow rate of dissolution is termed the carbonate compensation depth; it is kinetically controlled and can be as much as 2000 m below the saturation level (Skirrow, 1975). The compensation depth can be seen to correspond to the level at which there is an abrupt decrease in the abundance of $CaCO_3$ in sediments and, like the saturation level, it varies with location within the ocean system. It is clear that kinetic factors, briefly summarized here, have an important control on the accumulation of calcareous sediments on the sea-floor. Unfortunately, our understanding of these factors is still very limited and a fully quantitative approach is not yet possible. (Skirrow, 1975, gives a detailed exposition of our state of knowledge about dissolved carbon dioxide in seawater).

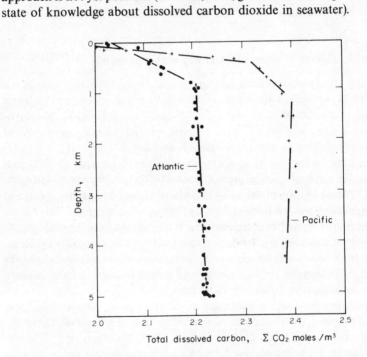

FIG. 11.5. Variation of total dissolved inorganic carbon content with depth in the Atlantic Ocean (36° N, 68° W) and in the Pacific Ocean (28° N, 122° W). (After Broecker, 1974)

The origin of ferro-manganese oxide deposits (including manganese nodules) is enigmatic. These deposits occur over vast areas of the sea-floor and have attracted much attention, not least because of their economic potential. In parts of the North Pacific, 75 % of the sea-floor is covered, but in the South Pacific, Atlantic and Indian Oceans, amounts are lower. In this regard, the rate of sedimentation of other materials is important, low rates being conducive to the development of oxide deposits. The deposits are precipitates of hydrous manganese and iron oxides, and occur mostly at the sediment surface. They have a highly variable morphology from nodular through to large slabs or as encrustations on other deposits. Their internal structures also vary and may exhibit concentric banding, globular, laminar, dendritic or cataclastic forms. The principal manganese minerals are two distinct phases named todorokite (a mixed $Mn^{2+}-Mn^{4+}$ hydrated oxide) and δ-MnO_2, one of the polymorphs of MnO_2. Iron is most probably present as some form of $FeO \cdot OH$., such as goethite or hydrogoethite. (There is, however, some confusion over the mineralogy of manganese nodules, see Cronan, 1976.)

Ferro-manganese oxide deposits have a wide range in composition; a 'world-ocean' average abundance is given in Table 11.12, together with the enrichment factor in relation to the crustal abundance of each element (see Chapter 4). There are many viewpoints on the possible sources of the manganese, iron and other metals; volcanic emanations, seawater, leaching of basalts, and hydrothermal solutions have been considered, but no clear consensus has emerged. However, some indication of the possible source(s) can come from consideration of the kinetics of manganese nodule growth. For example, Boudreau and Scott (1978) have shown that the pore-water from sediments cannot supply sufficient manganese when the oxidizing layer of the sedimentary column is greater than 40 cm, as is usually the case. Diffusion-controlled flux of manganese from seawater across the benthic boundary layer can account for the manganese in deep sea nodules and also matches the observed rates of nodule growth as determined by radiometric dating methods. These rates are of the order of a few millimetres per million years (or at a manganese deposition rate of about $1 \mu g \, cm^{-2} a^{-1}$).

Studies on the rate of oxidation of Mn^{2+} (the common oxidation state of manganese in seawater, see Table 11.3) to Mn^{4+} with concomitant precipitation have shown the importance of a catalytic reaction surface (Stumm and Morgan, 1970; Wilson, 1980). Once a deposit of ferric oxide has formed it becomes an adsorbent for Mn^{2+} ions which are then slowly oxidized to Mn oxides, possibly by reaction (11.11):

$$Mn^{2+} + 2OH^- + \tfrac{1}{2}O_2 = MnO_2 + H_2O. \qquad (11.11)$$

The MnO_2 so formed becomes in its turn a reactive surface for further oxidation. Wilson (1980) has shown that goethite and montmorillonite could

TABLE 11.12. *Average abundance of elements in ferro-manganese oxide deposits of the world's ocean, and enrichment factors for each element*

	Abundance (wt %)	Enrichment factor
Na	1.94	0.82
Mg	1.82	0.78
Al	3.10	0.38
Si	8.62	0.31
P	0.22	2.1
K	0.64	0.31
Ca	2.53	0.61
Ti	0.64	1.1
V	0.056	4.1
Cr	0.001	0.14
Mn	16.17	170
Fe	15.61	2.8
Co	0.30	120
Ni	0.49	65
Cu	0.26	47
Zn	0.071	10
Sr	0.083	2.2
Zr	0.065	3.9
Mo	0.041	270
Pd	5.5×10^{-7}	0.83
Cd	7.9×10^{-4}	40
Ba	0.20	4.7
Ir	9.4×10^{-7}	71
Au	2.5×10^{-7}	0.62
Hg	5.0×10^{-5}	6.2
Tl	1.3×10^{-2}	290
Pb	8.7×10^{-2}	70
Bi	8×10^{-4}	47

Based on data given in Cronan (1976).

be important catalysts. In his experiments, an increase in the amounts of these minerals (and, therefore, an increase in the concentration of adsorption sites) leads to the reaction being of a first-order form (i.e. the rate of oxidation of Mn^{2+} is directly proportional to the concentration of Mn^{2+}). The relatively slow rates of the reaction (at pH $\simeq$ 8) is commensurate with the growth rates of nodules on the sea-floor, and taken together with the diffusion work, suggests that seawater is the dominant source of the iron and manganese.

11.7 Constancy of seawater composition

The idea that the ocean system is at a 'steady state' has received greater support in the form of evidence concerning the constancy in sedimentary mineral assemblages over much of the Phanerozoic than from evidence of the exact balance between the input and removal of a number of elements (Section 11.4). Many authors have addressed themselves to the problem of identifying possible changes in seawater composition with time, and the classic work of Conway (1942, 1943) and of Rubey (1951) represents a major part of the earlier attempts.

Rubey had estimated the discharge of water from hot springs of the continent and of the ocean floor to be about 6.6×10^{16} g a^{-1}. If this rate is continued over 4 Ga then the total amount discharged is 2.6×10^{26} g. Only about 0.5 % of this amount need be newly emergent water to account for the present-day volume of the ocean. The question then arises: how fast was the initial outgassing of the Earth and to what extent has newly emergent water been added to the ocean during the Phanerozoic? In so far as the atmosphere and its evolution affects the chemical nature of seawater it is important to attempt to establish the changes in the atmosphere with time. Evidence on this seems to indicate that the composition has not undergone a steady change during the Phanerozoic, although large changes during the Proterozoic are evident. Oxygen is usually absent from the gaseous emanations of volcanoes and so the accumulation of atmospheric oxygen must have been by photo-synthesis (and possibly by some photodissociation of water in the upper atmosphere).The mineralogical and biological records suggest that the original reduced gases of the early atmosphere had been nearly completely oxidized by 3.0 Ga ago and that oxygen was a significant component of the atmosphere by the late Precambrian. Recent evidence on the occurrence of charcoal in rocks from the Lower Carboniferous onwards indicates that the level of oxygen has been above 6 % by volume during most of the Phanerozoic (Cope and Chaloner, 1980).

Studies of the compositions of pore waters extracted from sediments of different ages indicate some change in the Na:K concentration ratio, which was interpreted earlier as reflecting a change in seawater composition but must now be considered a result of diagenesis. For example, Sayles and Manheim (1975) have recorded significant changes in the concentration of Mg, Ca, Sr, K and Na of pore waters in a variety of marine sediments through pore water–sediment reaction. Also, systematic changes with time in the chemical and mineralogical compositions of sedimentary rocks may be explained by secondary processes. Mackenzie (1975) has carefully reviewed this evidence, including the data gathered and discussed by Russian geochemists, especially Ronov. Shales, for example, show a reduction in the ratio MO/Al_2O_3 for many elements (M) with increasing age. These changes are reflected in differences in the clay mineralogy of the shales of differing ages—recent ones have a diverse mineralogy but often are montmorillonite-rich, whereas older ones are characterized by the presence of illite and chlorite. Secondary reactions of the following generalized kind will affect the chemical composition:

$$2.5\,X^+_{0.05}\,K_{0.35}\,[(Mg_{0.34}\,Fe^{3+}_{0.17}\,Fe^{2+}_{0.04}\,Al_{1.50})(Al_{0.17}\,Si_{3.83})O_{10}(OH)_2]$$
montmorillonite-rich clay

$$+\,K^+ + AlO_4^-$$

$$=2.5\,X^+_{0.05}\,K_{0.75}\,[(Mg_{0.34}\,Fe^{3+}_{0.17}\,Fe^{2+}_{0.04}\,Al_{1.50})(Al_{0.57}\,Si_{3.43})O_{10}(OH)_2]$$
illite-rich clay

$$+\,SiO_2 \tag{11.12}$$

where X^+ represents exchangeable univalent cations.

The fact that carbonate rocks tend to be more dolomitic the older they are may also be interpreted as a diagenetic (i.e. secondary dolomitization) cause rather than a change in the Mg/Ca ratio of seawater with time.

Thus, the evidence from the sedimentary record discussed above is not a sensitive indicator of any possible changes in seawater composition. Rather, it appears as if the initial characteristics of the various sediment types have been remarkably constant throughout the Phanerozoic. It is, therefore, reasonable to ask: by how much can the composition of seawater change before producing any significant change in the nature of sediments and authigenic minerals? Holland (1972) asked this particular question with respect to marine evaporites, for they have shown the same depositional sequence (a carbonate section followed by an anhydrite–gypsum section, then a halite section) since late Precambrian time.

Holland (1972) considered the extent of variations in seven of the major components of seawater—Na, Mg, Ca, K, Cl, SO_4^{2-} and HCO_3^-—that would be possible without there being a change in the sequence of precipitated minerals during evaporation of seawater, and concluded that these constituents were never more than twice or less than half as abundant as in present-day seawater. Holland's approach to the problem can be illustrated using two major constituents—Ca^{2+} and HCO_3^-, as in Fig. 11.6. On this figure the composition of present-day seawater is marked with a circle, and Holland states that the composition must have remained within the stippled area during the Phanerozoic. An upper limit to a possible increase in calcium concentration (three times that of present-day seawater) is imposed by the solubility of gypsum. If seawater were saturated with respect to gypsum then this mineral would be associated with limestone in a slightly evaporative setting. This has not been observed. Similarly, if the calcium concentration were reduced by a factor of 30, seawater would become simultaneously saturated with halite and gypsum during evaporation. Such a phenomenon is not shown in evaporite sequences. The inclined left and right boundaries (see Fig. 11.6) are set by considerations of pH, which could not have been greater than about 9.0 without precipitation of brucite, and of the balance of Ca^{2+} and

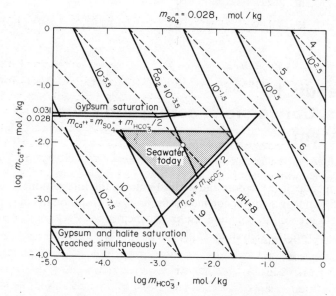

$m_{SO_4^=} = 0.028$, mol / kg

Gypsum saturation

$m_{Ca^{++}} = m_{SO_4^=} + m_{HCO_3^-}/2$

Seawater today

$m_{Ca^{++}} = m_{HCO_3^-}/2$

pH = 8

Gypsum and halite saturation reached simultaneously

$\log m_{HCO_3^-}$, mol / kg

$\log m_{Ca^{++}}$, mol / kg

FIG. 11.6. The permitted concentration range of calcium and bicarbonate in seawater, consistent with the mineralogy of marine evaporites when $m_{SO_4^{2-}}$ = 0.028 mol kg^{-1} and m_{NaCl} = 0.55 mol kg^{-1}. (After Holland, 1972)

HCO_3^- ions. In present-day seawater the number of Ca^{2+} is four times the number of HCO_3^- ions, so that during precipitation of calcite, the bicarbonate ions are used up, leaving some calcium ions for gypsum formation. If the number of calcium ions is equal to half the number of bicarbonate ions (i.e. lower right-hand boundary) then there would be no calcium available for gypsum. Further constraints of this above type can be imposed to restrict possible changes only to the stippled area and, more generally, to justify Holland's statement concerning the constancy of seawater composition. This constancy is a useful basis for studies of sedimentary geochemistry and of mass balances in the ocean system.

Suggested further reading

BERNER, R. A. (1971) *Principles of Chemical Sedimentology*. McGraw-Hill. 240 pp.
CRONAN, D. S. (1980) *Underwater Minerals*. Academic Press Inc. 364 pp.
HOLLAND, H. D. (1978) *The Chemistry of the Atmosphere and Oceans*. J. Wiley & Sons. 351 pp.
RILEY, J. P. and SKIRROW, G. (editors) (1975–1978) *Chemical Oceanography*. Second edition. Academic Press Inc. Seven volumes. These volumes cover a very wide range of topics relating to chemical oceanography. Volume 1 is particularly useful for it deals with the basic chemical principles of the chemistry of ocean water.
STUMM, W. and MORGAN, J. J. (1970) *Aquatic Chemistry. An Introduction Emphasizing Chemical Equilibria in Natural Waters*. J. Wiley & Sons. 583 pp.

Appendix I
Units, Conversion Factors and Physical Constants

THIS Appendix lists the base and derived units in the International System of Units (SI) and gives conversion factors for units which although not part of the SI system are still in common use in geochemistry. A selection of relevant physical constants is given in Table I.5.

TABLE I.1. *The SI base units*

Physical quantity	Name	Symbol
Length	metre	m
Mass	kilogram	kg
Time	second	s
Electric current	ampere	A
Thermodynamic temperature	kelvin	K
Amount of substance	mole	mol
Luminous intensity	candela	cd

TABLE I.2. *Some SI derived units*

Physical quantity	Name	Symbol	Definition*
Energy	joule	J	$m^2 \, kg \, s^{-2}$
Force	newton	N	$m \, kg \, s^{-2}$
Pressure	pascal	Pa	$m^{-1} \, kg \, s^{-2}$
Electric charge	coulomb	C	$s \, A$
Electric potential difference	volt	V	$m^2 \, kg \, s^{-3} \, A^{-1}$

*The definition is in terms of the SI base units.

304

TABLE I.3. *SI prefixes used to construct multiples of units*

Multiple	Prefix	Symbol	Multiple	Prefix	Symbol
10^{-1}	deci	d	10	deca	da
10^{-2}	centi	c	10^2	hecto	h
10^{-3}	milli	m	10^3	kilo	k
10^{-6}	micro	μ	10^6	mega	M
10^{-9}	nano	n	10^9	giga	G
10^{-12}	pico	p	10^{12}	tera	T
10^{-15}	femto	f	10^{15}	peta	P
10^{-18}	atto	a	10^{18}	exa	E

TABLE I.4. *Some other units and conversion factors*

Physical quantity	Name	Symbol	Conversion
Length	ångström	Å	10^{-10} m
Mass	tonne	t	10^3 kg $=$ Mg
Pressure	bar	bar	10^5 Pa
Pressure	atmosphere	atm	101,325 Pa
Energy	calorie (thermochemical)	cal	4.184 J
Energy	electronvolt	eV	$1\,\text{eV} \approx 1.602\,189 \times 10^{-19}$ J
Kinematic viscosity	stokes	St	10^{-4} m^2 s^{-1}
Dynamic viscosity	poise	P	10^{-1} Pa s

TABLE I.5. *Values of some physical constants*

Quantity	Symbol	Value
Planck constant	h	$6.626\,176 \times 10^{-34}$ J s
Avogadro constant	N_A	$6.022\,045 \times 10^{23}$ mol^{-1}
Faraday constant	F	$9.648\,456 \times 10^4$ C mol^{-1}
Boltzmann constant	k	$1.380\,662 \times 10^{-23}$ J K^{-1}
Gas constant	R	$8.314\,41$ J K^{-1} mol^{-1}

Appendix II
The Periodic Table of the Elements

Period	Group Ia	Group IIa	Group IIIa	Group IVa	Group Va	Group VIa	Group VIIa	Group VIII			Group Ib	Group IIb	Group IIIb	Group IVb	Group Vb	Group VIb	Group VIIb	Group O
1 $1s$	1 H																1 H	2 He
2 $2s2p$	3 Li	4 Be											5 B	6 C	7 N	8 O	9 F	10 Ne
3 $3s3p$	11 Na	12 Mg											13 Al	14 Si	15 P	16 S	17 Cl	18 Ar
4 $4s3d$ $4p$	19 K	20 Ca	21 Sc	22 Ti	23 V	24 Cr	25 Mn	26 Fe	27 Co	28 Ni	29 Cu	30 Zn	31 Ga	32 Ge	33 As	34 Se	35 Br	36 Kr
5 $5s4d$ $5p$	37 Rb	38 Sr	39 Y	40 Zr	41 Nb	42 Mo	43 Tc	44 Ru	45 Rh	46 Pd	47 Ag	48 Cd	49 In	50 Sn	51 Sb	52 Te	53 I	54 Xe
6 $6s$ $(4f)$ $5d$ $6p$	55 Cs	56 Ba	57* La	72 Hf	73 Ta	74 W	75 Re	76 Os	77 Ir	78 Pt	79 Au	80 Hg	81 Tl	82 Pb	83 Bi	84 Po	85 At	86 Rn
7 $7s$ $(5f)$ $6d$	87 Fr	88 Ra	89** Ac															

*Lanthanide series $4f$	58 Ce	59 Pr	60 Nd	61 Pm	62 Sm	63 Eu	64 Gd	65 Tb	66 Dy	67 Ho	68 Er	69 Tm	70 Yb	71 Lu
**Actinide series $5f$	90 Th	91 Pa	92 U	93 Np	94 Pu	95 Am	96 Cm	97 Bk	98 Cf	99 Es	100 Fm	101 Md	102 No	103 Lr

Appendix III
Electronic Configurations of the Elements

Element	1s	2s	2p	3s	3p	3d	4s	4p	4d	4f	5s	5p	5d	5f	5g
1. H	1														
2. He	2														
3. Li	2	1													
4. Be	2	2													
5. B	2	2	1												
6. C	2	2	2												
7. N	2	2	3												
8. O	2	2	4												
9. F	2	2	5												
10. Ne	2	2	6												
11. Na	2	2	6	1											
12. Mg	2	2	6	2											
13. Al	2	2	6	2	1										
14. Si	2	2	6	2	2										
15. P	2	2	6	2	3										
16. S	2	2	6	2	4										
17. Cl	2	2	6	2	5										
18. Ar	2	2	6	2	6										
19. K	2	2	6	2	6		1								
20. Ca	2	2	6	2	6		2								
21. Sc	2	2	6	2	6	1	2								
22. Ti	2	2	6	2	6	2	2								
23. V	2	2	6	2	6	3	2								
24. Cr	2	2	6	2	6	5	1								
25. Mn	2	2	6	2	6	5	2								
26. Fe	2	2	6	2	6	6	2								
27. Co	2	2	6	2	6	7	2								
28. Ni	2	2	6	2	6	8	2								
29. Cu	2	2	6	2	6	10	1								
30. Zn	2	2	6	2	6	10	2								
31. Ga	2	2	6	2	6	10	2	1							
32. Ge	2	2	6	2	6	10	2	2							
33. As	2	2	6	2	6	10	2	3							
34. Se	2	2	6	2	6	10	2	4							
35. Br	2	2	6	2	6	10	2	5							
36. Kr	2	2	6	2	6	10	2	6							
37. Rb	2	2	6	2	6	10	2	6			1				
38. Sr	2	2	6	2	6	10	2	6			2				
39. Y	2	2	6	2	6	10	2	6	1		2				
40. Zr	2	2	6	2	6	10	2	6	2		2				
41. Nb	2	2	6	2	6	10	2	6	4		1				
42. Mo	2	2	6	2	6	10	2	6	5		1				
43. Tc	2	2	6	2	6	10	2	6	6		1				
44. Ru	2	2	6	2	6	10	2	6	7		1				
45. Rh	2	2	6	2	6	10	2	6	8		1				
46. Pd	2	2	6	2	6	10	2	6	10						
47. Ag	2	2	6	2	6	10	2	6	10		1				
48. Cd	2	2	6	2	6	10	2	6	10		2				
49. In	2	2	6	2	6	10	2	6	10		2	1			
50. Sn	2	2	6	2	6	10	2	6	10		2	2			
51. Sb	2	2	6	2	6	10	2	6	10		2	3			
52. Te	2	2	6	2	6	10	2	6	10		2	4			
53. I	2	2	6	2	6	10	2	6	10		2	5			
54. Xe	2	2	6	2	6	10	2	6	10		2	6			

Appendix III (*Cont.*)

Element	K	L	M	4s	4p	4d	4f	5s	5p	5d	5f	5g	6s	6p	6d	6f	6g	6h	7
55. Cs	2	8	18	2	6	10		2	6				1						
56. Ba	2	8	18	2	6	10		2	6				2						
57. La	2	8	18	2	6	10		2	6	1			2						
58. Ce	2	8	18	2	6	10	2	2	6				2						
59. Pr	2	8	18	2	6	10	3	2	6				2						
60. Nd	2	8	18	2	6	10	4	2	6				2						
61. Pm	2	8	18	2	6	10	5	2	6				2						
62. Sm	2	8	18	2	6	10	6	2	6				2						
63. Eu	2	8	18	2	6	10	7	2	6				2						
64. Gd	2	8	18	2	6	10	7	2	6	1			2						
65. Tb	2	8	18	2	6	10	9	2	6				2						
66. Dy	2	8	18	2	6	10	10	2	6				2						
67. Ho	2	8	18	2	6	10	11	2	6				2						
68. Er	2	8	18	2	6	10	12	2	6				2						
69. Tm	2	8	18	2	6	10	13	2	6				2						
70. Yb	2	8	18	2	6	10	14	2	6				2						
71. Lu	2	8	18	2	6	10	14	2	6	1			2						
72. Hf	2	8	18	2	6	10	14	2	6	2			2						
73. Ta	2	8	18	2	6	10	14	2	6	3			2						
74. W	2	8	18	2	6	10	14	2	6	4			2						
75. Re	2	8	18	2	6	10	14	2	6	5			2						
76. Os	2	8	18	2	6	10	14	2	6	6			2						
77. Ir	2	8	18	2	6	10	14	2	6	7			2						
78. Pt	2	8	18	2	6	10	14	2	6	9			1						
79. Au	2	8	18	2	6	10	14	2	6	10			1						
80. Hg	2	8	18	2	6	10	14	2	6	10			2						
81. Tl	2	8	18	2	6	10	14	2	6	10			2	1					
82. Pb	2	8	18	2	6	10	14	2	6	10			2	2					
83. Bi	2	8	18	2	6	10	14	2	6	10			2	3					
84. Po	2	8	18	2	6	10	14	2	6	10			2	4					
85. At	2	8	18	2	6	10	14	2	6	10			2	5					
86. Rn	2	8	18	2	6	10	14	2	6	10			2	6					
87. Fr	2	8	18	2	6	10	14	2	6	10			2	6					1
88. Ra	2	8	18	2	6	10	14	2	6	10			2	6					2
89. Ac	2	8	18	2	6	10	14	2	6	10			2	6	1				2
90. Th	2	8	18	2	6	10	14	2	6	10			2	6	2				2
91. Pa	2	8	18	2	6	10	14	2	6	10	2		2	6	1				2
92. U	2	8	18	2	6	10	14	2	6	10	3		2	6	1				2
93. Np	2	8	18	2	6	10	14	2	6	10	5		2	6					2
94. Pu	2	8	18	2	6	10	14	2	6	10	6		2	6					2
95. Am	2	8	18	2	6	10	14	2	6	10	7		2	6					2
96. Cm	2	8	18	2	6	10	14	2	6	10	7		2	6	1				2
97. Bk	2	8	18	2	6	10	14	2	6	10	8		2	6	1				2
98. Cf	2	8	18	2	6	10	14	2	6	10	10		2	6					2
99. Es	2	8	18	2	6	10	14	2	6	10	11		2	6					2
100. Fm	2	8	18	2	6	10	14	2	6	10	12		2	6					2
101. Md	2	8	18	2	6	10	14	2	6	10	13		2	6					2
102. No	2	8	18	2	6	10	14	2	6	10	14		2	6					2
103. Lr	2	8	18	2	6	10	14	2	6	10	14		2	6	1				2

Source: Cotton and Wilkinson (1980).

Appendix IV
Atomic Weights ($^{12}C = 12.000$)

Element	Symbol	Atomic no.	Atomic weight	Element	Symbol	Atomic no.	Atomic weight
Aluminium	Al	13	26.98154	Neodymium	Nd	60	144.24
Antimony	Sb	51	121.7	Neon	Ne	10	20.179
Argon	Ar	18	39.948	Neptunium	Np	93	237.0482
Arsenic	As	33	74.9216	Nickel	Ni	28	58.70
Barium	Ba	56	137.33	Niobium	Nb	41	92.9064
Beryllium	Be	4	9.01218	Nitrogen	N	7	14.0067
Bismuth	Bi	83	208.9804	Osmium	Os	76	190.2
Boron	B	5	10.81	Oxygen	O	8	15.9994
Bromine	Br	35	79.904	Palladium	Pd	46	106.4
Cadmium	Cd	48	112.41	Phosphorus	P	15	30.97376
Calcium	Ca	20	40.08	Platinum	Pt	78	195.09
Carbon	C	6	12.011	Potassium	K	19	39.0983
Cerium	Ce	58	140.12	Praseodymium	Pr	59	140.0977
Caesium	Cs	55	132.9054	Protactinium	Pa	91	231.0395
Chlorine	Cl	17	35.453	Radium	Ra	88	226.0254
Chromium	Cr	24	51.996	Rhenium	Re	75	186.207
Cobalt	Co	27	58.9332	Rhodium	Rh	45	102.9055
Copper	Cu	29	63.546	Rubidium	Rb	37	85.4678
Dysprosium	Dy	66	162.50	Ruthenium	Ru	44	101.07
Erbium	Er	68	167.26	Samarium	Sm	62	150.4
Europium	Eu	63	151.96	Scandium	Sc	21	44.9559
Fluorine	F	9	18.9984	Selenium	Se	34	78.96
Gadolinium	Gd	64	157.25	Silicon	Si	14	28.0855
Gallium	Ga	31	69.72	Silver	Ag	47	107.868
Germanium	Ge	32	72.59	Sodium	Na	11	22.98977
Gold	Au	79	196.9665	Strontium	Sr	38	87.62
Hafnium	Hf	72	178.49	Sulphur	S	16	32.06
Helium	He	2	4.00260	Tantalum	Ta	73	180.9479
Holmium	Ho	67	164.9304	Tellurium	Te	52	127.60
Hydrogen	H	1	1.0079	Terbium	Tb	65	158.9254
Indium	In	49	114.82	Thallium	Tl	81	204.37
Iodine	I	53	126.9045	Thorium	Th	90	232.0381
Iridium	Ir	77	192.22	Thulium	Tm	69	168.9342
Iron	Fe	26	55.847	Tin	Sn	50	118.69
Krypton	Kr	36	83.80	Titanium	Ti	22	47.90
Lanthanum	La	57	138.906	Tungsten	W	74	183.85
Lead	Pb	82	207.2	Uranium	U	92	238.029
Lithium	Li	3	6.941	Vanadium	V	23	50.9414
Lutetium	Lu	71	174.97	Xenon	Xe	54	131.30
Magnesium	Mg	12	24.305	Ytterbium	Yb	70	173.04
Manganese	Mn	25	54.9380	Yttrium	Y	39	88.9059
Mercury	Hg	80	200.59	Zinc	Zn	30	65.38
Molybdenum	Mo	42	95.94	Zirconium	Zr	40	91.22

Appendix V
Ionic Radii (Å)

THE table gives three sets of ionic radii (Å) commonly used in geochemistry—those by: Shannon and Prewitt (1969, 1970) revised by Shannon (1976), abbreviated S & P; Whittaker and Muntus (1970), abbreviated W & M; Ahrens (1952), abbreviated Ah. The radii are given for different coordination numbers (CN) in roman numerals, and spin states (Sp) where appropriate, but only the common types of four coordination: square planar (IV sq), square pyramidal (IV py), and tetrahedral (IV) are differentiated. *L* or *H* denotes low-spin or high-spin state respectively. The list excludes ions with oxidation states that are unlikely to be of geochemical relevance, but all coordination numbers as listed by Shannon (1976) are included. In the W & M radius list, approximate values for anions other than those of O or F are given in parenthesis. Ah radius values given in parenthesis are doubtful.

Ion	CN and Sp	S & P radius	W & M radius	Ah radius
Ag$^+$	II	.67	.75	
	IV sq	1.02	1.10	
	V	1.09	1.20	
	VI	1.15	1.23	1.26
	VII	1.22	1.32	
	VIII	1.28	1.38	
Al^{3+}	IV	.39	.47	
	V	.48	.56	
	VI	.535	.61	.51
As^{5+}	IV	.335	.42	
	VI	.46	.58	.46
Au^{3+}	IV sq	.68	.78	
B^{3+}	III	.01	.10	
	IV	.11	.20	
	VI	.27		

Ion	CN and Sp	S & P radius	W & M radius	Ah radius
Ba^{2+}	VI	1.35	1.44	1.34
	VII	1.38	1.47	
	VIII	1.42	1.50	
	IX	1.47	1.55	
	X	1.52	1.60	
	XI	1.57		
	XII	1.61	1.68	
Be^{2+}	III	.16	.25	
	IV	.27	.35	
	VI	.45		
Bi^{3+}	V	.96	1.07	
	VI	1.03	1.10	(.96)
	VIII	1.17	1.19	
Br$^-$	VI	1.96	(1.88)	1.96
	VIII		(1.84)	
Br^{7+}	IV	.25	.34	

Appendix V (*Cont.*)

Ion	CN and Sp	S & P radius	W & M radius	Ah radius	Ion	CN and Sp	S & P radius	W & M radius	Ah radius
Ca²⁺	VI	1.00	1.08	.99	Cu²⁺	IV sq	.57	.70	
	VII	1.06	1.15			IV	.57		
	VIII	1.12	1.20			V	.65	.73	
	IX	1.18	1.26			VI	.73	.81	(.72)
	X	1.23	1.36						
	XII	1.34	1.43		Dy³⁺	VI	.912	.99	.92
						VII	.97		
Cd²⁺	IV	.78	.88			VIII	1.027	1.11	
	V	.87	.95			IX	1.083		
	VI	.95	1.03						
	VII	1.03	1.08		Er³⁺	VI	.890	.97	.89
	VIII	1.10	1.15			VII	.945		
	XII	1.31	1.39			VIII	1.004	1.08	
Ce³⁺	VI	1.01	1.09			IX	1.062		
	VII	1.07							
	VIII	1.143	1.22		Eu²⁺	VI	1.17	1.25	
	IX	1.196	1.23			VII	1.20		
	X	1.25				VIII	1.25	1.33	
	XII	1.34	1.37			IX	1.30		
Ce⁴⁺	VI	.87	.88	.9		X	1.35		
	VIII	.97	1.05		Eu³⁺	VI	.947	1.03	.98
	X	1.07				VII	1.01	1.11	
	XII	1.14				VIII	1.066	1.15	
						IX	1.120		
Cl⁻	IV		(1.67)						
	VI		(1.72)	1.81	F⁻	II	1.285	1.21	
	VIII		(1.65)			III	1.30	1.22	
Cl⁷⁺	IV	.08	.28			IV	1.31	1.23	
	VI			.27		VI	1.33	1.25	1.33
Co²⁺	IV*H*	.58	.65		Fe²⁺	IV*H*	.63	.71	
	V	.67				IV sq *H*	.64		
	VI*L*	.65	.73			VI*L*	.61	.69	
	H	.745	.83	.72		*H*	.780	.86	.74
	VIII	.90				VIII	.92		
Cr²⁺	VI*L*	.73	.81		Fe³⁺	IV*H*	.49	.57	
	H	.80	.90			V	.58		
Cr³⁺	VI	.615	.70	.63		VI*L*	.55	.63	
						H	.645	.73	.64
Cs⁺	VI	1.67	1.78	1.67		VIII*H*	.78		
	VIII	1.74	1.82						
	IX	1.78	1.86		Ga³⁺	IV	.47	.55	
	X	1.81	1.89			V	.55	.63	
	XI	1.85				VI	.620	.70	.62
	XII	1.88	1.96						
					Gd³⁺	VI	.938	1.02	.97
Cu⁺	II	.46	.54			VII	1.00	1.12	
	IV	.60				VII	1.053	1.14	
	VI	.77		.96		IX	1.107		

Appendix V (*Cont.*)

Ion	CN and Sp	S & P radius	W & M radius	Ah radius	Ion	CN and Sp	S & P radius	W & M radius	Ah radius
Ge⁴⁺	IV	.390	.48		Lu³⁺	VIII	.977	1.05	
	VI	.530	.62	.53		IX	1.032		
Hf⁴⁺	IV	.58			Mg²⁺	IV	.57	.66	
	VI	.71	.79	.78		V	.66	.75	
	VII	.76				VI	.720	.80	.66
	VIII	.83	.91			VIII	.89	.97	
Hg⁺	III	.97	1.05		Mn²⁺	IV*H*	.66		
	VI	1.19				VI*L*	.67	.75	
Hg²⁺	II	.69	.77			V*H*	.75		
	IV	.96	1.04			*H*	.830	.91	.80
	VI	1.02	1.10	1.10		VII*H*	.90		
	VIII	1.14	1.22			VIIĪ	.96	1.01	
					Mn⁴⁺	IV	.39		
Ho³⁺	VI	.901	.98	.91		VI	.530	.62	(.60)
	VIII	1.015	1.10						
	IX	1.072			Mo⁴⁺	VI	.650	.73	.70
	X	1.12			Mo⁶⁺	IV	.41	.50	
						V	.50	.58	
I⁻	VI		(2.13)	2.20		VI	.59	.68	.62
	VIII		(1.97)			VII	.73	.79	
In³⁺	IV	.62			Na⁺	IV	.99	1.07	
	VI	.800	.88	.81		V	1.00	1.08	
	VIII	.92	1.00			VI	1.02	1.10	.97
						VII	1.12	1.21	
Ir³⁺	VI	.68	.81			VIII	1.18	1.24	
Ir⁴⁺	VI	.625	.71	.68		IX	1.24	1.40	
						XII	1.39		
K⁺	IV	1.37			Nb⁵⁺	IV	.48	.40	
	VI	1.38	1.46	1.33		VI	.64	.72	.69
	VII	1.46	1.54			VII	.69	.74	
	VIII	1.51	1.59			VIII	.74		
	IX	1.55	1.63						
	X	1.59	1.67		Nd³⁺	VI	.983	1.06	1.04
	XII	1.64	1.68			VIII	1.109	1.20	
						IX	1.163	1.17	
La³⁺	VI	1.032	1.13	1.14		XII	1.27		
	VII	1.10	1.18						
	VIII	1.160	1.26		Ni²⁺	IV	.55		
	IX	1.216	1.28			IV sq	.49		
	X	1.27	1.36			V	.63		
	XII	1.36	1.40			VI	.690	.77	.69
Li⁺	IV	.590	.68		O²⁻	II	1.35	1.27	
	VI	.76	.82	.68		III	1.36	1.28	
	VIII	.92				IV	1.38	1.30	
						VI	1.40	1.32	1.40
Lu³⁺	VI	.861	.94	.85		VIII	1.42	1.34	

Appendix V (*Cont.*)

Ion	CN and Sp	S & P radius	W & M radius	Ah radius
Os⁴⁺	VI	.630	.71	.69
P⁵⁺	IV	.17	.25	
	V	.29		
	VI	.38		.35
Pa⁴⁺	VI	.90		.98
	VIII	1.01	1.09	
Pa⁵⁺	VI	.78		.89
	VIII	.91	.99	
	IX	.95	1.03	
Pb²⁺	IV py	.98	1.02	
	VI	1.19	1.26	1.20
	VII	1.23		
	VIII	1.29	1.37	
	IX	1.35	1.41	
	X	1.40		
	XI	1.45	1.47	
	XII	1.49	1.57	
Pb⁴⁺	IV	.65		
	V	.73		
	VI	.775	.86	.84
	VIII	.94	1.02	
Pd⁺	II	.59	.67	
Pd²⁺	IV sq	.64	.72	
	VI	.86	.94	(.80)
Pd⁴⁺	VI	.615	.70	.65
Pr³⁺	VI	.99	1.08	1.06
	VIII	1.126	1.22	
	IX	1.179		
Pt²⁺	IV sq	.60	.68	
Pt⁴⁺	VI	.625	.71	(.65)
Ra²⁺	VIII	1.48	1.56	
	XII	1.70	1.72	
Rb⁺	VI	1.52	1.57	(1.47)
	VII	1.56	1.64	
	VIII	1.61	1.68	
	IX	1.63		
	X	1.66	1.74	
	XI	1.69		
	XII	1.72	1.81	
	XIV	1.83		
Re⁴⁺	VI	.63	.71	(.72)

Ion	CN and Sp	S & P radius	W & M radius	Ah radius
Re⁵⁺	VI	.58	.60	
Re⁶⁺	VI	.55	.60	
Rh³⁺	VI	.665	.75	.68
Rh⁴⁺	VI	.60	.71	
Ru³⁺	VI	.68	.76	
Ru⁴⁺	VI	.620	.70	.67
S²⁻	IV		(1.56)	
	VI		(1.72)	
	VIII		(1.78)	
S⁶⁺	IV	.12	.20	
	VI	.29		
Sb³⁺	IV py	.76	.85	
	V	.80	.88	
Sb⁵⁺	VI	.60	.69	.62
Sc³⁺	VI	.745	.83	.81
	VIII	.870	.95	
Se²⁻	VI		(1.88)	
	VIII		(1.90)	
Se⁶⁺	IV	.28	.37	
	VI	.42		
Si⁴⁺	IV	.26	.34	
	VI	.400	.48	.42
Sm³⁺	VI	.958	1.04	1.00
	VII	1.02		
	VIII	1.079	1.17	
	IX	1.132		
	XIII	1.24		
Sn²⁺	VIII	1.22	1.30	
Sn⁴⁺	IV	.55		
	V	.62		
	VI	.690	.77	.71
	VII	.75		
	VIII	.81		
Sr²⁺	VI	1.18	1.21	1.12
	VII	1.21	1.29	
	VIII	1.26	1.33	
	IX	1.31		
	X	1.36	1.40	
	XII	1.44	1.48	

Appendix V (*Cont.*)

Ion	CN and Sp	S & P radius	W & M radius	Ah radius	Ion	CN and Sp	S & P radius	W & M radius	Ah radius
Ta^{5+}	VI	.64	.72	.68	U^{5+}	VI	.76	.84	
	VII	.69				VII	.84	1.04	
	VIII	.74	.77		U^{6+}	II	.45	.53	
						IV	.52	.56	
Tb^{3+}	VI	.923	1.00	.93		VI	.73	.81	.80
	VII	.98	1.10			VII	.81	.96	
	VIII	1.040	1.12			VIII	.86		
	IX	1.095							
Te^{4+}	III	.52	.60		V^{3+}	VI	.640	.72	
	IV	.66							
	VI	.97		(.70)					
					W^{4+}	VI	.66	.73	.70
Th^{4+}	VI	.94	1.08	1.02	W^{6+}	IV	.42	.50	
	VIII	1.05	1.12			V	.51		
	IX	1.09	1.17			VI	.60	.68	.62
	X	1.13							
	XI	1.18			Y^{3+}	VI	.900	.98	.92
	XII	1.21				VII	.96		
						VIII	1.019	1.10	
Ti^{3+}	VI	.670	.75	(.76)		IX	1.075	1.18	
Ti^{4+}	IV	.42							
	V	.51	.61						
	VI	.605	.69	.68	Yb^{3+}	VI	.868	.95	.86
	VIII	.74				VII	.925		
						VIII	.985	1.06	
Tl^+	VI	1.50	1.58	1.47		IX	1.042		
	VIII	1.59	1.68						
	XII	1.70	1.84		Zn^{2+}	IV	.60	.68	
Tl^{3+}	IV	.75				V	.68	.76	
	VI	.885	.97	.95		VI	.740	.83	.74
	VIII	.98	1.08			VIII	.90	.98	
Tm^{3+}	VI	.880	.96	.87	Zr^{4+}	IV	.59		
	VIII	.994	1.07			V	.66		
	IX	1.052				VI	.72	.80	.79
						VII	.78	.86	
U^{4+}	VI	.89		.97		VIII	.84	.92	
	VII	.95	1.06			IX	.89		
	VIII	1.00	1.08						
	IX	1.05	1.13						
	XII	1.17							

Appendix VI
Electronegativities (eV) of the Elements

THE following table, in order of atomic number, gives values estimated by Pauling's method (abbrev. P) and those calculated by Mulliken's method (abbrev. M) where available. For discussion see Chapter 6, Section 6.7.2. The Pauling-type values are given for the following oxidation states: elements of groups Ia and VIIb, 1; groups IIIa, IIIb and Vb, 3; group IV, 4; the rest, 2.

Element	P	M	Element	P	M	Element	P	M
3. Li	0.98	0.94	27. Co	1.88		55. Cs	0.79	
4. Be	1.57	1.46	28. Ni	1.91		56. Ba	0.89	
5. B	2.04	2.01	29. Cu	1.90	1.36	57. La		1.10
6. C	2.55	2.63	30. Zn	1.65	1.49	58. Ce		1.12
7. N	3.04	2.33	31. Ga	1.81	1.95	59. Pr		1.13
8. O	3.44	3.17	32. Ge	2.01		60. Nd		1.14
9. F	3.98	3.91	33. As	2.18	1.75	62. Sm		1.17
11. Na	0.93	0.93	34. Se	2.55	2.23	64. Gd		1.20
12. Mg	1.31	1.32	35. Br	2.96	2.76	66. Dy		1.22
13. Al	1.61	1.81	37. Rb	0.82		67. Ho		1.23
14. Si	1.90	2.44	38. Sr	0.95		68. Er		1.24
15. P	2.19	1.81	39. Y	1.22		69. Tm		1.25
16. S	2.58	2.41	40. Zr	1.33		71. Lu		1.27
17. Cl	3.16	3.00	42. Mo	2.16		74. W	2.36	
19. K	0.82	0.80	45. Rh	2.28		77. Ir	2.20	
20. Ca	1.00		46. Pd	2.20		78. Pt	2.28	
21. Sc	1.36		47. Ag	1.93	1.36	79. Au	2.54	
22. Ti	1.54		48. Cd	1.69	1.4	80. Hg	2.00	
23. V	1.63		49. In	1.78	1.80	81. Tl	2.04	
24. Cr	1.66		50. Sn	1.96		82. Pb	2.33	
25. Mn	1.55		51. Sb	2.05	1.65	83. Bi	2.02	
26. Fe	1.83		53. I	2.66	2.56	92. U		1.38

Source: values cited in Cotton and Wilkinson (1966).

Appendix VII
Geochemical Reference
Literature

At the present time the two principal reference works for general geochemical data are:

Handbook of Geochemistry, executive editor K. H. WEDEPOHL, published in stages between 1969 and 1978. Published by Springer-Verlag. It comprises two volumes; the first contains twelve chapters on topics such as crystal chemistry, thermodynamics, evaluation of geochemical data and average compositions of rocks. Volume II, in five parts is a systematic treatment of the chemical elements, with sections which cover crystal chemistry; isotopes; abundances in meteorites, common rock types and waters; and behaviour in magmatogenetic, metamorphic, weathering and alteration processes.

Data of Geochemistry, 6th edition, technical editor M. FLEISCHER. 1962 onwards. Published by United States Geological Survey as Professional Paper 440. This edition is being published in individual chapters; those already published are listed below.

Chapter B–1 Cosmochemistry. Part 1. Meteorites, by B. MASON (1979).

D. Composition of the earth's crust, by R, L. PARKER (1967).

F. Chemical composition of subsurface waters, by J. D. HEM and G. A. WARING (1963).

G. Chemical composition of rivers and lakes, by D. A. LIVINGSTONE (1963).

K. Volcanic emanations, by D. E. WHITE and G. A. WARING (1963).

L. Phase equilibrium relations of the common rock-forming oxides except water, by G. W. MOREY (1964).

N. Chemistry of igneous rocks, Part 1. The chemistry of the peralkaline oversaturated obsidians, by R. MACDONALD and D. K. BAILEY (1973).

S.   Chemical composition of sandstones—excluding carbonate and volcanic sands, by F. J. PETTIJOHN (1963).
T.   Nondetrital siliceous sediments, by E. R. CRESSMAN (1962).
W.   Chemistry of the iron-rich sedimentary rocks, by H. L. JAMES (1966).
Y.   Marine evaporites, by F. H. STEWART (1963).
JJ. Composition of fluid inclusions, by E. ROEDDER (1972).
KK. Compilation of stable isotope fractionation factors of geochemical interest, by I. FRIEDMAN and J. R. O'NEIL (1977).

The first edition of the *Data of Geochemistry*, by F. W. CLARKE, was published in 1908 as U.S. Geological Survey Bulletin 330. Later editions, also by Clarke, were published in 1911, 1916, 1920 and 1924 as Bulletins 491, 616, 695 and 770. These earlier editions contain information that is still useful today.

The following two books are also useful in giving general accounts of the geochemistry of the individual elements, but it should be remembered that they were published over 25 years ago:

GOLDSCHMIDT, V. M. (1954) *Geochemistry*, 730 pp. Oxford.
RANKAMA, K. and SAHAMA, Th. G. (1950) *Geochemistry*, 912 pp. University of Chicago.

The following is a selection of works which contain geochemical data on particular topics.

Meteorites
MASON, B. (1971) *Handbook of Elemental Abundances in Meteorites*, 555 pp. Gordon & Breach.

Thermodynamics
ROBIE, R. A., HEMINGWAY B. S. and FISHER, J. F. (1978) Thermodynamic properties of minerals and related substances at 298.15 K and 1 bar (10^5 Pascals) pressure and at higher temperatures. *U.S. Geol. Surv. Bull.* 1452, 456 pp.
HELGESON, H. C., DELANEY, J. M., NESBITT, H. W. and BIRD, D. K. (1978) Summary and critique of the thermodynamic properties of rock-forming minerals. *Am. J. Sci.* **278**-A, 1–229.

Electrode Potentials and Solubility Equilibria
CHARLOT, G. (1958) *Selected Constants and Oxidation-reduction Potentials.* Pergamon.
LATIMER, W. M. (1952) *Oxidation potentials*, 2nd edn. Prentice Hall.

SILLEN, L. G. and MARTELL, A. E. (1964) *Stability Constants of Metal-Ion Complexes*, Special Publication No. 17. The Chemical Society, London. Also, (1971) Supplement 1 to above. Special Publication No. 25. The Chemical Society, London.

References

AHRENS, L.H. (1952) The use of ionization potentials. Part 1. Ionic radii of the elements. *Geochim. Cosmochim. Acta* **2**, 155–169.

ALLÈGRE, C. J. and MINSTER, J. F. (1978) Quantitative models of trace element behavior in magmatic processes. *Earth Planet. Sci. Lett.* **38**, 1–25.

ALLER, L. H. (1961) *The Abundance of the Elements*, 283 pp. Interscience, Wiley.

ALLER, L. H. (1968) Chemical composition of diffuse nebulae In: *Origin and Distribution of the Elements* (ed. L. H. AHRENS), pp. 191–193. Pergamon Press.

ANDERS, E. (1963) Meteorite ages. In *The Solar System*. Vol. IV. *The Moon Meteorites and Comets* (eds. B. M. MIDDLEHURST and G. P. KUIPER), pp. 402–495. Univ. Chicago Press.

ANDERS, E. (1971) How well do we know 'cosmic' abundances? *Geochim. Cosmochim. Acta* **35**, 516–522.

ANDERS, E. (1977) Chemical compositions of the Moon, Earth, and eucrite parent body. *Phil. Trans. R. Soc. Lond.* A, **285**, 23–40.

ANDERSON, A. T. (1975) Some basaltic and andesitic gases. *Rev. Geophys. Space Phys.* **13**, 37–55.

ANDERSON, A. T., Jr., CLAYTON, R. N. and MAYEDA, T. K. (1971) Oxygen isotope thermometry of mafic igneous rocks. *J. Geol.* **79**, 715–729.

ANDERSON, D. L. and HANKS, T. C. (1972) Formation of the Earth's core. *Nature* **237**, 387–388.

ANDERSON, D. L., SAMMIS, C. and JORDAN, T. (1972) Composition of the mantle and core. In *The Nature of the Solid Earth* (ed. E. C. ROBERTSON), pp. 41–66. McGraw Hill.

APTED, M. J. and ROY, S. D. (1981) Corrections to the trace element fractionation equations of Hertogen and Gijbels (1976). *Geochim. Cosmochim. Acta* **45**, 777–778.

ARRHENIUS, G. and ALFVÉN, H. (1971) Fractionation and condensation in space. *Earth Planet. Sci. Lett.* **10**, 253–267.

ATHERTON, M. P. (1976) Crystal growth models in metamorphic tectonites. *Phil. Trans. R. Soc. Lond.* A, **283**, 255–270.

BAILEY, J. C. (1977) Fluorine in granitic rocks and melts: a review. *Chem. Geol.* **19**, 1–42.

BARTH, T. F. W. (1951) The feldspar geological thermometers. *Neues Jahrb. Mineral.* **82**, 143–154.

BARTH, T. F. W. (1952) *Theoretical Petrology*, 387 pp. Wiley.

BERMANN, B. L., VAN HEMERT, R. L. and BOWMAN, C. D. (1969) Threshold photoneutron cross section for Mg^{26} and a source of stellar neutrons. *Phys. Rev. Lett.* **23**, 386–389.

BERNAL, J. D. (1960) Geometry of the structure of monatomic liquids. *Nature* **185**, 68–70.

BERNAL, J. D. (1964) The structure of liquids. *Proc. R. Soc. Lond.* A, **280**, 299–322.

BERNER, R. A. (1971) *Principles of Chemical Sedimentology*, 240 pp. McGraw-Hill.

BERNER, R. A. and HOLDREN, G. R. (1979) Mechanism of feldspar weathering. II. Observations on feldspars from soils. *Geochim. Cosmochim. Acta* **43**, 1173–1186.

BIGNELL, R. D., CRONAN, D. S. and TOOMS, J. S. (1976) Red Sea metalliferous brine precipitates. *Geol. Assoc. Can. Spec. Publ.* **14**, 147–179.

BISCHOFF, J. L. and DICKSON, F. W. (1975) Seawater–basalt interaction at 200°C and 500 bars: implications for origin of sea-floor heavy-metal deposits and regulation of seawater chemistry. *Earth Planet. Sci. Lett.* **25**, 385–397.

BISCHOFF, J. L. and SEYFRIED, W. E. (1978) Hydrothermal chemistry of seawater from 25° to 350°C. *Am. J. Sci.* **278**, 838–860.

BITHER, T. A., BOUCHARD, R. J., CLOUD, W. H., DONAHUE, P. C. and SIEMONS, W. J. (1968)

Transition metal pyrite dichalcogenides. High pressure synthesis and correlation of properties. *Inorg. Chem.* **7**, 2208–2220.

BLACK, L. P., GALE, N. H., MOORBATH, S., PANKHURST, R. J. and McGREGOR, V. R. (1971) Isotopic dating of very early Precambrian amphibolite facies gneisses from the Godthaab district, West Greenland. *Earth Planet. Sci. Lett.* **12**, 245–259.

BLANDER, M. and ABDEL-GAWAD, M. (1969) The origin of meteorites and the constrained equilibrium condensation theory. *Geochim. Cosmochim. Acta* **33**, 701–716.

BLANK, H. R. and GETTINGS, M. E. (1973) Subsurface form and extent of the Skaergaard intrusion, east Greenland. *Trans. Am. Geophys. Union* **54**, 507.

BLAXLAND, A. B., VAN BREEMEN, O., EMELEUS, C. H. and ANDERSON, J. G. (1978) Age and origin of the major syenite centers in the Gardar province of south Greenland: Rb–Sr studies. *Bull. Geol. Soc. Am.* **89**, 231–244.

BOON, J. A. and FYFE, W. S. (1972) The co-ordination number of ferrous ions in silicate systems. *Chem. Geol.* **10**, 287–298.

BOTTINGA, Y., KUDO, A. and WEILL, D. (1966) Some observations on oscillatory zoning and crystallization of magmatic plagioclase. *Am. Mineral.* **51**, 792–806.

BOTTINGA, Y. and WEILL, D. F. (1972) The viscosity of magmatic silicate liquids: a model for calculation. *Am. J. Sci.* **272**, 438–475.

BOUDREAU, B. P. and SCOTT, M. R. (1978) A model for the diffusion-controlled growth of deep sea manganese nodules. *Am. J. Sci.* **278**, 903–929.

BREWER, P. G. (1975) Minor elements in sea water. In *Chemical Oceanography* (eds. J. P. RILEY and G. SKIRROW), vol. 1, pp. 415–496. Academic Press.

BROECKER, W. S. (1974) *Chemical Oceanography*, 214 pp. Harcourt Brace Jovanovich, Inc.

BROOKS, C. K. (1969) On the distribution of zirconium and hafnium in the Skaergaard intrusion, East Greenland. *Geochim. Cosmochim. Acta* **33**, 357–374.

BROOKS, C. K., HENDERSON, P. and RØNSBO, J. G. (1981) Rare-earth partition between allanite and glass in the obsidian of Sandy Braes, Northern Ireland. *Mineral. Mag.* **44**, 157–160.

BROWN, G. M., PECKETT, A., PHILLIPS, R. and EMELEUS, C. H. (1973) Mineral-chemical variations in the Apollo 16 magnesio-feldspathic highland rocks. *Proc. 4th Lunar Sci. Conf.*, pp. 505–518.

BUCHWALD, V. F. (1975) *Handbook of Iron Meteorites*, 3 vols. Univ. California Press.

BUDDINGTON, A. F. and LINDSLEY, D. H. (1964) Iron–titanium oxide minerals and synthetic equivalents. *J. Petrol.* **5**, 310–357.

BURBIDGE, E. M., BURBIDGE, G. R., FOWLER, W. A. and HOYLE, F. (1957) Synthesis of the elements in stars. *Rev. Mod. Phys.* **29**, 547–650.

BURNHAM, C. W. (1979) Magmas and hydrothermal fluids. In *Geochemistry of Hydrothermal Ore Deposits* (ed. H. L. BARNES), 2nd edn., pp. 71–136. Wiley.

BURNS, R. G. (1970) *Mineralogical Applications of Crystal Field Theory*, 224 pp. Cambridge.

BURNS, R. G. and FYFE, W. S. (1967) Trace element distribution rules and their significance. *Chem. Geol.* **2**, 89–104.

BURNS, R. G. and VAUGHAN, D. J. (1970) Interpretation of the reflectivity behavior of ore minerals. *Am. Mineral.* **55**, 1576–1586.

BURTON, J. A., PRIM, R. C. and SLICHTER, W. P. (1953) The distribution of solute in crystals grown from the melt. Part 1. Theoretical. *J. Chem. Physics* **21**, 1987–1999.

BURTON, J. D. and LISS, P. S. (1973) Processes of supply and removal of dissolved silicon in the oceans. *Geochim. Cosmochim. Acta* **37**, 1761–1773.

BUSECK, P. R. and GOLDSTEIN, J. I. (1969) Olivine compositions and cooling rates of pallasitic meteorites. *Bull. Geol. Soc. Am.* **80**, 2141–2158.

CAMERON, A. G. W. (1959) A revised table of the abundances of the elements. *Astrophys. J.* **129**, 676–699.

CAMERON, A. G. W. (1968) A new table of abundances of the elements in the solar system. In *Origin and Distribution of the Elements* (ed. L. H. AHRENS), pp. 125–143. Pergamon.

CAMERON, A. G. W. (1973) Abundances of the elements in the solar system. *Space Science Rev.* **15**, 121–146.

CARMICHAEL, D. M. (1969) On the mechanism of prograde metamorphic reactions in quartz-bearing pelitic rocks. *Contrib. Mineral. Petrol.* **20**, 244–267.

CARMICHAEL, D. M. (1977) Chemical equilibria involving pure crystalline compounds. In *Application of Thermodynamics to Petrology and Ore Deposits* (ed. H. J. GREENWOOD), pp. 47–65. Mineral Assoc. Can. Short Course Handbook, 2.

CARMICHAEL, I. S. E., NICHOLLS, J. and SMITH, A. L. (1970) Silica activity in igneous rocks. *Am. Mineral.* **55**, 246–263.

CARMICHAEL, I. S. E., TURNER, F. J. and VERHOOGEN, J. (1974) *Igneous Petrology*, 739 pp. McGraw-Hill.

CARRUTHERS, J. R. (1975) Crystal growth from the melt. In *Treatise on Solid State Chemistry* (ed. N. B. HANNAY), vol. 5. *Changes of State.* Plenum Press.

CHARLOT, G. (1958) *Selected Constants and Oxide-Reduction Potentials.* Pergamon.

CHESWORTH, W. (1973) The residua system of chemical weathering: a model for the chemical breakdown of silicate rocks at the surface of the earth. *J. Soil Sci.* **24**, 69–81.

CHRISTIAN, J. W. (1975) *The Theory of Transformations in Metals and Alloys. Part 1 Equilibrium and General Kinetic Theory*, 2nd edn. 586 pp. Pergamon.

CLARK, A. M. and LONG, J. V. P. (1971) The anisotropic diffusion of nickel in olivine. In *Thomas Graham Memorial Symposium on Diffusion Processes*, pp. 511–521. Gordon & Breach.

CLARK, S. P., TUREKIAN, K. K. and GROSSMAN, L. (1972) Early history of the earth. In *The Nature of the Solid Earth* (ed. E. C. ROBERTSON), pp. 3–18. McGraw-Hill.

CLARKE, F. W. (1908) The data of geochemistry. *U.S. Geol. Surv. Bull.* **330**, 716 pp.

CLARKE, F. W. and WASHINGTON, H. S. (1924) The composition of the earth's crust. *U.S. Geol. Surv. Prof. Pap.* **127**, 117 pp.

CLAYTON, D. D. (1968) *Principles of Stellar Evolution and Nucleosynthesis*, 612 pp. McGraw-Hill.

CLAYTON, D. D., FOWLER, W. A., HULL, T. E. and ZIMMERMAN, B. A. (1961) Neutron capture chains in heavy element synthesis. *Ann. Phys.* **12**, 331–408.

CLAYTON, R. N., ONUMA, N. and MAYEDA, T. K. (1976) A classification of meteorites based on oxygen isotopes. *Earth Planet. Sci. Lett.* **30**, 10–18.

CONWAY, E. J. (1942) Mean geochemical data in relation to oceanic evolution. *Proc. R. Irish Acad.* B, **48**, 119–159.

CONWAY, E. J. (1943) The chemical evolution of the ocean. *Proc. R. Irish Acad.* B, **48**, 161–212.

COPE, M. J. and CHALONER, W. G. (1980) Fossil charcoal as evidence of past atmospheric composition. *Nature* **283**, 647–649.

CORYELL, C. G., CHASE, J. W. and WINCHESTER, J. W. (1963) A procedure for geochemical interpretation of terrestrial rare-earth abundance patterns. *J. Geophys. Res.* **68**, 559–566.

COTTON, F. A. and WILKINSON, G. (1966) *Advanced Inorganic Chemistry. A Comprehensive Text*, 2nd edn., 1366 pp. Wiley.

COTTON, F. A. and WILKINSON, G. (1980) *Advanced Inorganic Chemistry. A Comprehensive Text*, 4th edn., 1396 pp. Wiley.

COX, K. G., BELL, J. D. and PANKHURST, R. J. (1979) *The Interpretation of Igneous Rocks*, 450 pp. George Allen & Unwin.

CRANK, J. (1975) *The Mathematics of Diffusion*, 2nd edn., 414 pp. Oxford.

CRONAN, D. S. (1976) Manganese nodules and other ferro-manganese oxide deposits. In *Chemical Oceanography*, 2nd edn. (eds. J. P. RILEY and R. CHESTER), vol. 5, pp. 217–263. Academic.

CRONAN, D. S. (1980) *Underwater Minerals*, 364 pp. Academic.

CURTIS, D., GLADNEY, E. and JURNEY, E. (1980) A revision of the meteorite based cosmic abundance of boron. *Geochim. Cosmochim. Acta* **44**, 1945–1953.

DALE, I. M. and HENDERSON, P. (1972) The partition of transition elements in phenocryst-bearing basalts and the implications about melt structure. *24th Int. Geol. Congress, Sect.* **10**, 105–111.

DARKEN, L. S. (1948) Diffusion, mobility and their interrelation through free energy in binary metallic systems. *Am. Inst. Mining Metall. Engineers Trans.* **175**, 184–201.

DAVIS, D. W., GRAY, J., CUMMING, G. L. and BAADSGAARD, H. (1977) Determination of the ^{87}Rb decay constant. *Geochim. Cosmochim. Acta* **41**, 1745–1749.

DEER, W. A., HOWIE, R. A. and ZUSSMAN, J. (1966) *An Introduction to the Rock Forming Minerals*, 528 pp. Longmans.

322   *Inorganic Geochemistry*

DEGENS, E. T. and ROSS, D. A., eds. (1969) *Hot Brines and Recent Heavy Metal Deposits in the Red Sea*, 600 pp. Springer.
DEINES, P. (1977) On the oxygen isotope distribution among mineral triplets in igneous and metamorphic rocks. *Geochim. Cosmochim. Acta* **41**, 1709–1730.
DENBIGH, K. (1971) *The Principles of Chemical Equilibrium*, 3rd edn., 494 pp. Cambridge.
DE PIERI, R. and QUARENI, S. (1978) Partition coefficients of alkali and alkaline-earth elements between alkali feldspar phenocrysts and their lava matrix. *Mineral. Mag.* **42**, 63–67.
DISSANAYAKE, C. B. and VINCENT, E. A. (1972) Zinc in rocks and minerals from the Skaergaard intrusion, east Greenland. *Chem. Geol.* **9**, 285–297.
DISSANAYAKE, C. B. and VINCENT, E. A. (1975) Mercury in rocks and minerals of the Skaergaard intrusion, east Greenland. *Mineral. Mag.* **40**, 33–42.
DONALDSON, C. H. (1975) Calculated diffusion coefficients and the growth rate of olivine in a basalt magma. *Lithos* **8**, 163–174.
DOWTY, E. (1980) Crystal growth and nucleation theory and the numerical simulation of igneous crystallization In: *Physics of Magmatic Processes* (ed. R. B. HARGRAVES), Princeton Univ. Press, Chapter 10, pp. 419–485.
DRAKE, M. J. and HOLLOWAY, J. R. (1978) 'Henry's Law' behaviour of Sm in a natural plagioclase/melt system: Importance of experimental procedure. *Geochim. Cosmochin. Acta* **42**, 679–683.
DRAKE, M. J. and HOLLOWAY, J. R. (1981) Partitioning of Ni between olivine and silicate melt: the 'Henry's Law problem' reexamined. *Geochim. Cosmochim. Acta* **45**, 431–437.
DREVER, J. I. (1974) The magnesium problem. In *The Sea* vol. 5: *Marine Chemistry* (ed. E. D. GOLDBERG), pp. 337–357.
DUCHESNE, J. -C. (1972) Iron–titanium oxide minerals in the Bjerkrem–Sogndal massif, south-western Norway. *J. Petrol*, **13**, 57–81.
DUDAS, M. J., SCHMITT, R. A. and HARWARD, M. E. (1971) Trace element partitioning between volcanic plagioclase and dacitic pyroclastic matrix. *Earth Planet. Sci. Lett.* **11**, 440–446.
DU FRESNE, E. R. and ANDERS, E. (1962) On the chemical evolution of the carbonaceous chondrites. *Geochim. Cosmochim. Acta* **26**, 1085–1114.
DUNITZ, J. D. and ORGEL, L. E. (1957) Electronic properties of transition—metal oxides. II. Cation distribution amongst octahedral and tetahedral sites. *J. Phys. Chem. Solids* **3**, 318–323.
DUPUY, C. (1972) Coefficients de partage du strontium entre phases leucocrates des ignimbrites de Toscane (Italie). *Bull. Soc. Fr. Mineral. Cristallogr.* **95**, 322–329.
EDMUNDS, W. M. (1975) Geochemistry of brines in the Coal Measures of northeast England. *Trans. Instn. Min. Metall. (Sect. B: Appl. Earth Sci.)* **84**, B39–B52.
EDMUNDS, W. M. and ATHERTON, M. P. (1971) Polymetamorphic evolution of garnet in the Fanad aureole, Donegal, Eire. *Lithos* **4**, 147–161.
EGGLER, D. H. and ROSENHAUER, M. (1978) Carbon dioxide in silicate melts: II. Solubilities of CO_2 and H_2O in Ca $MgSi_2O_6$(diopside) liquids and vapors at pressures to 40 kb. *Am. J. Sci.* **178**, 64–94.
ELDERFIELD, H. (1977) Authigenic silicate minerals and the magnesium budget in the oceans. *Phil. Trans. R. Soc. Lond.* A, **286**, 273–281.
EWART, A. and TAYLOR, S. R. (1969) Trace element geochemistry of the rhyolitic volcanic rocks, Central North Island, New Zealand. Phenocryst data. *Contrib. Mineral. Petrol*, **22**, 127–146.
FAURE, G. (1977) *Principles of Isotope Geology*, 464 pp. J. Wiley & Sons.
FAURE, G. and POWELL, J. L. (1972) *Strontium Isotope Geology*, 188 pp. Springer–Verlag, New York.
FEGLEY, B., JR. and LEWIS, J. S. (1980) Volatile element chemistry in the solar nebula: Na, K, F, Cl, Br and P. *Icarus* **41**, 439–455.
FENN, P. M. (1977) The nucleation and growth of alkali feldspars from hydrous melts. *Can. Mineral.* **15**, 135–161.
FISHER, G. W. (1970) The application of ionic equilibria to metamorphic differentiation: an example. *Contrib. Mineral. Petrol.* **29**, 91–103.
FISHER, G. W. (1977) Nonequilibrium thermodynamics in metamorphism. In *Thermodynamics in Geology* (ed. D. G. FRASER), pp. 381–403. Reidel.

FLEISCHER, M., ed. (1962 onwards) *Data of Geochemistry*, 6th edn. *U.S. Geol. Surv. Prof. Pap.* 440.

FLOOD, H. and KNAPP, W. J. (1968) Structural characteristics of liquid mixtures of feldspar and silica. *J. Am. Ceram. Soc.* **51**, 259–263.

FLORY, P. J. (1953) *Principles of Polymer Chemistry*, 688 pp. Cornell University Press.

FOLAND, K. A. (1974) Alkali diffusion in orthoclase. In *Geochemical Transport and Kinetics* (eds. A. W. HOFMANN, B. J. GILETTI, H. S. YODER, Jr. and R. A. YUND), pp. 77–98. Carnegie Inst. Washington, Publ. 634.

FORESTER, R. W. and TAYLOR, H. P., Jr. (1976) ^{18}O-depleted igneous rocks from the Tertiary Complex of the Isle of Mull, Scotland. *Earth Planet. Sci. Lett.* **32**, 11–17.

FRASER, D. G., ed. (1977) *Thermodynamics in Geology*. D. REIDEL Publ. Co. 410 pp.

FREER, R. (1979) An experimental measurement of cation diffusion in almandine garnet. *Nature* **280**, 220–222.

FREER, R. and O'REILLY, W. (1980) The diffusion of Fe^{2+} ions in spinels with relevance to the process of maghemitization. *Mineral. Mag.* **43**, 889–899.

FRIEDMAN, I. and O'NEIL, J. R. (1977) Compilation of stable isotope fractionation factors of geochemical interest. In *Data of Geochemistry* (ed. M. FLEISCHER), 6th edn, *U.S. Geol. Surv. Prof. Pap.* 440–KK.

FROESE, E. (1976) Applications of thermodynamics in metamorphic petrology. *Pap. Geol. Surv. Can.* 75–43, 1–37.

FRONDEL, J. W. (1975) *Lunar Mineralogy*, 323 pp. Wiley.

FUGE, R. (1977) On the behavior of fluorine and chlorine during magmatic differentiation. *Contrib. Mineral. Petrol.* **61**, 245–249.

FUMI, F. G., and TOSI, M. P. (1964) Ionic sizes and Born repulsive parameters in the NaCl-type alkali halides. I. *J. Phys. Chem. Solids* **25**, 31–43.

FYFE, W. S., PRICE, N. J. and THOMPSON, A. B. (1978) *Fluids in the Earth's Crust*, 383 pp. Elsevier.

GALE, N. H., ARDEN, J. W. and HUTCHISON, R. (1975) The chronology of the Nakhla achondritic meteorite. *Earth Planet. Sci. Lett.* **26**, 195–206.

GANAPATHY, R. and ANDERS, E. (1974) Bulk compositions of the moon and earth, estimated from meteorites. *Proc. 5th Lunar Sci. Conf.*, pp. 1181–1206.

GARRELS, R. M. (1954) Mineral species as functions of pH and oxidation–reduction potentials, with special reference to the zone of oxidation and secondary enrichment of sulphide ore deposits. *Geochim. Cosmochim. Acta* **5**, 153–168.

GARRELS, R. M. and CHRIST, C. L. (1965) *Solutions, Minerals and Equilibria*, 450 pp. Harper.

GARRELS, R. M. and MACKENZIE, F. T. (1967) Origin of the chemical compositions of some springs and lakes. In *Equilibrium Concepts in Natural Water Systems*. Am. Chem. Soc. *Advances in Chem. Series No. 67* (ed. R. F. GOULD), pp. 222–242.

GARRELS, R. M. and MACKENZIE, F. T. (1971) *Evolution of the Sedimentary Rocks*, 397 pp. W. W. Morton & Co. Inc.

GARRELS, R. M. and MACKENZIE, F. T. (1972) A quantitative model for the sedimentary rock cycle. *Marine Chem.* **1**, 27–41.

GARRELS, R. M. and THOMPSON, M. E. (1962) A chemical model for sea water at 25° C and one atmosphere total pressure. *Am. J. Sci.* **260**, 57–66.

GAST, P. W. (1968) Trace element fractionation and the origin of tholeiitic and alkaline magma types. *Geochim. Cosmochim. Acta* **32**, 1057–1086.

GHOSE, S. (1965) Mg^{2+}–Fe^{2+} order in an orthopyroxene $Mg_{0.93}Fe_{1.07}Si_2O_6$. *Zeit. Krist.* **125**, 1–6.

GOLDICH, S. S. (1938) A study in rock weathering. *J. Geol.* **46**, 17–58.

GOLDSCHMIDT, V. M. (1933) Grundlagen der quantitativen Geochemie. *Fortschr. Mineral. Kristall. Petrogr.* **17**, 112–156.

GOLDSCHMIDT, V. M. (1937) The principles of distribution of chemical elements in minerals and rocks. *J. Chem. Soc. Lond.* 655–673.

GOLDSCHMIDT, V. M. (1954) *Geochemistry*, 730 pp. Clarendon Press, Oxford.

GOLDSTEIN, J. I. and OGILVIE, R. E. (1965) The growth of the Widmanstätten pattern in metallic meteorites. *Geochim. Cosmochim. Acta* **29**, 893–920.

GOLDSTEIN, J. I. and SHORT, J. M. (1967a) Cooling rates of 27 iron and stony-iron meteorites. *Geochim. Cosmochim. Acta* **31**, 1001–1023.

GOLDSTEIN, J. I. and SHORT, J. M. (1967b) The iron meteorites, their thermal history and parent bodies. *Geochim. Cosmochim. Acta* **31**, 1733–1770.

GOLES, G. G. (1969) Cosmic abundances, nucleosynthesis and cosmic chronology. In *Handbook of Geochemistry*, Part 1 (ed. K. H. WEDEPOHL), pp. 116–133. Springer–Verlag.

GOODMAN, R. J. (1972) The distribution of Ga and Rb in coexisting groundmass and phenocryst phases of some basic volcanic rocks. *Geochim. Cosmochim. Acta* **36**, 303–318.

GOVOROV, I. N. and STUNZHAS, A. A. (1963) Mode of transport of beryllium in alkali metasomatism. *Geochemistry* No. 4, 402–409. Transl. from *Geokhimiya* (Publ. Acad. Sci. USSR), 1963, no. 4, 383–390.

GRAY, C. M., PAPANASTASSIOU, D. A. and WASSERBURG, G. J. (1973) The identification of early condensates from the solar nebula. *Icarus* **20**, 213–239.

GREEN, D. H. and RINGWOOD, A. E. (1967) The genesis of basalt magmas. *Contrib. Mineral. Petrol.* **15**, 103–190.

GREEN, T. H., GREEN, D. H. and RINGWOOD, A. E. (1967) The origin of high-alumina basalts and their relationship to other basaltic magma types. *Earth Planet. Sci. Lett.* **2**, 41–51.

GREENLAND, L. P. (1970) An equation for trace element distribution during magmatic crystallization. *Am. Mineral.* **55**, 455–465.

GREENWOOD, H. J., ed. (1977) *Application of Thermodynamics to Petrology and Ore Deposits.* Mineral. Assoc. Can., Short Course Handbook, 2 (Vancouver).

GREENWOOD, N. N. (1968) *Ionic Crystals, Lattice Defects and Nonstoichiometry*, 194 pp. Butterworths.

GROSSMAN, L. (1972) Condensation in the primitive solar nebula. *Geochim. Cosmochim. Acta* **36**, 579–619.

GROSSMAN, L. and LARIMER, J. W. (1974) Early chemical history of the solar system. *Rev. Geophys. Space Sci.* **12**, 71–101.

GROVER, J. (1977) Chemical mixing in multicomponent systems: an introduction to the use of Margules and other thermodynamic excess functions to represent non-ideal behaviour. In *Thermodynamics in Geology* (ed. D. G. FRASER), NATO Advanced Study Institutes Series, pp. 67–97. Reidel Publ. Co.

GUGGENHEIM, E. A. (1952) *Mixtures*, Clarendon Press, Oxford.

GUNNER, J. D. (1974) Investigations of lower Palaeozoic granites in the Beardmore Glacier region. *Ant. J. U.S.* **9**, 76–81.

HAGGERTY, S. E. (1972) Apollo 14: Subsolidus reduction and compositional variations of spinels. *Proc. 3rd Lunar Sci. Conf.*, pp. 305–332.

HAJASH, A. (1975) Hydrothermal processes along mid-ocean ridges: an experimental investigation. *Contrib. Mineral. Petrol.* **53**, 205–226.

HAKLI, T. and WRIGHT, T. L. (1967) The fractionation of nickel between olivine and augite as a geothermometer. *Geochim. Cosmochim. Acta* **31**, 877–884.

HAMILTON, D. L., BURNHAM, C. W. and OSBORN, E. F. (1964) The solubility of water and effects of oxygen fugacity and water content on crystallization of mafic magmas. *J. Petrol.* **5**, 21–39.

HANOV, J. S. (1969) Barite saturation in sea water. *Geochim. Cosmochim. Acta* **33**, 894–898.

HANSON, G. N. (1978) The application of trace elements to the petrogenesis of igneous rocks of granitic composition. *Earth Planet. Sci. Lett.* **38**, 26–43.

HARGRAVES, R. B., ed. (1980) *Physics of Magmatic Processes*, 585 pp. Princeton University Press.

HARRISS, R. C. and ADAMS, J. A. S. (1966) Geochemical and mineralogical studies on the weathering of granitic rocks. *Am. J. Sci.* **264**, 146–173.

HARRISON, W. J. and WOOD, B. J. (1980) An experimental investigation of the partitioning of REE between garnet and liquid with reference to the role of defect equilibria. *Contrib. Mineral. Petrol.* **72**, 145–155.

HART, R. (1970) Chemical exchange between sea water and deep ocean basalts. *Earth Planet. Sci. Lett.* **9**, 269–279.

HART, S. R. (1981) Diffusion compensation in natural silicates. *Geochim. Cosmochim. Acta* **45**, 279–291.

HART, S. R. and BROOKS, C. (1974) Clinopyroxene-matrix partitioning of K, Rb, Cs, Sr and Ba. *Geochim. Cosmochim. Acta* **38**, 1799–1806.

HART, S. R., ERLANK, A. J. and KABLE, J. D. (1974) Sea floor basalt alteration: some chemical and Sr isotopic effects. *Contrib. Mineral. Petrol.* **44**, 219–230.

HASKIN, L. A., SHIK, C.-Y., BANSAL, B. M., RHODES, J. M., WIESMANN, H. and NYQUIST, L. E. (1974) Chemical evidence for the origin of 76535 as a cumulate. *Proc. 5th Lunar Sci. Conf.*, pp. 1213–1225.

HAUGHTON, D. R., ROEDER, P. L. and SKINNER, B. J. (1974) Solubility of sulfur in mafic magmas. *Econ. Geol.* **69**, 451–467.

HELGESON, H. C. (1968) Evaluation of irreversible reactions in geochemical processes involving minerals and aqueous solutions – I. Thermodynamic relations. *Geochim. Cosmochim. Acta* **32**, 853–877.

HELGESON, H. C. (1971) Kinetics of mass transfer among silicates and aqueous solutions. *Geochim. Cosmochim. Acta* **35**, 421–469.

HELGESON, H. C. (1972) Kinetics of mass transfer among silicates and aqueous solutions: Correction and clarification. *Geochim. Cosmochim. Acta* **36**, 1067–1070.

HELGESON, H. C., DELANEY, J. M., NESBITT, H. W. and BIRD, D. K. (1978) Summary and critique of the thermodynamic properties of rock-forming minerals. *Am. J. Sci.* **278**-A, 1–229.

HELGESON, H. C., GARRELS, R. M. and MACKENZIE, F. T. (1969) Evaluation of irreversible reactions in geochemical processes involving minerals and aqueous solutions—II. Applications. *Geochim. Cosmochim. Acta* **32**, 455–482.

HENDERSON, L. M. and KRACEK, F. C. (1927) The fractional precipitation of barium and radium chromates. *J. Am. Chem. Soc.* **49**, 739–749.

HENDERSON, P. (1975) Geochemical indicator of the efficiency of fractionation of the Skaergaard intrusion, East Greenland. *Mineral. Mag.* **40**, 285–291.

HENDERSON, P. (1979) Irregularities in patterns of element partition. *Mineral. Mag.* **43**, 399–404.

HENDERSON, P. and DALE, I. M. (1969) The partitioning of selected transition element ions between olivine and groundmass of oceanic basalts. *Chem. Geol.* **5**, 267–274.

HENDERSON, P. and WILLIAMS, C. T. (1979) Variation in trace element partition (crystal magma) as a function of crystal growth rate. In *Origin and Distribution of the Elements* (ed. L. H. AHRENS), pp. 191–198. Pergamon.

HERTOGEN, J. and GIJBELS, R. (1976) Calculation of trace element fractionation during partial melting. *Geochim. Cosmochim. Acta* **40**, 313–322.

HESS, P. C. (1977) Structure of silicate melts. *Can. Mineral.* **15**, 162–178.

HESS, P. C. (1980) Polymerization model for silicate melts. In *Physics of Magmatic Processes* (ed. R. B. HARGRAVES), pp. 3–48. Princeton Univ. Press.

HIGUCHI, H. and NAGASAWA, H. (1969) Partition of trace elements between rock-forming minerals and the host volcanic rocks. *Earth Planet. Sci. Lett.* **7**, 281–287.

HILDEBRAND, J. H. (1929) Solubility. XII. Regular solutions. *J. Am. Chem. Soc.* **51**, 66–80.

HILDEBRAND, J. H. and SCOTT, R. H. (1962) *Regular Solutions*. Prentice Hall.

HOEFS, J. (1973) *Stable Isotope Geochemistry*, 140 pp. Springer-Verlag.

HOFMANN, A. W. (1980) Diffusion in natural silicate melts: a critical review. In *Physics of Magmatic Processes* (ed. R. B. HARGRAVES), pp. 385–417. Princeton Univ. Press.

HOLDREN, G. R., Jr. and BERNER, R. A. (1979) Mechanism of feldspar weathering—1. Experimental studies. *Geochim. Cosmochim. Acta* **43**, 1161–1171.

HOLLAND, H. D. (1972) The geologic history of sea water—an attempt to solve the problem. *Geochim. Cosmochim. Acta* **36**, 637–651.

HOLLAND, H. D. (1978) *The Chemistry of the Atmosphere and Oceans*, 351 pp. Wiley.

HOLLAND, J. G. and LAMBERT, R. St. J. (1972) Major element chemical composition of shields and the continental crust. *Geochim. Cosmochim. Acta* **36**, 673–683.

HUMPHRIS, S. E. and THOMPSON, G. (1978a) Hydrothermal alteration of oceanic basalts by seawater. *Geochim. Cosmochim. Acta* **42**, 107–125.

HUMPHRIS, S. E. and THOMPSON, G. (1978b) Trace element mobility during hydrothermal alteration of oceanic basalts. *Geochim. Cosmochim. Acta* **42**, 127–136.

HURLEY, P. M. (1968a) Absolute abundances and distribution of Rb, K and Sr in the earth. *Geochim. Cosmochim. Acta* **32**, 273–284.

HURLEY, P. M. (1968b) Correction to: Absolute abundances and distribution of Rb, K and Sr in the earth. *Geochim. Cosmochim. Acta* **32**, 1025–1030.

HUTCHISON, R. (1974) The formation of the Earth. *Nature* **250**, 556–568.

HUTCHISON, R., BEVAN, A. W. R. and HALL, J. M. (1977) *Appendix to the Catalogue of Meteorites*, 297 pp. British Museum (Natural History).

HUTCHISON, R., BEVAN, A. W. R., EASTON, A. J. and AGRELL, S. O. (1981) Mineral chemistry and relations among H-group chondrites. *Proc. R. Soc. Lond.* A, **374**, 159–178.

IRVINE, T. N. and SMITH, C. H. (1967) The ultramafic rocks of the Muskox intrusion Northwest Territories, Canada: In *Ultramafic and Related Rocks* (ed. P. J. WYLLIE), pp. 38–49. Wiley.

IRVING, A. J. (1978) A review of experimental studies of crystal/liquid trace element partitioning. *Geochim. Cosmochim. Acta* **42**, 743–770.

IRVING, A. J. and FREY, F. A. (1978) Distribution of trace elements between garnet megacrysts and host volcanic liquids of kimberlitic to rhyolitic composition. *Geochim. Cosmochim. Acta* **42**, 771–787.

JACOBS, J. A., RUSSEL, R. D. and WILSON, J. T. (1974) *Physics and Geology*, 2nd edn., 622 pp. McGraw Hill.

JÄGER, E. and HUNZIKER, J. C., eds. (1979) *Lectures in Isotope Geology*, 329 pp. Springer-Verlag.

JAMBON, A. and CARRON, J-P. (1978) Étude experimentale de la diffusion cationique dans un verre basaltique: alcalins et alcalino-terreux. *Bull. Minéral.* **101**, 22–26.

JAMBON, A., CARRON, J-P. and DELBOVE, F. (1978) Donnees preliminaires sur la diffusion dans les magmas hydratés: le cesium dans un liquide granitique a 3 kbar. *C.R. Acad. Sci. Paris* D, **287**, 403–406.

JENSEN, B. B. (1973) Patterns of element partitioning. *Geochim. Cosmochim. Acta* **37**, 2227–2242.

JESSBERGER, E. K., HUNEKE, J. C. and WASSERBURG, G. J. (1974) Evidence for a ~ 4.5 aeon age of plagioclase clasts in a lunar highland breccia. *Nature* **248**, 199–202.

JOHANNES, W. (1967) Experimente zur metasomatischen Magnesitbildung. *Neues Jahrb. Mineral. Monatsh.* 321–333.

KASTNER, M. (1974) The contribution of authigenic feldspars to the geochemical balance of alkali metals. *Geochim. Cosmochim. Acta* **38**, 650–653.

KATSURA, T. and NAGASHIMA, S. (1974) Solubility of sulfur in some magmas at 1 atmosphere. *Geochim. Cosmochim. Acta* **38**, 517–531.

KEIL, K. (1969) Meteorite composition. In *Handbook of Geochemistry* (ed. K. H. WEDEPOHL), **1**, pp. 78–115. Springer.

KIRKPATRICK, R. J. (1975) Crystal growth from the melt: a review. *Am. Mineral.* **60**, 798–814.

KIRKPATRICK, R. J. (1977) Nucleation and growth of plagioclase, Makaopuki and Alae lava lakes, Kilauea Volcano, Hawaii. *Geol. Soc. Am. Bull.* **88**, 78–84.

KLEIN, C., Jr. (1972) Lunar materials: their mineralogy, petrology and chemistry. *Earth Sci. Rev.* **8**, 169–204.

KRAUSKOPF, K. (1967) *Introduction to Geochemistry*, 721 pp. McGraw Hill, New York.

KUSHIRO, I. (1969) The system forsterite–diopside–silica with and without water at high pressures. *Am. J. Sci.* **267**-A, 269–294.

KUSHIRO, I. (1972) Effect of water on the compositions of magmas formed at high pressures. *J. Petrol.* **13**, 311–334.

KUSHIRO, I. (1975) On the nature of silicate melt and its significance in magma genesis: regularities in the shift of the liquidus boundaries involving olivine, pyroxene, and silica minerals. *Am. J. Sci.* **275**, 411–431.

KUSHIRO, I., YODER, H. S., Jr. and MYSEN, B. O. (1976) Viscosities of basalt and andesite melts at high pressures. *J. Geophys. Res.* **81**, 6351–6356.

LACY, E. D. (1965) A statistical model of polymerisation/depolymerisation relationships in silicate melts and glasses. *Phys. Chem. Glasses* **6**, 171–180.

LACY, E. D. (1967) The Newtonian flow of simple silicate melts at high temperature. *Phys. Chem. Glasses* **8**, 238–246.

LAGACHE, M. (1965) Contribution a l'etude de l'alteration des feldspaths dans l'eau, entre 100 et

200°C, sous diverses pressions de CO_2, et application à la synthèse des minéraux argileux. *Bull. Soc. Fr. Mineral. Cristallogr.* **88**, 223–253.

LAGACHE, M. (1976) New data on the kinetics of the dissolution of alkali feldspars at 200°C in CO_2 charged water. *Geochim. Cosmochim. Acta* **40**, 157–161.

LARIMER, J. W. (1967) Chemical fractionations in meteorites—I. Condensation of elements. *Geochim. Cosmochim. Acta* **31**, 1215–1238.

LARIMER, J. W. (1971) Composition of the earth: Chondritic or achondritic? *Geochim. Cosmochim. Acta* **35**, 769–786.

LARIMER, J. W. and ANDERS, E. (1967) Chemical fractionations in meteorites—II. Abundance patterns and their interpretations. *Geochim. Cosmochim. Acta* **31**, 1239–1270.

LATIMER, W. M. (1952) *Oxidation Potentials*, 2nd edn. Prentice Hall.

LEEMAN, W. P. (1979) Partitioning of Pb between volcanic glass and coexisting sanidine and plagioclase feldspars. *Geochim. Cosmochim. Acta* **43**, 171–175.

LEEMAN, W. P. and SCHEIDEGGER, K. F. (1977) Olivine/liquid distribution coefficients and a test for crystal-liquid equilibrium. *Earth Planet. Sci. Lett.* **35**, 247–257.

LERBEKMO, J. F. and PLATT, R. L. (1962) Promotion of pressure-solution of silica in sandstones. *J. Sediment. Petrol.* **32**, 514–519.

LERMAN, A. (1970) Chemical equilibria and evolution of chloride brines. *Spec. Pap. Mineral. Soc. Am.* **3**, 291–306.

LERMAN, A. (1979) *Geochemical Processes. Water and Sediment Environments*, 481 pp. Wiley.

LEVIN, B. J. (1979) On the core of the moon. *Proc. 10th Lunar Sci. Conf.*, pp. 2321–2323.

LIVINGSTONE, D. A. (1963) Chemical composition of rivers and lakes. In *Data of Geochemistry* (ed. M. FLEISCHER), 6th edn. U.S. Geol. Surv. Prof. Pap. 440-G. 64 pp.

LOOMIS, T. P. (1977) Kinetics of a garnet granulite reaction. *Contrib. Mineral. Petrol.* **62**, 1–22.

LONG, D. T. and ANGINO, E. E. (1977) Chemical speciation of Cd, Cu, Pb and Zn in mixed freshwater, seawater and brine solutions. *Geochim. Cosmochim. Acta* **41**, 1183–1191.

LORD, H. C. III (1965) Molecular equilibria and condensation in a solar nebula and cool stellar atmospheres. *Icarus* **4**, 279–288.

LOUGHNAN, F. C. (1969) *Chemical Weathering of the Silicate Minerals*, 154 pp. American Elsevier.

LOVERING, J. F., NICHIPORUK, W., CHODOS, A. and BROWN, H, (1957) The distribution of gallium, germanium, cobalt, chromium, and copper in iron and stony-iron meteorites in relation to nickel content and structure. *Geochim. Cosmochim. Acta* **11**, 263–278.

LOWRY, R. K., HENDERSON, P. and NOLAN, J. (1982) Tracer diffusion of some alkali, alkaline-earth and transition element ions in a basaltic and an andesitic melt, and the implications concerning melt structure. (In preparation.)

LOWRY, R. K., REED, S. J. B., NOLAN, J., HENDERSON, P. and LONG, J. V. P. (1981) Lithium tracer-diffusion in an alkali-basaltic melt—an ion-microprobe determination. *Earth Planet. Sci. Lett.* **53**, 36–40.

McCLURE, D. S. (1957) The distribution of transition metal cations in spinels. *J. Phys. Chem. Solids* **3**, 311–317.

McELHINNEY, M. W., ed. (1976) *The Earth: Its Origin, Structure and Evolution*, 597 pp. Academic Press.

McINTIRE, W. L. (1963) Trace element partition coefficients—a review of theory and applications to geology. *Geochim. Cosmochim. Acta* **27**, 1209–1264.

MACKENZIE, F. T. (1975) Sedimentary cycling and the evolution of sea water. In *Chemical Oceanography* (eds. J. P. RILEY and G. SKIRROW), 2nd edn. **1**, 309–364.

MACKENZIE, F. T. and GARRELS, R. M. (1966) Chemical mass balance between rivers and oceans. *Am. J. Sci.* **264**, 507–525.

MACKENZIE, J. D. (1962) Oxide melts. *Adv. Inorg. Chem. Radiochem.* **4**, 293–318.

McKIE, D. and McKIE, C. (1974) *Crystalline Solids*, 628 pp. Nelson, London.

MARTIN, R. F., WHITLEY, J. E. and WOOLLEY, A. R. (1978) An investigation of rare-earth mobility: fenitized quartzites, Borrelan Complex, N.W. Scotland. *Contrib. Mineral. Petrol.* **66**, 69–73.

MASON, B. (1962) *Meteorites*, 274 pp. Wiley.

MASON, B. (1966) *Principles of Geochemistry*, 3rd edn., 329 pp. Wiley.

MASON, B. (1971) *Handbook of Elemental Abundances* in Meteorites, 555 pp. Gordon & Breach.

MASON, B. (1975) The Allende meteorite—Cosmochemistry's Rosetta Stone? *Accounts Chem. Res.* **8**, 217–224.

MASON, B. (1979) Cosmochemistry. Part 1, Meteorites. In *Data of Geochemistry* (ed. M. FLEISCHER), 6th edn. U.S. Geol. Surv. Prof. Pap. 440-B-1, 132 pp.

MASON, B. and JAROSEWICH, E. (1973) The Barea, Dyarrl Island, and Emery meteorites, and a review of the mesosiderites. *Mineral. Mag.* **39**, 204–215.

MASON, B. and WIIK, H. B. (1964) The amphoterites and meteorites of similar composition. *Geochim. Cosmochim. Acta* **28**, 533–538.

MASSON, C. R., SMITH, I. B. and WHITEWAY, S. G. (1970) Activities and ionic distributions in liquid silicates: application of polymer theory. *Can. J. Chem.* **48**, 1456–1464.

MASUDA, A. (1962) Regularities in variation of relative abundances of lanthanide elements and an attempt to analyse separation—index patterns of some minerals. *J. Earth Sci. Nagoya Univ.* **10**, 173–187.

MATSUI, A., ONUMA, N., NAGASAWA, H., HIGUCHI, H. and BANNO, S. (1977). Crystal structure control in trace element partition between crystal and magma. *Bull. Soc. Fr. Mineral. Cristallogr.* **100**, 315–324.

MEDARIS, L. G. (1969) Partitioning of Fe^{2+} and Mg^{2+} between coexisting synthetic olivine and orthopyroxene. *Am. J. Sci.* **267**, 945–968.

MILNE-THOMSON, L. M. and COMRIE, L. J. (19) *Standard four-figure mathematical tables*. Macmillan, London.

MISENER, D. J. (1974) Cationic diffusion in olivine to 1400°C and 35 kbar. In *Geochemical Transport and Kinetics* (eds. A. W. HOFMANN, B. J. GILETTI, H. S. YODER, Jr. and R. A. YUND), pp. 117–129. Carnegie Inst. Washington, Publ. 634.

MOORBATH, S., O'NIONS, R. K., PANKHURST, R. J., GALE, N. H. and McGREGOR, V. R. (1972) Further rubidium–strontium age determinations on the very early Precambrian rocks of the Godthaab district, West Greenland. *Nature Phys. Sci.* **240**, 73–82.

MORRIS, R. V. and HASKIN, L. A. (1974) EPR measurement of the effect of glass composition on the oxidation state of europium. *Geochim. Cosmochim. Acta* **38**, 1435–1445.

MOTTL, N. J. and HOLLAND, H. D. (1978) Chemical exchange during hydrothermal alteration of basalt by seawater—I. Experimental results for major and minor components of seawater. *Geochim. Cosmochim. Acta* **42**, 1103–1115.

MOTTL, M. J., HOLLAND, H. D. and CORR, R. F. (1979) Chemical exchange during hydrothermal alteration of basalt by seawater-II. Experimental results for Fe, Mn and sulfur species. *Geochim. Cosmochim. Acta* **43**, 869–884.

MUEHLENBACHS, K. and CLAYTON, R. N. (1972) Oxygen isotope studies of fresh and weathered submarine basalts. *Can. J. Earth Sci.* **9**, 172–184.

MUEHLENBACHS, K. and CLAYTON, R. N. (1976) Oxygen isotope composition of the oceanic crust and its bearing on sea water. *J. Geophys. Res.* **81**, 4365–4369.

MUELLER, R. F. (1969) Kinetics and thermodynamics of intracrystalline distributions. *Mineral. Soc. Am. Spec. Pap.* **2**, 83–93.

MURTHY, V. R. and HALL, H. T. (1970) The chemical composition of the Earth's core: possibility of sulphur in the core. *Phys. Earth Planet. Inter.* **6**, 125–130.

MYSEN, B. O. (1976a) The role of volatiles in silicate melts: solubility of carbon dioxide and water in feldspar, pyroxene and feldspathoid melts to 30 kb and 1625°C. *Am. J. Sci.* **276**, 969–996.

MYSEN, B. O. (1976b) Partitioning of samarium and nickel between olivine, orthopyroxene, and liquid; preliminary data at 20 kbar and 1025°C. *Earth Planet. Sci. Lett.* **31**, 1–7.

MYSEN, B. O. (1977) The solubility of H_2O and CO_2 under predicted magma genisis conditions and some petrological and geophysical implications. *Rev. Geophys. Space Phys.* **15**, 351–361.

MYSEN, B. O, (1979) Nickel partitioning between olivine and silicate melt: Henry's law revisited. *Am. Mineral.* **64**, 1107–1114.

MYSEN, B. O. and VIRGO, D. (1980) Trace element partitioning and melt structure: an experimental study at 1 atm. pressure. *Geochim. Cosmochim. Acta* **44**, 1917–1930.

MYSEN, B. O., VIRGO, D., HARRISON, W. J. and SCARFE, C. M. (1980) Solubility mechanisms of H_2O in silicate melts at high pressures and temperatures: a Raman spectroscopic study. *Am. Mineral.* **65**, 900–914.

NAFZIGER, R. H. and MUAN, A. (1967) Equilibrium phase compositions and thermodynamic properties of olivines and pyroxenes in the system $MgO-'FeO'-SiO_2$. *Am. Mineral.* **52**, 1364–1385.

NAGASAWA, H. (1970) Rare earth concentrations in zircons and apatites and their host dacites and granites. *Earth Planet. Sci. Lett.* **9**, 359–364.

NAGASAWA, H. and SCHNETZLER, C. C. (1971) Partitioning of rare earth, alkali and alkaline earth elements between phenocrysts and acidic igneous magma. *Geochim. Cosmochim. Acta* **35**, 953–968.

NAGASAWA, H. and WAKITA, H. (1968) Partition of uranium and thorium between augite and host lavas. *Geochim. Cosmochim. Acta* **32**, 917–921.

NAGY, B. (1975) *Carbonaceous Meteorites*, 747 pp. Elsevier.

NAVROTSKY, A. (1976) Silicates and related minerals: solid state chemistry and theormodynamics applied to geothermometry and geobarometry. *Prog. Solid State Chem.* **11**, 203–264.

NAVROTSKY, A. (1978) Thermodynamics of element partitioning: (1) Systematics of transition metals in crystalline and molten silicates and (2) Defect chemistry and 'the Henry's Law problem'. *Geochim. Cosmochim. Acta* **42**, 887–902.

NAVROTSKY, A. and KLEPPA, O. J. (1967) The thermodynamics of cation distribution in simple spinels. *J. Inorg. Nucl. Chem.* **29**, 2701–2714.

NELSON, L. S., BLANDER, M., SKAGGS, S. R. and KEIL, K. (1972) Use of a CO_2 laser to prepare chondrule like spherules from supercooled molten oxide and silicate droplets. *Earth Planet. Sci. Lett.* **14**, 338–344.

NESBITT, H. W. (1980) A consistency test for single ion activity coefficients in electrolytic solutions, including seawater. *Chem. Geol.* **29**, 107–116.

NICHOLLS, J., CARMICHAEL, I. S. E. and STORMER, J. C., Jr. (1971) Silica activity and P_{total} in igneous rocks. *Contrib. Mineral. Petrol.* **33**, 1–20.

NOBLE, D. C. and HEDGE, C. E. (1970) Distribution of rubidium between sodic sanidine and natural silicic liquid. *Contrib. Mineral. Petrol.* **29**, 234–241.

OHMOTO, H. (1972) Systematics of sulfur and carbon isotopes in hydrothermal ore deposits. *Econ. Geol.* **67**, 551–578.

OHMOTO, H. and RYE, R. O. (1974) Hydrogen and oxygen isotopic compositions of fluid inclusions in the Kuroko Deposits, Japan. *Econ. Geol.* **69**, 947–953.

OLLIER, C. D. (1975) *Weathering*, 304 pp. Longmans.

O'NIONS, R. K. and SMITH, D. G. W. (1973) Bonding in silicates: an assessment of bonding in orthopyroxene. *Geochim. Cosmochim. Acta* **37**, 249–257.

ONUMA, N., HIGUCHI, H., WAKITA, H. and NAGASAWA, H. (1968) Trace element partition between two pyroxenes and the host lava. *Earth Planet. Sci. Lett.* **5**, 47–51.

ONUMA, N., CLAYTON, R. N. and MAYEDA, T. K. (1972a) Oxygen isotope temperatures of "equilibrated" ordinary chondrites. *Geochim. Cosmochim. Acta* **36**, 157–168.

ONUMA, N., CLAYTON, R. N. and MAYEDA, T. K. (1972b) Oxygen isotope cosmothermometer. *Geochim. Cosmochim. Acta* **36**, 169–188.

OWEN, D. and MCCONNELL, J. D. C. (1974) Spinodal unmixing in an alkali feldspar. In *The Feldspars* (Proc. NATO Advanced Study Inst. July 1972) (eds. W. S. MACKENZIE and J. ZUSSMAN), pp. 424–439. Manchester Univ. Press.

PAČES, T. (1973) Steady-state kinetics and equilibrium between ground water and granitic rock. *Geochim. Cosmochim. Acta* **37**, 2641–2663.

PAKISER, L. C. and ROBINSON, R. (1967) Composition of the continental crust as estimated from seismic observations. In *The Earth Beneath the Continents* (ed. J. S. STEINHART and T. J. SMITH). *Am. Geophys. Union, Geophys. Mon.* **10**, 620–626.

PAPANASTASSIOU, D. A. and WASSERBURG, G. J. (1969) The determination of small time differences in the formation of planetary objects. *Earth Planet. Sci. Lett.* **5**, 361–376.

PARKER, R. L. (1967) Composition of the Earth's crust. In *Data of Geochemistry* (ed. M. FLEISCHER), 6th edn., 19 pp. U. S. Geol. Surv. Prof. Pap. 440-D.

PATCHETT, P. T. and TATSUMOTO, M. (1980) Lu-Hf total-rock isochron for the eucrite meteorites. *Nature* **288**, 571–574.

PAULING, L. (1960) *The Nature of the Chemical Bond*, 3rd edn., 644 pp. Cornell University Press.

PAULING, L. (1970) Crystallography and chemical bonding of sulfide minerals. *Mineral. Soc. Am. Spec. Pap.* **3**, 125–131.

PETROVIĆ, R., BERNER, R. A. and GOLDHABER, M. B. (1976) Rate control in dissolution of alkali feldspars—I. Study of residual feldspar grains by X-ray photoelectron spectroscopy. *Geochim. Cosmochim. Acta* **40**, 537–548.

PHILLIPS, C. S. G. and WILLIAMS, R. J. P. (1965/66) *Inorganic Chemistry*, 2 vols., 685 pp. and 683 pp. Clarendon Press, Oxford.

PHILPOTTS, J. A. and SCHNETZLER, C. C. (1970) Phenocryst–matrix partition coefficients for K, Rb, Sr and Ba with applications to anorthosite and basalt genesis. *Geochim. Cosmochim. Acta* **34**, 307–322.

PINCKNEY, D. M. and RAFTER, T. A. (1972) Fractionation of sulfur isotopes during ore deposition in the Upper Mississippi Valley zinc–lead deposit. *Econ. Geol.* **67**, 315–328.

POLDERVAART, A. (1955) Chemistry of the earth's crust. In *Crust of the Earth—a symposium* (ed. A. POLDERVAART), pp. 119–144. Geol. Soc. Am. Spec. Pap. 62.

POWELL, R. (1974) A comparison of some mixing models for crystalline silicate solid solutions. *Contrib. Mineral. Petrol.* **46**, 265–274.

POWELL, R. (1977) Activity–composition relationships for crystalline solutions. In *Thermodynamics in Geology* (ed. D. G. FRASER), pp. 57–65. D. Reidel Publ. Co.

POWELL, R. (1978a) *Equilibrium Thermodynamics in Petrology. An Introduction*, 284 pp. Harper & Row.

POWELL, R. (1978b) The thermodynamics of pyroxene geotherms. *Phil. Trans. R. Soc. Lond.* A, **288**, 459–469.

POWELL, M. and POWELL, R. (1977) Plagioclase—alkali-feldspar geothermometry revisited. *Mineralog. Mag.* **41**, 253–256.

PRESTON, M. A. (1962) *Physics of the Nucleus*, 671 pp. Addison-Wesley Publ. Co.

PUTNIS, A. and McCONNELL, J. D. C. (1980) *Principles of Mineral Behaviour*, 257 pp. Blackwells.

PYTKOWICZ, R. M. (1975) Some trends in marine chemistry and geochemistry. *Earth Sci. Rev.* **11**, 1–46.

RANKAMA, K. and SAHAMA, Th. G. (1950) *Geochemistry*, 912 pp. University of Chicago Press.

RAYLEIGH, J. W. S. (1896) Theoretical considerations respecting the separation of gases by diffusion and similar processes. *Philos. Mag.* 5th ser. **42**, 493–498.

RAYLEIGH, J. W. S. (1902) On the distillation of binary mixtures. *Philos. Mag.* 6th ser. **4**, 521–537.

REEVES, H. (1971) *Nuclear Reactions in Stellar Surfaces and their Relation with Stellar Evolution*, 87 pp. Gordon & Breach Science Publ.

RICHARD, P., SHIMIZU, N. and ALLEGRE, C. J. (1976) $^{143}Nd/^{146}Nd$, a natural tracer: an application to oceanic basalts. *Earth Planet. Sci. Lett.* **31**, 269–278.

RICHARDSON, F. D. (1974) *Physical Chemistry of Melts in Metallurgy*, 2 vols., 559 pp. Academic Press.

RILEY, J. P. and SKIRROW, G., eds. (1975–1978) *Chemical Oceanography*, 2nd edn., 7 vols. Academic Press.

RINGWOOD, A. E. (1966a) The chemical composition and origin of the earth. In *Advances in Earth Sciences* (ed. P. M. HURLEY), pp. 287–356. MIT Press.

RINGWOOD, A. E. (1966b). Chemical evolution of the terrestrial planets. *Geochim. Cosmochim. Acta* **30**, 41–104.

RINGWOOD, A. E. (1977) Composition of the core and implications for origin of the earth. *Geochem. J.* **11**, 111–135.

RINGWOOD, A. E. (1979a) *Origin of the Earth and Moon*, 295 pp. Springer-Verlag.

RINGWOOD, A. E. (1979b) Composition and origin of the Earth. In *The Earth: Its Origin, Structure and Evolution* (ed. M. W. McELHINNEY), pp. 1–58. Academic Press.

ROBERTSON, D. E. (1970) The distribution of cobalt in oceanic waters. *Geochim. Cosmochim. Acta* **34**, 553–567.

ROBIE, R. A., HEMINGWAY, B. S. and FISHER, J. R. (1978) Thermodynamic properties of minerals and related substances at 298.15 K and 1 bar (10^5 Pascals) pressure and at higher temperatures. *U.S. Geol. Surv. Bull.* **1452**, 456 pp.

ROCK, P. A. (1969) *Chemical Thermodynamics: Principles and Applications.* Macmillan.

ROEDDER, E. (1972) Composition of fluid inclusions. In *Data of Geochemistry* (ed. M. FLEISCHER), U.S. Geol. Surv. Prof. Pap. 440-JJ.

RONOV, A. B. and YAROSHEVSKIY, A. A. (1976) A new model for the chemical structure of the Earth's crust. *Geochem. Int.* **13**, No. 6, 89–121.

ROSS, J. E. and ALLER, L. H. (1976) The chemical composition of the Sun. *Science* **191**, 1223–1229.

RUBEY, W. M. (1951) Geologic history of sea water. An attempt to state the problem. *Geol. Soc. Am. Bull.* **62**, 1111–1148.

RUCKLIDGE, J. (1972) Chlorine in partially serpentinized dunite. *Econ. Geol.* **67**, 38–40.

RUXTON, B. P. (1968) Measures of the degree of chemical weathering of rocks. *J. Geol.* **76**, 518–527.

RYE, R. O. and OHMOTO, H. (1974) Sulfur and carbon isotopes and ore genesis: a review. *Econ. Geol.* **69**, 826–842.

RYERSON, F. J. and HESS, P. C. (1978) Implications of liquid–liquid distribution coefficients to mineral-liquid partitioning. *Geochim. Cosmochim. Acta* **42**, 921–932.

SATO, M., HICKLING, N. L. and MCLANE, J. E. (1973) Oxygen fugacity values of Apollo 12, 14 and 15 lunar samples and reduced state of lunar magmas. *Proc. 4th Lunar Sci. Conf.*, pp. 1061–1079.

SAMFORD, R. F. (1981) Three Fortran programs for finite-difference solutions to binary diffusion in one and two phases with composition- and time-dependent diffusion coefficient. *Computers Geosci.* **7**.

SAXENA, S. K. and GHOSE, S. (1971) Mg^{2+}–Fe^{2+} order—disorder in orthopyroxenes and the thermodynamics of the orthopyroxene crystalline solution. *Am. Mineral.* **56**, 532–559.

SAXENA, S. K. and RIBBE, P. H. (1972) Activity–composition relations in feldspars. *Contrib. Mineral. Petrol.* **37**, 131–138.

SAYLES, F. L. and MANHEIM, F. T. (1975) Interstitial solutions and diagenesis in deeply buried marine sediments: results from the Deep Sea Drilling Project. *Geochim. Cosmochim. Acta* **39**, 103–127.

SCARFE, C. M. (1973) Viscosities of basic magmas at varying pressures. *Nature Phys. Sci.* **241**, 101–102.

SCHMITT, H. H. (1975) Apollo and the geology of the Moon. *J. Geol. Soc. Lond.* **131**, 103–119.

SCHNETZLER, C. C. and PHILPOTTS, J. A. (1968) Partition coefficients of rare-earth elements and barium between igneous matrix material and rock-forming mineral phenocrysts—I. In *Origin and Distribution of the Elements* (ed. L. H. AHRENS), pp. 929–938. Pergamon.

SCHNETZLER, C. C. and PHILPOTTS, J. A. (1970) Partition coefficients of rare-earth elements between igneous matrix material and rock-forming mineral phenocrysts—II. *Geochim. Cosmochim. Acta* **34**, 331–340.

SCHOCK, H. H. (1977) Trace element partitioning between phenocrysts of plagioclase, pyroxenes and magnetite and the host pyroclastic matrix. *J. Radioanal. Chem.* **38**, 327–340.

SCHWARCZ, H. P., SCOTT, S. D. and KISSIN, S. A. (1975) Pressures of formation of iron meteorites from sphalerite compositions. *Geochim. Cosmochim. Acta* **39**, 1457–1466.

SCOTT, E. R. D. and WASSON, J. T. (1975) Classification and properties of iron meteorites. *Rev. Geophys. Space Phys.* **13**, 527–546.

SCOTT, S. D. (1976) Application of the sphalerite geobarometer to regionally metamorphosed terrains. *Am. Mineral.* **61**, 661–670.

SCOTT, S. D. and BARNES, H. L. (1971) Sphalerite geothermometry and geobarometry. *Econ. Geol.* **66**, 653–669.

SEARS, D. W. (1978) *The Nature and Origin of Meteorites*, 187 pp. Adam Hilger Ltd. (Bristol).

SEIFERT, F. A. and VIRGO, D. (1975) Kinetics of the Fe^{2+}–Mg, order–disorder reaction in anthophyllites: Quantitative cooling rates. *Science* **188**, 1107–1109.

SELBIN, J. (1973a) The origin of the chemical elements. I. *J. Chem. Ed.* **50**, 306–310.

SELBIN, J. (1973b) The origin of the chemical elements. II. *J. Chem. Ed.* **50**, 380–387.

SEYFRIED, W. E. and BISCHOFF, J. L. (1979) Low temperature basalt alteration by seawater: an experimental study at 70° C and 150° C. *Geochim. Cosmochim. Acta* **43**, 1937–1947.

SHANNON, R. D. (1976) Revised effective ionic radii and systematic studies of interatomic distances in halides and chalcogenides. *Acta Cryst.* A, **32**, 751–767.

SHANNON, R. D. and PREWITT, C. T. (1969) Effective ionic radii in oxides and fluorides. *Acta Cryst.* B, 25, 925–946.

SHANNON, R. D. and PREWITT, C. T. (1970) Revised values of effective ionic radii. *Acta Cryst.* B, **26**, 1046–1048.

SHAPIRO, M. M., SILBERRERG, R. and TSAO, C. H. (1970) Relative abundances of cosmic rays at their sources. *Acta Physica Academiae Scientiarum, Hungaricae* **29**, Suppl. 1, 479–484.

SHAW, D. M. (1953) The camouflage principle and trace element distribution in magmatic minerals. *J. Geol.* **61**, 142–151.

SHAW, D. M. (1970) Trace element fractionation during anatexis. *Geochim. Cosmochim. Acta* **34**, 237–243.

SHAW, D. M. (1978) Trace element behaviour during anatexis in the presence of a fluid phase. *Geochim. Cosmochim. Acta* **42**, 933–943.

SHIDO, F., MIYASHIRO, A. and EWING, M. (1974) Compositional variation in pillow lavas from the mid-Atlantic ridge. *Marine Geol.* **16**, 177–190.

SHIEH, Y. N. and TAYLOR, H. P., Jr. (1969) Oxygen and carbon isotope studies of contact metamorphism of carbonate rocks. *J. Petrol.* **10**, 307–331.

SILLÉN, L. G. (1961) The physical chemistry of sea water. In *Oceanography* (ed. M. SEARS), pp. 549–581. Am. Assoc. Advance Sci. Publ. No. 67.

SILLÉN, L. G. and MARTELL, A. E. (1964) *Stability Constants of Metal-Ion Complexes*, Spec. Publ. No. 17, The Chemical Society, London.

SILLÉN, L. G. and MARTELL, A. E. (1971) *Stability Constants of Metal-Ion Complexes* (Suppl. 1 to Spec. Publ. No. 17) Spec. Pub. No. 25, The Chemical Society, London.

SKIRROW, G. (1975) The dissolved gases—carbon dioxide. In *Chemical Oceanography* (eds. J. P. RILEY and G. SKIRROW), 2nd edn, **2**, 1–192.

SMITH, B. A. and GOLDSTEIN, J. I. (1977) The metallic microstructures and thermal histories of severely reheated chondrites. *Geochim. Cosmochim. Acta* **41**, 1061–1072.

SMITH, J. V. (1977) Possible controls on the bulk composition of the earth: implications for the origin of the earth and moon. *Proc. 8th Lunar Sci. Conf.*, pp. 333–369.

SMITH, J. V. (1979) Mineralogy of the planets: a voyage in space and time. *Mineral. Mag.* **43**, 1–89.

SPOONER, E. T. C., BECKINSALE, R. D., FYFE, W. S. and SMEWING, J. D. (1974) O^{18} enriched ophiolitic metabasic rocks from E. LIGURIA (Italy), PINDOS (Greece), and TROODOS (Cyprus). *Contrib. Mineral. Petrol.* **47**, 41–62.

STORMER, J. C., Jr. (1975) A practical two-feldspar geothermometer. *Am. Mineral.* **60**, 667–674.

STORZER, D. and SELO, M. (1976) Uranium content and fission track ages of some basalts from the FAMOUS area. *Bull. Soc. Geol. France* XVIII, 807–810.

STUMM, W. and BRAUNER, P. A. (1975) Chemical speciation. In *Chemical Oceanography* (eds. J. P. RILEY and G. SKIRROW), 2nd edn., **1**, 173–239.

STUMM, W. and MORGAN, J. J. (1970) *Aquatic Chemistry. An Introduction Emphasizing Chemical Equilibria in Natural Waters*, 583 pp. Wiley.

SUESS, H. E. and UREY, H. C. (1956) Abundances of the elements. *Rev. Mod. Phys.* **28**, 53–74.

SUN, C-O., WILLIAMS, R. J. and SUN, S-S. (1974) Distribution coefficients of Eu and Sr for plagioclase–liquid and clinopyroxene–liquid equilibria in oceanic ridge basalt: an experimental study. *Geochim. Cosmochim. Acta* **38**, 1415–1433.

TAKAHASHI, E. (1978) Partitioning of Ni^{2+}, Co^{2+}, Fe^{2+}, Mn^{2+} and Mg^{2+} between olivine and silicate melts: compositional dependence of partition coefficient. *Geochim. Cosmochim. Acta* **42**, 1829–1844.

TAYLOR, G. F. and HEYMANN, D. (1971) The formation of clear taenite in ordinary chondrites. *Geochim. Cosmochim. Acta* **35**, 175–188.

TAYLOR, H. P., Jr. (1968) The oxygen isotope geochemistry of igneous rocks. *Contrib. Mineral. Petrol.* **19**, 1–71.

TAYLOR, H. P., Jr. (1974) The application of oxygen and hydrogen isotope studies to problems of hydrothermal alteration and ore deposition. *Econ. Geol.* **69**, 843–883.

TAYLOR, H. P., Jr. (1978) Oxygen and hydrogen isotope studies of plutonic granitic rocks. *Earth Planet. Sci. Lett.* **38**, 177–210.

TAYLOR, H. P., Jr. and EPSTEIN, S. (1962) Relationship between O^{18}/O^{16} ratios in coexisting minerals of igneous and metamorphic rocks. Part I. Principles and experimental results *Geol. Soc. Amer. Bull.* **73**, 461–480.

TAYLOR, H. P., Jr. and FORESTER, R. W. (1971) Low-O^{18} igneous rocks from the intrusive complexes of Skye, Mull, and Ardnamurchan, Western Scotland. *J. Petrol.* **12**, 465–497.

TAYLOR, H. P., Jr. and TURI, B. (1976) High-^{18}O igneous rocks from the Tuscan magmatic province, Italy. *Contrib. Mineral. Petrol.* **55**, 33–54.

TAYLOR, S. R. (1964) The abundance of chemical elements in the continental crust—a new table. *Geochim. Cosmochim. Acta* **28**, 1273–1285.

TAYLOR, S. R. (1973) Tektites: a post-Apollo view. *Earth Sci. Rev.* **9**, 101–123.

TAYLOR, S. R. (1975) *Lunar Science: A Post-Apollo View*, 372 pp. Pergamon.

THOMPSON, G. (1973) A geóchemical study of the low-temperature interaction of sea-water and oceanic igneous rocks. *Trans. Am. Geophys. Un.* **54**, 1015–1019.

TICKLE, R. E. (1967) The electrical conductance of molten alkali silicates. Part 1. Experiments and results. *Phys. Chem. Glasses* **8**, 101–102.

TILLER, W. A., JACKSON, K. A., RUTTER, J. W. and CHALMERS, B. (1953) The redistribution of solute atoms during the solidification of metals. *Acta Met.* **1**, 428–437.

TÓMASSON, J. and KRISTMANNSDÓTTIR, H. (1972) High temperature alteration minerals and thermal brines, Reykjanes, Iceland. *Contrib. Mineral. Petrol.* **36**, 123–134.

TOOP, G. W. and SAMIS, C. S. (1962) Activities of ions in silicate melts. *Trans. Met. Soc. AIME* **224**, 878–887.

TOSSELL, J. A., VAUGHAN, D. J. and JOHNSON, K. H. (1974) The electronic structure of rutile, wustite, and hematite from molecular orbital calculations. *Am. Mineral.* **59**, 319–334.

TOULMIN, P. and BARTON, P. B., Jr. (1964) A thermodynamic study of pyrite and pyrrhotite. *Geochim. Cosmochim. Acta* **28**, 641–671.

TREUIL, M. and VARET, J. (1973) Critères volcanologiques, pétrologiques et géochimiques de la genèse et de la différenciation des magmas basaltiques: exemple de l'Afar. *Bull. Soc. Geol. France* 7th ser., **15**, 506–540.

TURNER, F. J. (1968) *Metamorphic Petrology. Mineralogical and Field Aspects*, 403 pp. McGraw Hill.

UREY, H. C., LOWENSTAM, H. A., EPSTEIN, S. and McKINNEY, C. R. (1951) Measurement of paleotemperatures and temperatures of the Upper Cretaceous of England, Denmark, and the southeastern United States. *Bull. Geol. Soc. Am.* **62**, 399–416.

VAN SCHMUS, W. R. (1969) Mineralogy and petrology of chondritic meteorites. *Earth Sci. Rev.* **5**, 145–184.

VAN SCHMUS, W. R. and HAYES, J. M. (1974) Chemical and petrographic correlations among carbonaceous chondrites. *Geochim. Cosmochim. Acta* **38**, 47–64.

VAN SCHMUS, W. R. and WOOD, J. A. (1967) A chemical-petrologic classification for the chondritic meteorites. *Geochim. Cosmochim. Acta* **31**, 747–765.

VANIMAN, D. T. and PAPIKE, J. J. (1977) Very low Ti (VLT) basalts: a new mare rock type from the Apollo 17 drill core. *Proc. 8th Lunar Sci. Conf.*, pp. 1443–1471.

VAUGHAN, D. J., BURNS, R. G. and BURNS, V. M. (1971) Geochemistry and bonding of thiospinel minerals. *Geochim. Cosmochim. Acta* **35**, 365–381.

VINCENT, E. A. and NIGHTINGALE, G. (1974) Gallium in rocks and minerals of.the Skaergaard intrusion. *Chem. Geol.* **14**, 63–73.

VIRGO, D. and HAFNER, S. S. (1970) Fe^{2+}, Mg order–disorder in natural orthopyroxenes. *Am. Mineral.* **55**, 201–223.

WAFF, H. F. (1975) Pressure-induced coordination changes in magmatic liquids. *Geophys. Res. Lett.* **2**, 193–196.

WAGER, L. R. (1960) The major element variation of the layered series of the Skaergaard intrusion and a re-estimation of the average composition of the hidden layered series and of the successive residual magmas. *J. Petrol.* **1**, 364–398.

WAGER, L. R. and BROWN, G. M. (1968) *Layered Igneous Rocks*, 588 pp. Oliver & Boyd.

WAGER, L. R. and DEER, W. A. (1939) Geological investigations in east Greenland, Pt. III. The petrology of the Skaergaard intrusion, Kangerdlugssuaq, east Greenland, *Medd. Grønland* 105, No. 4, 1–352.

WAGER, L. R. and MITCHELL, R. L. (1951) The distribution of trace elements during strong fractionation of basic magma—a further study of the Skaergaard intrusion, east Greenland. *Geochim. Cosmochim. Acta* 1, 129–208.

WALSH, D., DONNAY, G. and DONNAY, J. D. H. (1976) Ordering of transition metal ions in olivine. *Can. Mineral.* 14, 149–150.

WÄNKE, H., BADDENHAUSEN, H., DREIBUS, G., JAGOUTZ, E., KRUSE, H., PALME, H., SPETTEL, B. and TESCHKE, F. (1973) Multielement analyses of Apollo 15, 16 and 17 samples and the bulk composition of the moon. *Proc. 4th Lunar Sci. Conf.*, pp. 1461–1481.

WASSON, J. T. (1974) *Meteorites: Classification and Properties*, 316 pp. Springer-Verlag.

WASSON, J. T. and WAI, C. M. (1970) Composition of the metal, schreibersite and perryite of enstatite achondrites and the origin of enstatite chondrites and achondrites. *Geochim. Cosmochim. Acta* 34, 169–184.

WATSON, E. B. (1976) Two-liquid partition coefficients: experimental data and geochemical implications. *Contrib. Mineral. Petrol.* 56, 119–134.

WATSON, E. B. (1977) Partitioning of manganese between forsteite and silicate liquid. *Geochim. Cosmochim. Acta* 41, 1363–1374.

WEDEPOHL, K. H., executive editor (1969–1978) *Handbook of Geochemistry*, 2 vols. Springer.

WEI, G. C. T. and WUENSCH, B. J. (1976) Tracer concentration gradients for diffusion coefficients exponentially dependent on concentration. *J. Am. Ceram. Soc.* 59, 295–299.

WELLS, A. F. (1975) *Structural Inorganic Chemistry*, 4th edn., 1095 pp. Clarendon Press, Oxford.

WELLS, P. R. A. (1977) Pyroxene thermometry in simple and complex systems. *Contrib. Mineral. Petrol.* 62, 129–139.

WHITE, D. E. (1968) Environments of generation of some base-metal ore deposits. *Econ. Geol.* 63, 301–335.

WHITE, D. E. and WARING, G. A. (1963) Volcanic emanations. In *Data on Geochemistry*, 6th edn. (ed. M. FLEISCHER), *U.S. Geol. Surv. Bull.* 440-K.

WHITE, D. E., HEM, J. D. and WARING, G. A. (1963) Chemical composition of sub-surface waters. In *Data of Geochemistry*, 6th edn. (ed. M. FLEISCHER), *U.S. Geol. Surv. Bull.* 440-F.

WHITEWAY, S. G., SMITH, I. B. and MASSON, C. R. (1970) Theory of molecular size distribution in multichain polymers. *Can. J. Chem.* 48, 33–45.

WHITTAKER, E. J. W. (1967) Factors affecting element ratios in the crystallization of minerals. *Geochim. Cosmochim. Acta* 31, 2275–2288.

WHITTAKER, E. J. W. (1971) Madelung energies and site preferences in amphiboles, I. *Am. Mineral.* 56, 980–996.

WHITTAKER, E. J. W. (1978) The cavities in a random close-packed structure. *J. Non-Crystal. Solids* 28, 293–304.

WHITTAKER, E. J. W. and MUNTUS, R. (1970) Ionic radii for use in geochemistry. *Geochim. Cosmochim. Acta* 34, 945–956.

WIIK, H. B. (1956) The chemical composition of some stony meteorites. *Geochim. Cosmochim. Acta* 9, 279–289.

WILCOX, R. E. (1979) The liquid line of descent and variation diagrams. In *The Evolution of the Igneous Rocks. Fiftieth Anniversary Perspectives* (ed. H. S. YODER, Jr.), pp. 204–232. Princeton Univ. Press.

WILLIAMS, R. J. P. (1959) Deposition of trace elements in basic magma. *Nature* 184, 44.

WILSON, A. F., GREEN, D. C. and DAVIDSON, L. R. (1970). The use of oxygen isotope geothermometry on the granulites and related intrusives, Musgrave Ranges, Central Australia. *Contrib. Mineral. Petrol.* 27, 166–178.

WILSON, D. E. (1980) Surface and complexation effects on the rate of Mn(II) oxidation in natural waters. *Geochim. Cosmochim. Acta* 44, 1311–1317.

WILSON, T. R. S. (1975) Salinity and the major elements of sea water. In *Chemical Oceanography* (eds. J. P. RILEY and G. SKIRROW), 2nd edn., 1, 365–413.

WINCHELL, P. (1969) The compensation law for diffusion in silicates. *High Temp. Sci.* **1**, 200–215.

WINDLEY, B. F., ed. (1976) *The Early History of the Earth*, 619 pp. Wiley.

WOLLAST, R. (1967) Kinetics of the alteration of K-feldspar in buffered solutions at low temperature. *Geochim. Cosmochim. Acta* **31**, 635–648.

WOLLAST, R. (1974) The silica problem. In *The Sea*, vol. 5, *Marine Chemistry* (eds. E. D. GOLDBERG), pp. 359–392. Wiley.

WOOD, B. J. (1974) Crystal field spectrum of Ni^{2+} in olivine. *Am. Mineral.* **59**, 244–248.

WOOD, B. J. (1975) The application of thermodynamics to some subsolidus equilibria involving solid solutions. *Fortsch. Mineral.* **52**, 21–45.

WOOD, B. J. (1976) Samarium distribution between garnet and liquid at high pressure. *Carnegie Inst. Washington Yearb.* **75**, 659–662.

WOOD, B. J. (1977) The activities of components in clinopyroxene and garnet solid solutions and their application to rocks. *Phil. Trans. R. Soc. London Ser.* A, **286**, 331–342.

WOOD, B. J. and BANNO, S. (1973) Garnet–orthopyroxene and orthopyroxene–clinopyroxene relationships in simple and complex systems. *Contrib. Mineral. Petrol.* **42**, 109–124.

WOOD, B. J. and FRASER, D. G. (1976) *Elementary Thermodynamics for Geologists*, 303 pp. Oxford Univ. Press.

WOOD, J. A. (1964) The cooling rates and parent planets of several iron meteorites. *Icarus* **3**, 429–459.

WOOD, J. A. (1967) Chondrites: their metallic minerals, thermal histories, and parent planets. *Icarus* **6**, 1–49.

WOOD, J. A. (1968) *Meteorites and the Origin of Planets*, 117 pp. McGraw Hill.

WOOD, M. I. and HESS, P. C. (1980) The structural role of Al_2O_3 and TiO_2 in immiscible silicate liquids in the system SiO_2–MgO–CaO–FeO–TiO_2–Al_2O_3. *Contrib. Mineral. Petrol.* **72**, 319–328.

WYLLIE, P. J. (1971) *The Dynamic Earth: Textbook in Geosciences*, 416 pp. Wiley.

YODER, H. S., Jr. (1976) *Generation of Basaltic Magma*, 265 pp. National Academy of Sciences, Washington, DC.

YORK, D. and FARQUHAR, R. M. (1972) *The Earth's Age and Geochronology*, 178 pp. Pergamon.

YUND, R. A. and McCALLISTER, R. H. (1970) Kinetics and mechanisms of exsolution. *Chem. Geol.* **6**, 5–30.

ZACHARIASEN, W. H. (1932) The atomic arrangement in glasses. *J. Am. Chem. Soc.* **54**, 3841–3851.

Name Index

Subject Index